AF709532

A GUIDE TO THE BRITISH —FOOD— MANUFACTURING INDUSTRY

A Guide to the British Food Manufacturing Industry

Fourth edition

Revised and updated by
Di Amor
Freelance training and information broker
Reading
UK

Edited by **Peter Sheard**
University of Bristol
Division of Food Animal Science
Bristol
UK

CHAPMAN & HALL
London · Glasgow · Weinheim · New York · Tokyo · Melbourne · Madras

Published by Chapman & Hall, 2–6 Boundary Row, London SE1 8HN, UK

Chapman & Hall, 2–6 Boundary Row, London SE1 8HN, UK

Blackie Academic & Professional, Wester Cleddens Road, Bishopbriggs, Glasgow G64 2NZ, UK

Chapman & Hall GmbH, Pappelallee 3, 69469 Weinheim, Germany

Chapman & Hall USA, One Penn Plaza, 41st Floor, New York NY 10119, USA

Chapman & Hall Japan, ITP-Japan, Kyowa Building, 3F, 2-2-1 Hirakawacho, Chiyoda-ku, Tokyo 102, Japan

Chapman & Hall Australia, Thomas Nelson Australia, 102 Dodds Street, South Melbourne, Victoria 3205, Australia

Chapman & Hall India, R. Seshadri, 32 Second Main Road, CIT East, Madras 600 035, India

First edition 1986
Second edition 1988
Third edition 1991
Fourth edition 1995

© 1991 Nova Press, 1995 Di Amor

Typeset in 9/10 pt Times by Best-set Typesetter Ltd, Hong Kong

Printed in Great Britain by T.J. Press Ltd, Padstow, Cornwall

ISBN 0 412 57360 1

Apart from any fair dealing for the purposes of research or private study, or criticism or review, as permitted under the UK Copyright Designs and Patents Act, 1988, this publication may not be reproduced, stored, or transmitted, in any form or by any means, without the prior permission in writing of the publishers, or in the case of reprographic reproduction only in accordance with the terms of the licences issued by the Copyright Licensing Agency in the UK, or in accordance with the terms of licences issued by the appropriate Reproduction Rights Organization outside the UK. Enquiries concerning reproduction outside the terms stated here should be sent to the publishers at the London address printed on this page.

The publisher makes no representation, express or implied, with regard to the accuracy of the information contained in this book and cannot accept any legal responsibility or liability for any errors or omissions that may be made.

A catalogue record for this book is available from the British Library

Library of Congress Catalog Card Number: 94-72451

∞ Printed on permanent acid-free text paper, manufactured in accordance with ANSI/NISO Z39.48-1992 and ANSI/NISO Z39.48-1984 (Permanence of Paper).

TABLE OF CONTENTS

FOREWORD

Since Peter Sheard introduced the Third Edition of this guide in January 1991, the recession has been worse than predicted by the politicians and economists and much longer. Job losses in the food industry, often regarded as 'recession proof' have been very substantial, with numbers employed by the industry as a whole dropping from about 550 000 in the late 1980s to a figure of about 420 000 now. The themes of company annual reports are those of cost-cutting, rationalization and restructuring, and workforce reduction. Medium-sized companies seem to have been the worst affected. Companies covered in Section 2.2 of the last edition – Corton Beach, Alma Holdings, the Hughes Food Group, Polly Peck International, Clarke Foods – all ended up in the hands of receivers. Berisford, who had a turnover of £5bn in 1985, is now reduced to a shell company following disposal of most of its assets.

EC legislation has also seriously affected the food industry. This has ranged from changes in subsidy payments to the meat industry, to meat hygiene legislation affecting the operation of slaughterhouses, and the more general effects of the Food Safety Act 1990. These forced many companies into capital expenditure on premises, equipment and staff hygiene training which they might otherwise not have made. Hillsdown Holdings, for instance, had 28 slaughterhouses in 1992; by November 1993, it had disposed of or closed all of them.

1994 will bring the end to the statutory milk marketing scheme operated by the five Milk Marketing Boards in the UK for the last 60 years. The dairy industry is consequently awash with conflicting views on whether these fundamental changes will be better or worse for farmers and make milk cheaper or more expensive for the consumer. Its proposed replacement 'Milk Marque' will have to compete with other milk-buying co-operatives set up by major manufacturers trying to ensure secure milk supplies for their manufacturing operations. Northern Foods, Nestlé, Unigate and MD Foods have all been active in this arena.

Other interesting developments in the food industry include the increase in the number of companies producing so called 'value-added' products. Many of the commodity-based parts of the industry like dairy and meat see the likelihood of much greater returns on their investments from the production of ready-meals rather than fresh meat or milk. Manufacture of vegetarian, ethnic, and 'healthy' foods is also on the increase as well as the use of frozen or chilled 'bake-off' goods by in-store bakeries, delicatessens and the catering industry. Pizza toppings, sandwiches and sandwich filling manufacture comprise other growth areas.

SCOPE OF THE GUIDE

The guide concentrates on companies engaged in food manufacturing. These include those quoted on the London Stock Exchange, the Unlisted Securities Market and some of the largest private companies in England, Scotland, Wales and Northern Ireland. In the Third Edition several retailers with food manufacturing subsidiaries were mentioned – Asda, J. Sainsbury, Co-operative Retail Society (CRS) and the Northern Co-operative Society (NORCO). Of these only J. Sainsbury continues to have manufacturing interests and is the parent company of NewMarket Foods (formerly Haverhill Meat Products) covered in Section 2.2. Asda closed Lofthouse Foods following poor returns for several years and the failure to find a buyer for it. CRS withdrew from all its dairy activities, and NORCO called in the receivers when the Co-operative Wholesale Society (CWS) declined to enter into a rescue deal.

Companies registered in Eire are not included unless the parent company has operating subsidiaries in the UK, such as Avonmore Foods, Kerry Group, Kerrygold, Waterford Foods and Golden Vale. Food brokers, importers, wholesalers and retailers, suppliers of raw materials, ingredients or processing equipment are not generally included unless the company is also involved in food manufacturing. An exception to this is Ess-Food UK which disposed of its food manufacturing interest to Tulip International within the period covered by the guide.

The word 'food' is used to mean food and non-alcoholic beverages; companies whose major interests are alcoholic beverages are not featured.

Di Amor, January 1994

DISCLAIMER

The completion of this publication has been undertaken with great care and every effort has been made to ensure that it is accurate and without error. However, its author and editor cannot accept any responsibility whatsoever for loss of income or any other inconvenience caused by errors in the text.

In matters of great importance it is suggested that readers refer directly to primary sources of information such as a company's annual report and accounts which can be obtained as a microfiche (£3.00/company) from Companies House, Crown Way, Cardiff CF4 3UZ (tel. (0222) 380124; Fax (0222) 380900; DX 33050) or Companies House, 55/71 City Road London (tel. 071-253 9393; Fax 071-608 0435).

CORRECTIONS AND UPDATES

Collection of material for the guide is an ongoing process and companies listed in the current edition are warmly invited to bring any inaccuracies to the author's or editor's attention or submit material for the next edition.

They can be contacted as follows:

Dr Peter Sheard,
Division of Food Animal Science,
University of Bristol,
Churchill Building,
Langford,
Bristol BS18 7DY Tel (0272) 852581; Fax (0272) 852741

Mrs Di Amor,
8 Marlborough Court
Coley Avenue
Reading RG1 6PP Tel (0734) 502918

THE EDITOR

Dr Peter Sheard studied Food Processing Technology at Loughborough University. After completing a PhD at the University of Nottingham in 1986, he joined AFRC and worked for 5 years at the Institute of Food Research, Bristol Laboratory (formerly the Meat Research Institute). He is currently working at the University of Bristol in the Division of Food Animal Science where he is Course Organizer for an MSc Meat Science Course.

THE AUTHOR

Di Amor has a Biochemistry Degree from the University of Bristol. She has worked as an Information Scientist at both the Leatherhead Food Research Association and the International Food Information Service (part of the Commonwealth Agricultural Bureaux) at Reading. She was Short Course Development Officer in the Department of Food Science and Technology at the University of Reading from 1988 to 1992. She now works as a freelance training and information broker and is Chairperson of the Western Branch of the Institute of Food Science and Technology.

ABBREVIATIONS

A number of common abbreviations are used throughout this book:

£m	million £ sterling
£bn	billion £ sterling
pa	per annum
ppl	pence per litre
pw	per week
t	tonnes
ABF	Associated British Foods
AFRC	Agricultural and Food Research Council
BCCCA	Biscuit, Cake, Chocolate & Confectionery Alliance
BMMA	British Meat Manufacturers' Association
CAP	Controlled atmosphere packaging
CTN	Confectioners, tobacconists and newsagents
CRS	Co-operative Retail Society
CWS	Co-operative Wholesale Society
EC	European Community
Emps	Employees
GrandMet	Grand Metropolitan plc
MAFF	Ministry of Agriculture, Fisheries and Food
MAP	Modified atmosphere packaging
MLC	Meat and Livestock Commission
MMB	Milk Marketing Board
MMC	Monopolies and Mergers Commission
n/a	not available
RHM	Ranks Hovis McDougall
T/O	Turnover
T/O £–	Turnover not disclosed. (Applies to small companies which are exempt from disclosing turnover and number of employees under the Companies Act.)
UHC	Ultimate holding company

Part One

GENERAL INFORMATION

SECTION 1.1 TABLE OF FOOD PRODUCTS MANUFACTURED

The following table lists the main companies involved in each product area. For the large and medium-sized companies it may be a subsidiary or associate company of a different name which actually manufactures the product. To keep the table simple, subsidiaries have not been included but further information can be gained from the relevant company section and the index.

Table 1

Food products	Large companies [2.1]	Medium-sized companies [2.2]	Small companies [2.3]
Baby foods	Reckitt & Colman	H.J. Heinz SmithKline Beecham	Farley Health Products Cow & Gate Nutricia John Wyeth & Brother
Beverages – hot drinking (coffee, tea, ovaltine, horlicks, cocoa, etc)	Allied-Lyons Associated British Foods Cadbury Schweppes CWS Hillsdown Nestle Tomkins (RHM) Unilever	James Finlay Kraft General Foods SmithKline Beecham	Gold Crown Group Jacksons Foods Paulig UK Ringtons Wander
Biscuits	Allied-Lyons Associated British Foods CWS Nestlé Northern Foods United Biscuits	BSN Thomas Tunnock Weetabix	Furniss Group J.J. Lees Northumbrian Fine Foods Robertson & Baxter Speciality Foods Thomas Tunnock Walkers Shortbread
Bread, bakery mixes and products, including 'bake-off' frozen and chilled products	Allied-Lyons Associated British Foods Dalgety Northern Foods Tomkins (RHM)	Brake Bros Carr's Milling Ind. Greggs Wm Jackson & Sons William Price & Sons Warburtons Whitworths	D.S. Crawford Fletchers Bakeries Kavli Le Pain Croustillant MacPhie of Glenbervie John Millar & Sons Pilgrim Foods (croutons) J.F. Renshaw A.C. Skelton & Sons William Sword R. & K. Wise

Breakfast cereals	Nestlé CWS	Harrison & Crosfield Kellogg's Quaker Oats Weetabix Yorkshire Food Group	Cheshire Wholefoods Haldane Foods Group Jordan's Cereals Morning Foods Northumbrian Fine Foods	
Cakes, gateaux and desserts	Allied-Lyons Albert Fisher Grand Metropolitan Hazlewood Foods Northern Foods Tomkins (RHM) United Biscuits	CPC (United Kingdom) Kitchens of Sara Lee Pride Foods	Aulds (Foods) C & D Continental Patisserie Conrad & Essenhigh B.G. Foods Dairyfresh Desserts Elisabeth the Chef Giles Foods	Heritage Foods J.J. Lees Livera Foods Northumbrian Fine Foods Paradise Cuisine Strathmore Foods Sweet 'n Savoury R. & K. Wise
Coffee	Nestlé Tomkins (RHM) Unilever	Kraft General Foods	Gold Crown Group Paulig UK Ringtons	
Confectionery (sugar and chocolate)	Cadbury Schweppes Mars Nestlé	Kraft General Foods James Finlay Portfolio Foods Swizzels Matlow Thorntons Warner-Lambert	Bendicks (Mayfair) Blue Bird Confectionery Brandway Group (Beech's) Bristow's of Devon Duncans of Scotland Dunhills (Pontefract) Halo Foods Kinnerton (Confectionery) Leaf United Kingdom J.J. Lees Lofthouse of Fleetwood Lovell's Confectionery Meltis John Millar & Sons Needlers Northumbrian Fine Foods R. & W. Scott Taveners Walkers Nonsuch Wrigley & Co. Ltd	

Table 1 *Continued*

Food products	Large companies [2.1]	Medium-sized companies [2.2]	Small companies [2.3]
Dairy products	Booker MMB–Dairy Crest Nestlé Northern Foods Unigate	Avonmore Cavaghan & Gray Express Foods Golden Vale Kennerty Farm Kerrygold Kerry Group Kraft General Foods MD Foods MMBs (N. Ireland and Scottish) Raines Dairy Foods Waterford Foods	Aberdeen & District MMB Anchor Foods A.J. & R.G. Barber Castle Dairies Bridge Farm Dairies Central Midlands Co-op T.W. Clothier Coombe Farm Foods Golden Vale Horlicks Farms & Dairies Ilchester Cheese Co. Ingle Smokehouse Kavli J.G. Quicke & Partners Millway Foods Muller Dairy Long Clawson Dairy Lord Rayleigh's Dairy Mackie's Aberdeen Dairy MacPhie of Glenverbie North of Scotland MMB North Downs Dairy Co. L.E. Pritchitt ULN (UK) Yeo Valley Farms
Delicatessen products and ready meals (see also meat, fish, poultry, ethnic and vegetarian products)	Booker CWS Hazlewood Foods Nestlé Northern Foods Unilever	Brake Bros Campbell's UK Cavaghan & Gray Geest H.J. Heinz Hygrade Foods	Devon Quality Foods Giles Foods Heritage Food Group International Cuisine Jugg Foods Kitchen Range Foods Paradise Cuisine Puritan Maid Strathmore Foods Sweet 'n Savoury Unique Cuisine Waldens Wiltshire Foods

		Kraft General Foods Lloyd Maunder Perkins Foods	Lion Foods	
Dried foods – including fruits, nuts, pulses, rice, etc.		Cargill JLI Group Whitworths Yorkshire Food Group	Edward Billington Brandway Group James Budgett Haldane Foods Jacksons Foods Phoenix Foods	
Ethnic foods	CWS Hazlewood Foods Nestlé Tomkins (RHM) Unigate	Cavaghan & Gray H.J. Heinz C. Shippam	Abel Eastern Foods Bart Spices Beacon Foods Daloon Foods Noon Foods Pataks (Spices)	S & A Foods Tillery Valley Foods Westler Foods
Fats and oils	Unilever	Acatos & Hutcheson Borthwicks	Matthew Foods	
Fish and fish products (including shellfish, surimi, etc.)	Allied-Lyons Booker Albert Fisher Hazlewood Foods Hillsdown Unilever	Cavaghan & Gray Perkins Foods C. Shippam	Abacus Frozen Foods Allan & Dey Ken Bell International Cox Seafoods Dawnfresh Seafoods Faroe Seafood Co. Heritage Foods Group Icelandic Freezing Plant Kibun (UK) Macfish Moray Seafoods Oak Lodge Salmon Pinegain Princes Ltd Rahbek-Food Sco-Fro Foods F. Smales & Son R. & R. Spinks	

Table 1 ***Continued***

Food products	Large companies [2.1]	Medium-sized companies [2.2]	Small companies [2.3]
Frozen foods (see also meat, fish and poultry products)	Booker Albert Fisher Grand Metropolitan Nestlé Unilever United Biscuits	Brake Bros Campbell's UK Dalepak William Jackson Kitchens of Sara Lee McCain Foods Perkins Foods Pride Foods	Abacus Frozen Foods Everest Foods Heritage Foods Allan & Dey Glendale Frozen Foods F. Smales & Son Tendafrost Frozen Foods Sweet 'n Savoury Frozen Foods Walden Wiltshire Foods
Fruit, vegetables and salads (fresh, frozen and canned)	Booker CWS Dalgety Albert Fisher Hazlewood Foods Hillsdown Nestlé Northern Foods Unilever	Fyffes Geest Perkins Foods Pride Foods	Belgravia Salads Del Monte Everest Foods Hitchen Foods M & W Mack Princes Ltd Top Hat Holdings Sco-Fro Shearway Solway Tendafrost Frozen Foods Tinsley
Herbs, spices and flavourings	Dalgety Hazlewood Foods Tomkins (RHM) Unilever	Borthwicks McCormick	Bart Spices Beacon Foods British Pepper & Spice Co. Lion Foods Pataks (Spices) Zest Foods

Ice cream	Associated British Foods Booker Grand Metropolitan Hazlewood Foods Mars Nestlé Unilever	Pride Foods	Berrydales De Roma Ice Cream Fredericks Dairies Granelli McDermott Mackie's Aberdeen Dairy J.G. Quicke & Partners New England Food Co. Sidoli Ice Cream Stella Products Thayers Ice Cream Treats Ice Cream Tudor Dairies	
Jams and preserves	Associated British Foods Hazlewood Foods Hillsdown Tomkins (RHM)		Baxter's of Speyside F. Duerr & Sons Elesenham Quality Foods Merrydown plc Rayner & Co Robertson & Baxter R. & W. Scott Wilkin & Sons	
Margarines (including low-fat, dairy-based and non-dairy spreads)	Associated British Foods MMB–Dairy Crest Unigate Unilever	Avonmore Golden Vale Kerry Kerrygold Kraft General Foods	Anchor Foods Matthews Foods	
Meat and meat products (includes bacon, cooked meats and value-added products)	Booker Grand Metropolitan Hazlewood Foods Hillsdown Holdings Nestlé Northern Foods Unigate Unilever Union International	Geo. Adams & Sons Anglo Beef Processors Avonmore Foods F.W. Baker Brake Bros B. Brooks Campbell's UK Cavaghan & Gray Dalepak Dalehead	Aberdeen Meat Marketing John C. Bain A.J. & R.G. Barber G.N. Barlow Beck Food Group Beni Foods Ltd Bennetts of Doncaster Edward Billington G.D. Bowes Sarah Brownridge	Kays Cooked Meats Lagan Meat Lark Valley Foods Laxgate Mr Lazenby's Mathers (Inverurie) E.C. Matheson McIntosh Donald McIntoch of Dyce

Table 1 ***Continued***

Food products	Large companies [2.1]	Medium-sized companies [2.2]	Small companies [2.3]	
		Grampian Food Group	Brown Brothers	McKey Food Services
		David A Hall	Manufacturing	McLaren Foods
		Hygrade Foods	Buchan Meat Producers	McQuillan Meats
		Wm. Jackson	Buitelaar (Kilmarnock)	Meadow Vale Foods
		Kerry Group	Butcher & Baker Foods	Henry Newbould
		Lloyd Maunder	Case & Sons	Nimmo Holdings
		NewMarket Foods	Cheale Meat	Oak Forest Foods
		Perkins Foods	Chitty Group	Paradise Cuisine
		Pigs Marketing Board	T.W. Clothier	Perimax Meat
		Samworth Brothers	Cranswick plc	Plumrose
		C. Shippam	Crook of Harleston	Plumtree Farms
		Sims Food Group	Jeffrey Davies & Davies	Pukka Pies
		Tulip International	DMJ Ltd	Puredrive
			Eryri Foods	Rea Valley
			Farmers Boy	Roach Foods
			Fribo Foods	Saxby Bros
			J.W. Galloway	Scotch Premier Meat
			F.A. Gill	Simpson Ready Foods
			Gillard Holdings	A.C. Skelton
			Glendale Frozen Foods	Walter Smith
			Global Group	A.G. Stewart
			Halal Meat Co.	St Merryn Meat
			H. Hargrave	Thornley & Son
			Peter Hollingsworth	Westlake Foods
			Hyman Foods	Westler Foods
			Inverurie Scotch Meat	

Milk, liquid	CWS Northern Foods MMB–Dairy Crest Unigate	Avonmore Foods Kennerty Farm Kerry Group Kerrygold MD Foods MMB for N. Ireland Waterford Foods Scottish MMB	Aberdeen & District MMB Castle Dairies Central Midlands Co-op Golden Vale Dairies Lancashire Dairies Long Clawson Mackie's Aberdeen Dairy North of Scotland MMB
Pasta and pizza	Associated British Foods Hazlewood Foods Mars Nestlé Tomkins (RHM) Unilever United Biscuits	Brake Bros Campbell's UK CPC (United Kingdom) Geest H.J. Heinz McCain Foods Perkins William Price & Sons Pride Foods	Canadian Pizza Gianova Foods Haldane Foods Pasta Reale Schwan's Europe
Pet foods	Dalgety Mars Nestlé	F.W. Baker Harrisons & Crosfield H.J. Heinz Quaker Oats	Armitage Bros Town & Country Petfoods
Pickles & Sauces (savoury, wet and dry, cook-in, etc.)	Associated British Foods CWS Dalgety Albert Fisher Hillsdown Hazlewood Foods Nestlé Reckitt & Colman Tomkins (RHM) Unilever	CPC United Kingdom BSN H.J. Heinz C. Shippam	Baxter's of Speyside Beacon Foods Elsenham Quality Foods Golden West Foods International Cuisine Lion Foods MacPhie of Glenverbie Merrydown Rayner & Co. Robertson & Baxter R. & W. Scott Wilkin & Sons Zest Foods

Table 1 *Continued*

Food products	Large companies [2.1]	Medium-sized companies [2.2]	Small companies [2.3]
Poultry and poultry products	Hillsdown Northern Foods	Cargill Faccenda Favour Parker Grampian Country Food Kerry Group Lloyd Maunder Marshall Food Group Bernard Matthews Moy Park Nickerson Group G.W. Padley Perkins Foods John Rannoch C. Shippam Webbs Country Foods	Attwell Farms Crown Chickens Dove Valley Food for Thought Lion Mark Holdings Parkham Foods Pershore Poultry Webbs Poultry & Meat Group
Sandwiches and sandwich fillings and toppings	Booker Hillsdown Northern Foods	G.W. Padley Geest Golden Vale	Dell Foods Emley Farm Foods Everest Foods Food for Thought Fresh-pak Eggs Solway Foods Surrey Hills Top Hat Holdings
Snacks	Dalgety United Biscuits	Bensons Crisps JLI Group Portfolio Foods Walkers Smiths Snack Foods	Highlander Snacks The Snack Factory Northumbrian Fine Foods

Soft drinks and fruit juices	Cadbury Schweppes CWS Grand Metropolitan Reckitt & Colman Unigate	Bass plc A.G. Barr Pride Foods SmithKline Beecham	Alpine Soft Drinks Ltd G. Barraclough Boxford (Suffolk) Cambrian Soft Drinks Carters Drinks Group Cooke Bros Crystal Drinks Hall & Woodhouse (Panda) Hey UK Merrydown plc J.N. Nichols (Vimto) Orchard House Princes Rubicon Products Benjamin Shaw & Son Silver Spring Wells Soft Drinks	
Soups (canned, dried and fresh chilled)	Nestlé Unilever	Avonmore Campbell's UK CPC United Kingdom Geest H.J. Heinz Harrison & Crosfield	Baxter's of Speyside Haldane Foods Group New Covent Garden Soup Co.	
Sugars, sugar products	Associated British Foods Tate & Lyle	Cargill Yorkshire Food Group	James Budgett Edward Billington	
Tea	Allied-Lyons Associated British Foods CWS Hillsdown Unilever	James Finlay	Gold Crown Group Jackson Foods Paulig UK Ringtons	
Vegetarian and healthy foods	Hazlewood Foods Nestlé Unilever	Campbell's UK Dalepak Pride Foods	Berrydales Cauldron Foods Fribo Foods Goodlife Foods Haldane Foods Group Kitchen Range Foods	Marlow Foods Meridian Foods Northumbrian Fine Foods Simpson Ready Foods Speciality Foods Tillery Valley Foods

SECTION 1.2 TOP 100 FOOD COMPANIES BY TURNOVER

This table ranks the UK's largest food manufacturers by turnover. It has been compiled from the companies listed in Section 2, but excludes companies registered in Eire.

Note 1. For a few of the companies listed a substantial proportion of turnover is generated from pharmaceuticals, toiletries or other non-food activities or from operations outside the UK. Further details are given in individual company entries in Section 2.

Note 2. The figures should be used as a general guide only. They are not all for the same financial year and some have been restated in accordance with FRS No. 3.

Table 2

Company	Turnover £m	Pre-tax profit (loss) £m	Employees	Year end
1. Unilever plc[1] & NV	24 700	2029	287 000	31/12/92
2. Grand Metropolitan plc	7913	925	102 405	30/9/92
3. Allied-Lyons plc	5266	505	71 713	6/3/93
4. SmithKline Beecham plc	5219	1115	53 700	31/12/92
5. Dalgety plc	4470	112.2	15 417	30/6/93
6. Bass plc	4450	508	84 095	30/9/92
7. Hillsdown Holdings plc	4373	154.1	44 196	31/12/92
8. Associated British Foods plc	3954	297	51 724	12/9/92
9. Booker plc	3395	91.0	22 734	26/12/92
10. Cadbury Schweppes plc	3372	332.7	36 579	2/1/93
11. Tate & Lyle plc	3366	189.5	17 004	26/9/92
12. Co-op. Wholesale Society Ltd	3253	(0.4)	36 400	9/1/93
13. United Biscuits (Holdings) plc	3118	162.3	38 698	2/1/93
14. MMB of England & Wales	2573	–	2396	31/3/93
15. Northern Foods plc	2026	153.2	30 219	31/3/93
16. Harrisons & Crosfield plc	1942	85	32 177	31/12/92
17. Unigate plc	1925	104.1	27 257	31/3/93
18. Reckitt & Colman plc	1904	181.7	20 800	2/1/93
19. Nestlé Holdings (UK) Ltd	1677	22.6	19 870	31/12/92
20. Ranks Hovis McDougall (Tomkins)	1560	92.6	28 551	5/9/92
21. Mars UK & GB (Food Manufacturers GB Company) Ltd	1430	4.59	6009	26/12/92
22. Cargill plc	1301	15.5	6020	31/5/92
23. Albert Fisher Group plc	1284	31.5	6564	31/8/93
24. Dairy Crest	1156	28.2	9287	31/3/93
25. Union International plc	1091	32.5	17 903	31/12/92
26. Hazlewood Foods plc	762	55.0	9749	31/3/93
27. Geest plc	625	20.4	6998	2/1/93
28. Kellogg Company of GB Ltd	579	90.7	2870	31/12/92
29. Christian Salvesen plc	528	75.9	11 098	31/3/93
30. H.J. Heinz Co. Ltd	486	45.6	3486	24/4/93
31. Perkins Foods plc	354	22	1980	31/12/92
32. Express Foods Ltd (11 months only)	290	7.48	1700	Oct. 1993

Table 2 ***Continued***

Company	Turnover £m	Pre-tax profit (loss) £m	Employees	Year end
33. BSN (UK) Ltd	281	19.0	3559	31/12/92
34. Brake Bros plc	281	16.0	2718	31/12/92
35. General Foods Ltd (see Kraft General Foods)	267	39.3	1546	14/12/91
36. Princes Ltd	260	5.66	579	31/12/92
37. CPC (United Kingdom) Ltd	251	9.55	1958	30/9/92
38. MMB for N. Ireland	248	–	814	31/3/93
39. Scottish MMB	226	–	401	31/3/93
40. Acatos & Hutcheson plc	222	10.2	895	27/9/93
41. MD Foods plc	218	(0.6)	2201	26/9/92
42. McCain Foods (GB) Ltd	211	22.6	1969	30/6/92
43. Kraft Foods Ltd (see Kraft General Foods)	206	25.4	758	14/12/92
44. Golden West Foods Ltd	202	3.10	485	2/1/93
45. Weetabix Ltd	184	24.0	2466	31/7/92
46. Anchor Foods Ltd	182	1.4	410	31/5/93
47. Scottish Pride	167	3.4	974	31/3/93
48. Quaker Oats Ltd	166	6.3	981	30/6/92
49. Pride Foods Ltd	165	6.4	1193	28/12/93
50. James Finlay plc	165	10.1	33 290	31/12/92
51. William Jackson & Son plc	151	0.89	3600	24/4/93
52. Warner Lambert (UK) Ltd	151	(2.3)	1967	30/11/91
53. Campbell's UK Ltd	150	(1.55)	2241	2/8/92
54. G.W. Padley (Holdings) Ltd	150	3.66	3334	1/8/92
55. Warburtons Ltd	149	11.6	5430	30/9/92
56. Marshall Food Group Ltd	147	2.55	3764	30/9/92
57. Bernard Matthews plc	144	3.3	2478	3/1/93
58. Grampian Country Food Group	140	(0.42)	1734	31/5/93
59. Kerrygold Co.	128	(0.13)	405	28/12/91
60. F.W. Baker Ltd	122	2.8	859	31/7/92
61. NewMarket Foods Ltd	120	(1.44)	1865	14/3/92
62. Coombe Farm Foods Holdings Ltd	117	1.51	395	31/3/92
63. Favour Parker Ltd	116	(4.1)	2073	1/2/92
64. Faccenda Holdings Ltd	114	1.45	1077	30/4/92
65. Moy Park Ltd	113	2.42	2317	31/3/92
66. Cranswick plc	110	2.2	243	31/3/93
67. Whitworth's Holdings Ltd	109	(0.4)	1937	31/3/92
68. McCormick (UK) plc	105	4.5	664	31/10/92
69. JLI Group plc	105	5.2	1148	31/3/93
70. Edward Billington & Son Ltd	103	1.21	483	31/4/92
71. Tulip International UK Bacon Division Ltd	102	(5.84)	874	26/9/92
72. Greggs plc	101	7.1	6075	26/12/92
73. Dalehead Foods Holdings Ltd	97	1.05	817	16/5/92
74. A.G. Barr plc	95	6.0	1228	31/10/92
75. Thorntons plc	92	(4.8)	2775	26/6/93
76. Robertson & Baxter Ltd	91	27.35	472	31/12/92
77. Cow & Gate Nutricia Ltd	91	9.05	405	2/1/93

Table 2 ***Continued***

Company	Turnover £m	Pre-tax profit (loss) £m	Employees	Year end
78. Wrigley & Co. Ltd	89	12.93	622	31/12/92
79. Raines Dairy Foods Ltd	86	2.79	138	26/9/92
80. Portfolio Foods	80	(1.3)	1300*	31/3/93
81. H. Hargrave & Co. (Wholesale Pork) Ltd	79	(1.14)	564	25/9/92
82. Dromona Quality Foods	78	3.36	522	31/3/93
83. Müller Dairy (UK) Ltd	77	1.05	57	31/12/92
84. Webbs Country Foods Ltd	75	4.1	1477	30/9/92
85. Cavaghan & Gray Ltd	75	2.3	1734	28/3/92
86. Global Group plc	75	1.4	558	31/12/92
87. Del Monte Foods (UK) Ltd	74	4.19	182	29/11/91
88. B. Brooks (Norwich) Ltd	73	2.2	733	30/1/93
89. Carr's Milling Industries plc	73	1.12	805	28/8/93
90. Laxgate Ltd	72	0.75	231	28/2/93
91. Benjamin Shaw & Sons Ltd	66	(1.09)	383	31/1/93
92. McIntosh Donald Ltd	63	–	260	n/a
93. Yorkshire Food Group plc	63	2.7	722	31/12/92
94. Samworth Brothers Ltd	62	5.67	1372	31/12/92
95. Lord Rayleigh's Dairies Ltd	60	0.36	519	31/3/93
96. Lion Mark Holdings Ltd	60	3.51	392	31/12/91
97. Thames Valley Eggs Ltd	60	4.11	445	26/9/92
98. Lloyd Maunder Ltd	60	1.92	861	26/9/92
99. Hygrade Foods Ltd	59	0.71	951	28/3/92
100. Kitchen's of Sara Lee – UK Ltd	59	(4.98)	888	26/6/92

[1] Unilver plc: T/O £7517m; pre-tax profit £594m
* Estimated figure

SECTION 1.3 COMPANY CHANGES

Acquisitions, disposals, mergers and other changes, January 1991–December 1993

Figures in square parentheses denote the Section of the book where information about the company is given:
[2.1] Large companies >5000 employees
[2.2] Medium-sized companies 500–5000 employees
[2.3] Small companies 500 employees or less

Acatos & Hutcheson plc [2.2]

1991 Food packaging subsidiary P&W Plastics Ltd was sold.

Settlement reached on an outstanding loan from the joint venture in Spain with Macisa.

1992 In 1992, all the group's edible oils and fats businesses were merged into one company: Pura Food Products Ltd.

Allied-Lyons plc [2.1]

1991 Olympia & York sold its 9% stake in Allied-Lyons for £400m at an estimated profit of £32m.

Cantrell & Cochrane (part of Hiram-Walker) acquired a majority interest in Italian drinks company Barbieri SpA, based near Venice. Barbieri has a staff of 90 and a turnover of £16.3m.

Italian biscuit and confectionery subsidiary Sapori sold to Dutch company, Chocolate Groder.

1992 Lyons Maid ice cream business sold to Clarke Foods for £12m including unspecified debts. Clarke's had recently acquired the ice cream division of Hillsdown Holdings.

Initiated moves to merge UK brewing and wholesaling interests with Carlsberg A/S.

Sold Showerings, Vine Products and Whiteways and Warninck's businesses to a management buyout team for £140m. This is now the Gaymer Group.

Lyons-Tetley sold its Classic home baking range to the Brandway Group [2.3] (owner of the Supercook brand).

Acquired Modern Maid Food Products Inc. (USA), a manufacturer of prepared mixes and speciality coatings for £23m.

Acquired the majority of the Dawn Donuts chain of stores in the USA.

Acquired Ets Picard SA, a French bakery group which manufactures long-life toasts and biscottes.

Golden Dipt's (USA) retail range of sauces, seasonings and packaged coatings sold to McCormick & Co. Inc.

Bought Soutra Foods from Pauls Food Group.

Disposed of its investment in Coca-Cola Beverages Nederland BV.

Acquired the French distribution group Perrier-Jouet/Barton & Guestier Distribution.

Disposed of Harvey's brandy and distribution business in Spain to Pedro Domecq SA in exchange for cash and additional shares.

Established a joint fruit freezing and storage operation in Poland, intended to supply Margetts Foods Ltd.

1993 The Baskin-Robbins operation is involved in a joint venture with Kwality in India to open a store in Bombay. Kwality holds about 40% of the Indian ice cream market.

The Carlsberg-Tetley joint venture entered a complex distribution agreement with Eldridge Pope Brewery in Dorset.

Alma Holdings Ltd

1992 Called in the receivers KMPG Peat Marwick in February 1992 following trading losses running at £1.0m/month. The company had factories in Dundee, Glenrothes and Kirkaldy employing 775 people and held 7% in value of the UK sugar confectionery market. Subsequently, Portfolio Foods bought Barker & Dobson, Keiller's confectionery interests, Bensons and Milady for £3m. Keiller's preserves plant, equipment and brand names went to James Robertson in Manchester. Hacks and Victory V brands were acquired for over £3m by Ernest Jackson in Devon, part of Cadbury Schweppes.

Albrighton plc

1991 Management at Lovell's Confectionery [2.3] bought the company back from Albrighton for an undisclosed sum. Albrighton's interests were focused mainly on the construction industry.

Alpine Soft Drinks Ltd [2.3]

During 1991, Alpine Group plc purchased the Farringford Hotel, changed its name to Farringford plc and entered negotiations to dispose of its loss-making soft drinks business to International Media Communications. The deal was completed in January 1992. IMC is primarily concerned with in-flight entertainments and video recording.

Anglo Beef Processors Ltd (ABP) [2.2]

1991 ABP was set up in 1991 to handle the UK interests of Goodman International, the major Irish meat processor.

Asda

1993 Closed Lofthouse Foods, following several unprofitable years, and the failure to find a buyer for the business.

Associated British Foods plc [2.1]

1991 Acquired British Sugar plc from Berisford International in January 1991 for £880m.

1991 Sold plant and equipment for frozen gateaux manufacture from the Silbury Frozen Foods factory at Calne, Wiltshire to Sweet 'n Savoury Frozen Foods of Hartlepool, Cleveland [2.3].

Associated Fisheries plc

1992 In September 1992 Associated Fisheries sold its 50% interest in Macfish, the group's fish processing businesses, to Foodmark A/S of Denmark. This disposal brought to an end Associated Fisheries' investment in food processing, allowing it to concentrate on the import, export and sale of fish and fish products through its subsidiaries in the Billingsgate fish market, in the caviar trade and in surimi products. (See also Geest, who disposed of their 50% share at the same time.)

Associated Fresh Foods Ltd

Acquired by MD Foods of Denmark in December 1990. The group took on the name of MD Foods plc [2.2] in March 1992.

Avonmore Foods plc [2.2]

1991 Golden Foods Products acquired the cheese business of Caterpak Foods, Coventry for IR£0.35m. Caterpak supplies grated cheese for sandwich fillings to manufacturers and food service outlets.

Joint venture agreement with Belgian dairy co-operative called Kempico. Under an agreement mozzarella cheese will be produced, marketed and distributed on the Continent.

1992 Meat products division acquired certain assets and trade names including 'Murphy Evergreen' of John Murphy & Sons.

Through its subsidiary T.H. Goodwin & Son Ltd, the group purchased the remaining 25% interest in Goodwins Cheese Ltd, a profitable manufacturer of cheddar and mozzarella cheeses for £730 000 (sterling).

Under an agreement reached in April 1992, Avonmore took a 60% holding in Barretts and Baird, an independent, family owned meat processing company. The remaining 40% was acquired in March 1993. Barretts and Baird, with a turnover of about £110m made a loss of £800 000 in 1990, but returned a small profit (£180 000) in 1991.

Paid the receiver IR£5.5m for certain assets, including four factories, of United Meat Packers in Ireland.

Bought 49% of shares in the Master Pork Group (Ireland). Master Pork had pretax losses of IR£2.8m on a turnover of IR£16m in the year to December 1991. Avonmore has the option to purchase the other 51% on or after May 1995. The purchase increased Avonmore's share of Irish pig slaughtering to 33%.

Bought 100% of Harzland Fleisch-Service GmbH, Salzgitter in Germany for IR£6.5m.

Avonmore Dairies bought Whitecroft Dairies, Shropshire for £4.1m. This gave Avonmore its 16th milk depot in the UK and contributed 2.9 million gallons to total annual sales of milk. Bennetts Dairy Worcester, also joined the group.

Churchfields Dairy of Salisbury and Tom Parker Dairies of Fareham were bought for a total of £11.4m. These two dairies have always worked closely, hence the joint acquisition. Considerable investment was subsequently made in the Salisbury plant.

1993 In July 1993, Avonmore bought Midlands and Western Divisions of Dairy Crest's dairy business for IR£21.62m. Avonmore's UK liquid milk volume is now 85 million gallons, giving it a 7% share of the market.

Barretts & Baird (Holdings) Ltd

1992 Bought by Avonmore Foods plc [2.2].

Bart Spices Ltd [2.3]

1992 Bought by West Trust, which also owns Veeraswarmy's Ethnic Foods, for £2.5m.

Bass plc [2.2]

1993 In August 1993, Bass Taverns sold 44 of its pubs in London and Surrey to Greene King for £17.6m. Later the same month it also announced the sale of a further 46 pubs to Marston, Thompson & Evershed for £10.8m. The pubs sold to Marston are mainly in Bristol and the South West and in Yorkshire and Lincolnshire. These deals left Bass with 4300 pubs. Under the MMC's Beer Orders it is entitled to

4500 pubs and during 1993 it invested £34m in opening 22 pubs and developing a further 16.

Bass also sold its Augustus Barnett chain of off-licences to Allied-Lyons for £40m. Allied already owns Victoria Wine.

Beck Food Group Ltd [2.3]

1993 The Beck Food Group formed from the management buyout of Frans Buitelaar, bought Hillsdown's Perth Abattoir, renaming it the Perth Fresh Meat Company.

Belgravia Salads Ltd [2.3]

1991 Acquired Yours Tastefully, a distributor of chilled and frozen foods to the catering trade.

Bensons Crisps plc [2.2]

1991 Bought Dominator Ltd, Newbury (trading as A.K. Snacks) for an undisclosed sum. Also acquired was Dominator's subsidiary Five Star Snack Foods Ltd. This formed the basis of a new Retail Van Sales Division. The purchase of Gerald Harvey Ltd in 1992 futher expanded this division.

1992 Sudbury Snack, Wessex Wholesale, Golden Glow Nuts and Wilcox – small snack food companies – were also acquired during 1992.

Berisford International plc [2.2]

1991 Scotia Haven Foods sold to the Yorkshire Food Group [2.2] for £11.1m.

H.E. Daniel sold to CPL Group of Bishop's Stortford. Both companies make fragrances and flavourings.

Merged its sugar trading interests with the activities of Czarnikow Holdings into a new independently financed venture C. Czarnikow Sugars Ltd, in which Berisford retains a 15% interest.

Cocoa trading interests were sold to Bencroft Ltd, a company controlled by six managers of the cocoa business, in a deal which leaves Berisford with a 49% holding.

1992 The Group's financial obligations to Rayner Coffee International (RCI) were restructured during 1992 which freed Berisford from a considerable drain on its resources. RCI has now ceased to trade although certain assets of RCI were retained.

British Sugar plc sold to Associated British Foods plc for £880m.

Treehouse Farms Inc., a California-based almond processor was sold to a US subsidiary of the Yorkshire Food Group.

Sold its 71% interest in Enzymatix Ltd to Dr C.T. Evans and his associates.

1993 Made a bid for the Clark's Shoes business, which was turned down by the shareholders.

Blue Bird Confectionery [2.3]

1991 Bought by the Jack Chia Group, based in Singapore. Jack Chia has interests in confectionery, pharmaceuticals, publishing and property.

Booker plc [2.1]

1991 Acquired the remaining 10.3% of Arbor Acres for £25m from the Rockefeller family. Arbor Acres is the world's largest chicken breeding business.

Acquired the whole issued share capital of Atlantic Lobster Inc. for £3.2m. Ibec Aquaculture, the Canadian salmon farming operation, performed poorly and units were closed or sold.

Bought Wm Bennett Frozen Foods Ltd, Sheffield for £2.2m and acquired the business and assets of Ross Young Ltd's wholesale catering depots from United Biscuits for £5.75m. Both now form part of Pullman Foods.

Scottish Seafoods (fish division) bought the Inlaks scampi processing plant at Kilkeel, Northern Ireland.

Initiated the expansion of the distribution business in Continental Europe by acquiring in instalments a 40% interest in Recheio Distribuicao SA (a subsidiary of Jeromono Martins Holdings) in Portugal.

Completed its withdrawal from retailing in the UK by selling Holland & Barrett and the Kingswood-GK Chemist's chain to Lloyd's Chemists for about £55m.

Continued to withdraw from operations which do not fit the strategic focus of the group by selling:

L. Noel & Sons Ltd (herbs and spices) to Dublin-based investment company Alen McGuire & Partners for £8.5m.

Dixon Pork & Bacon Company (abattoir, pork processing and bacon plant) to Dalehead Foods.

The Foodfinders Ltd (speciality food marketing and distribution).

P. Leiner Nutritional Products (60% interest), (US manufacturer of vitamins, health and beauty products).

Booker Nutritional Products Ltd (supplier and manufacturer of vitamins and dietary supplements) to Ferrosan for £11.4m.

L'Herbier de Provence SA (herbal skincare, toiletry products and nutritional supplements business based in France).

1992 Bluecrest Foods (fish division) acquired Findus's 11-acre site/six-factory site at Humberstone Road, Grimsby.

Whitworth's Fresh Produce was sold to a management team (PBE Investments) for £7.6m. The deal included Bravi which has chilled potato chip operations in Holland and Belgium.

Brewhurst Health Food Supplies was sold.

Buckingham Foods bought the fixed assets (book value £1.7m) and goodwill of Eatfresh sandwich and chilled snack business from Grand Metropolitan.

Borthwicks plc [2.2]

1991 Completed the sale of its EC-approved abattoir on a 10-acre site at Colne, Lancs, for £1.8m. The purchaser is a new company Neerock Ltd, trading as Woodhead Bros Meat Co.

Barnett & Foster acquired Cooke, Tweedale & Lindsay, a Manchester-based flavour business, strong in the confectionery sector, for £0.85m.

1992 Bought a Hong Kong flavour company from its parent, F & C International Inc. of the USA, for £265 000.

1993 In February 1993, its Consumer Products business (prepared meals) was sold to the Hygrade Foods Group for £275 000 in cash.

After year-end in March 1993, Borthwicks sold its remaining abattoir at Ringmer to

Anglo Dutch Meats (Charing) for £0.29m and the Burton Son & Saunders bakery business.

Barnett & Foster and Globe flavours were merged to form Borthwick Flavours.

Brake Bros plc [2.2]

1991 Spent a total of £3.4m on acquiring Midfish Ltd, W.H. Hooper and Sons (both specialist suppliers of fresh and frozen fish and seafoods), Everfresh Frozen Foods Ltd and London Larder Ltd. Everfresh has four Midlands depots supplying frozen foods and chilled meat products to caterers. London Larder concentrates on fine foods for restaurants and hotels.

1992 Acquired several French frozen food distributors to the catering trade: Anjou Surgelation SA at Angers; Cogeshalles SA in the Paris area; Picardie Volailles SA, St Quentin North of Paris; Dorcier SA, Montmoreau, South West France.

1993 Paid £10m in shares for Country Choice Foods which has a staff of 250 and operates from eight depots supplying frozen bakery products to bakeries, fish and chips shops and snack bars.

Brandway Group Ltd [2.3]

1992 Purchased the Classics home baking range from Allied-Lyons which offered synergy with its own Supercook range.

British Pepper and Spice

1993 The Australian Burns Philp Group bought British Pepper and Spice, together with the Dutch Euroma Business, from Albert Fisher for £25m. The herb and spice companies had been acquired by Fisher as part of Hunter Saphir.

B. Brooks (Norwich) Ltd [2.2]

1991 Bought St Ives Meats from Dalgety.

BSN (UK) Ltd [2.2]

1991 Made bid for Grand Metropolitan subsidiary Express Foods.

Acquired Amoy Industries of Hong Kong, a company jointly owned by Pillsbury and the Hong Kong Development Co.

1992 Increased its stake in Galbani, the Italian cheese company first to 60% in 1991, and further to 70% in January 1993.

1993 Acquired a 40% stake in B.E. International (Bombay Emporium) a London-based company which imports a range of Indian and Chinese foods.

Cadbury Schweppes plc [2.1]

1991 Ernest Jackson bought the Zubes brand of throat lozenges from Roberts Laboratories.

1992 Paid the receivers of Alma Holdings £3m for Hacks and Victory V brands.

Acquired Aguas Minerales, the leading mineral water business in Mexico for £184m.

Took a 70% stake in Piasten Schokoladen-fabrik Hofmann, a Bavarian confectionery business with a turnover of £43m.

1993 Investing £20m in a joint venture to build a factory near Beijing in China.

Paid £154m for an additional 20.2% stake in Dr Pepper/Seven-Up taking its share of the third-largest US soft drinks group to 25.9%. Cadbury's plans to take over Dr Pepper were subsequently thwarted when Dr Pepper adopted a 'poison pill' approach, a device designed to protect companies against hostile takeovers.

Made a bid of £217m for A & W Brands, the leading American root beer brand.

Campbell's UK Ltd [2.2]

1991 Sold Slaters (Port Talbot) Ltd, its chilled ethnic food business, to Crestwish, a new company formed to carry out the purchase. The new company was named Madigan Foods. (Madigan Foods was subsequently taken over by Roy's Quality Foods, part of Unigate [2.1].)

1992 Sold catering and bakery distribution business to Mill House Foods, a privately owned company. (Mill House subsequently merged with Country Choice Foods which became part of Brake Bros.)

As part of the restructuring programme, acquired the business and assets of Freshbake Foods from Campbell Foods plc.

1993 Bought the Fray Bentos brand from Brooke Bond Foods for an undisclosed sum.

Canadian Pizza [2.3]

1992 A UK management team bought out their Canadian partners for £15.5m. The company supplies fresh, frozen and chilled pizzas and pizza crusts to supermarkets, caterers and industrial customers. It had a turnover of £11m in 1991.

1993 In November 1993, the company was floated on the stock market with a view to raising extra cash to reduce debt and fund an expansion programme.

Carr's Milling Industries plc

1991 For nil consideration, Carr's acquired the remaining 40% issued share capital of Caltech Biotechnology Ltd.

1992 James Reive & Sons Ltd, closed in 1991, was subsequently sold to Solway Agriculture for £150 000 cash in 1992.

Acquired the remaining 25% of the issued share capital of Dickensian Coffee Houses Ltd.

1993 In February 1993, egg producing and packing activities of North Country Eggs were discontinued.

Cavaghan & Gray Ltd [2.2]

1991 Acquired Calthwaite Dairy Products for £1.

1992 Bought the Emile Tissot ready meals business from Hunter Saphir for £3.75m in September 1992.

Clarke Foods

1990 Clarke's bought the Fiesta and Horton ice cream businesses from Hillsdown.

1992 Paid Allied-Lyons approximately £13m for the Lyons Maid ice cream business. The deal was partly funded by an £8m rights issue. Following acquisition Clarke closed the Greenford site and invested heavily in new plant and equipment, particularly at the Liverpool ice lolly factory.

Following a poor summer and problems with commissioning new equipment, Clarke's went into receivership with £40m in debts in October 1992.

In November 1992, Nestlé bought the trade and assets of Clarke Foods (UK) from the receivers.

Cliffords Foods plc

1993 Bought by Unigate [2.1] in March 1993 for £50.4m.

Copella Fruit Juices

(see Boxford (Suffolk) Holdings [2.3])

1992 Copella Fruit Juices, which was bought by Taunton Cider in 1986, was sold back to the original owners (the Peake family) for an undisclosed sum.

1993 Copella acquired the Cawston Vale brand name from Robert Marshall.

Co-operative Retail Services Ltd

Disposed of its dairy businesses to MD Foods, leaving it with no food manufacturing operations.

Co-operative Wholesale Society Ltd [2.1]

1992 Societies transferred to the CWS were Greater Nottingham (for £380m), Cumbrian, South Midlands, Enfield and St Albans and Kilsyth. Transfer of the Greater Nottingham Co-op meant that CWS became the largest retail co-op in Europe.

1993 Liquid milk businesses in London and Cornwall were sold to Unigate, and in Northern Ireland to Leckpatrick.

A possible merger with Norco, to solve Norco's financial problems was turned down by CWS on the grounds that all its best assets had already been sold.

Corton Beach plc

1990 Called in the receivers in October 1990.

CPC United Kingdom Ltd [2.2]

1993 CPC International has bought 85% of the shares of German potato products manufacturer Pfanni-Werke Otto Eckart.

Cranswick plc [2.3]

(formerly Cranswick Mill Group plc)

1992 Acquired the Charter Pork Cuts abattoir of R.W. Charter & Sons, Yorkshire for £1.2m.

Paid £3m for Yorkshire Country Pork Ltd of Preston, which operates a meat processing and packaging plant with a weekly capacity of 7000 pigs. It has since been integrated with the group's other butcheries in Yorkshire and Humberside to form Cranswick Country Foods Ltd.

Bought F.T. Sutton & Son (Rossendale) Ltd for £1m. Sutton manufactures and wholesales cooked meats and fresh pork products and supports Cranswick's plan to broaden its activities from pig marketing and feed production into pork and pork products.

1993 Acquired George Buckton (1985) Ltd, a producer of pigeon corn and bird food for a maximum of £1m.

Crown Chicken Ltd [2.3]

1993 Formed following a £6m management buyout of part of Harvest Poultry from Hillsdown Holdings.

Dairy Crest Ltd

(see Milk Marketing Board [2.1])

Dalgety plc [2.1]

1991 Sold its US fresh produce business to an investment group including senior management for £16.5m.

Dalgety Meat Products at St Ives (bacon slicing business) sold to B. Brooks (Norwich) [2.2]

Sold Hammonds sauce business to Albert Fisher for £12.24m in December 1991.

1992 Acquired Sooner Snacks, based in Scunthorpe from US company Border for £43m. Before being owned by Border, Sooner was part of Rowntree.

Sold Deans Farm Eggs to a management team.

Sold Federal Bakeries to RHM for £22m in October 1992.

Sold Modern Maid Food Products, its US food ingredients business, for £23m.

Reduced its holding in Dalgety Farmers Ltd (Australia) from 41 to 16%.

1993 Bought W.J. Oldacre (animal feeds) and Mortons Foods (crumbs and coatings) from Unigate for a total of £24.7m. Mortons became part of Unigate following its acquisition of Cliffords Foods plc.

Bought Priday's of Gloucester for an undisclosed sum. Priday's (formerly Priday Metford) will retain its name and customer base. It has a modern flour mill which was refurbished in 1990.

Dalgety paid £42m in cash for Paragon Petcare, part of BP Nutrition. The deal should give Dalgety 24% of the petfood market.

Del Monte Foods International (DMFI) (Del Monte Foods UK [2.3])

1992 Formed from Del Monte Foods Europe (of which Del Monte Foods UK is a part) following a management and investor buyout from Merrill Lynch, who bought Del Monte Foods from the takeover specialists Kohl Kravis Roberts.

In October 1992, a South African-based group, Royal Foods and Anglo-American Corporation of South Africa took over DMFI for £360m. Further organizational changes followed.

Del Monte Produce

1992 Bought from the receivers of Polly Peck International by Grupo Cabal, a Mexican investment group, for approx £260m, in August 1992.

Ess-Food (UK) Group Ltd [2.2]

1991 Ess-Food UK sold its bacon processing, slicing and packing plant at Thetford to a major new Danish pigmeat processing operation, Tulip International A/S. The sale left Ess-Food UK as a major shareholder in the Danish Bacon Co. plc, whose activities are distribution and retailing.

Ess-Food UK merged its operations with those of Danish Viking Foods, closing the Viking depot in Aylesbury and centralizing buying and sales at Ess-Food's head office at Welwyn Garden City.

Express Foods Ltd [2.2]

1992 Formed as an independent company in November 1992, following a £96m management buyout of GrandMet's cheese and foodservice operations. Express has a turnover of about £270m and employs 1700 people. It currently has six cheese factories and two distribution depots. Its new brand name is The Cheese Company.

Albert Fisher Group plc [2.1]

1991 Acquired two seafood suppliers: Peter Vassallo, Newcastle-upon-Tyne and Rokerij La Couronne in Belgium.

Bought Fruco, a German importer, packer and distributor of fresh fruit and vegetables.

Acquired the Hammonds Sauces business from Dalgety for £12.24m. Hammonds operates from plants in Bradford and Barrowfield in Lancashire.

1992 Acquired 75% interest in Spanish fruit and vegetable canner, Conservas Fernandez SA which has a turnover of about £16m and is based in Murcia.

Bought Jacques Onona & Co. a specialist fruit and vegetable importer.

Acquired Branston Produce, a supplier of fresh vegetables based at Branston near Lincoln.

1993 Took over the Hunter Saphir Group for £29.3m.

Sold the business and assets of Holco BV and Holco Konserven Vertriebs GmbH, Dutch mushroom growers and processors, to Hillsdown for £11.7m.

Sold British Pepper and Spice and Euroma Holdings BV to the Australian Burns Philip Group for £25m.

Sold the bakery mix division of Evedale Foods to Dawn Foods (UK), but retained the major part of the Evedale Food business in sauces and dressings.

Acquired Reddi-Made Foods of Tampa, Florida in a cash deal worth £5.3m.

Entered the primary fish processing sector with the purchase of Peterhead-based Forman Fish.

The Unimerchants business, importing dried pasta and other Italian foods, was sold to a management buyout team for £2m.

The small frozen desserts business Guanaria was sold to Tudor Dairies for £2m.

Forte plc

1991 Changed its name from Trusthouse Forte plc to Forte plc.

1992 Sold its contract catering subsidiary, Gardner Merchant Ltd and its associated businesses Ring & Brymer (out-door caterer) and Catering Partnerships (in-store restaurants). The sale netted £257m for Forte largely in the form of goodwill.

Puritan Maid Ltd [2.3], the group's wholesaler, manufacturer and distributor of food products reduced its manufacturing operations during 1992. It sold its sausage and burger manufacturing division to the Sims Food Group in July 1992, and closed its ice cream, bakery products and gateaux division. This leaves it with a frozen entrée unit, making airline meals at Humphreys Road in Dunstable. Its primary function is now the purchasing and distribution of foods for the Forte Group.

Fyffes plc [2.2]

1991 Acquired James Lindsay for IR£11.9m.

1992 Negotiations with Saba Trading AB of Sweden resulted in a trading agreement with Saba to supply a major proportion of their banana requirements.

1993 Acquired BRDR Lembcke A/S for IR£51.7m. Lembcke is Denmark's largest importer of fresh produce and pot plants and is a subsidiary of Axel Johnson AB.

Geest plc [2.2]

1991 Bought B.J. Marshall, a fresh produce business specializing in supplying the catering industry.

Bought Kent Salad Holdings for an undisclosed sum. Kent Salads is a major pre-packer of salads and iceberg lettuce with a turnover of £8.5m.

1992 In September 1992, as part of its plan to dispose of non-core businesses, Geest sold its 50% holding in the assets, stock and trade of Macfish Ltd to Foodmark A/S of Denmark for £9.7m. The other 50% stockholder was Associated Fisheries plc, which also disposed of its holding to Foodmark.

Acquired a 60% interest in AVM Cargo NV, which on acquisition changed its name to Geest Belgium NV.

Sold two US companies: Manor Hill Food Corporation (purchased in 1989) and Mayfield Importing Company Inc. (which trades as Sidari's Italian Foods) for a total consideration of $5.7m.

1993 Acquired Shieldness Produce Ltd of Bo'ness near Edinburgh, to support and broaden the produce range and the leadership position of Caledonian Foods. Shieldness has a turnover of £7m and 150 employees. It occupies a leasehold site of 7.25 acres and a 70 000-sq.ft factory.

Gerber Foods

(see Pride Foods [2.2])

Giles Foods Ltd [2.3]

(formerly Farmer Giles Foods)

1991 The company was the subject of management buy-in, and subsequently changed its name to Giles Foods.

Goodman International

(see Anglo Beef Processors [2.2])

1991 Following collapse of Middle Eastern markets and allegations of illegal dealing in intervention beef, the Irish Courts appointed an examiner to investigate the affairs of Goodman International. The result was a debt restructure agreement which left Larry Goodman with a 40% holding and a consortium of banks with 60%.

Golden Vale plc [2.2]

1991 Purchased two doorstep milk delivery businesses in South Wales for a total of £3.5m: E. Ladbrooke & Sons (trading as Bridgend Creamery) and Collins Dairies of Llantrisant.

1993 In April 1993, Golden Vale took over Leckpatrick for £22m.

It acquired a 33% stake in Vejle Margainefabrik, a Danish margarine manufacturer.

Purchased Vonk BV, a Dutch processed cheese manufacturer for a maximum of

£8.4m. Vonk has a turnover of £45m and debts of £20m. The deal included Vonk's 61% share of a Polish cheesemaker, Vonkpol.

In November 1993, GV bought the Haslington Cheese Co. and Haslington Food Ingredients for £5.3m.

Acquired the share capital of Cheese Express and Vonk UK in Tewkesbury. Cheese Express sells grated cheese and acts as agent for Vonk BV.

Grampian Country Food Group [2.2]

1993 Acquired Cymru Country Chickens, Llangefni, North Wales from Hillsdown Holdings in October 1993. The deal included 22 breeding and rearing farms, a hatchery, a feed mill and an oven-ready frozen chicken operation.

Grand Metropolitan plc [2.1]

1991 Sold four flour mills (acquired with the Pillsbury takeover) to Cargill for about $100m.

Somerwest, the West Country and Wales-based pork and bacon products company was sold to Dalehead Foods for an undisclosed sum. Somerwest had a turnover of about £20m in 1990.

Pillsbury bought Aunt Nellie's of Wisconsin, a producer of glass-jar packed vegetables. The company has a turnover of about $67m.

1992 Sold the Express Food business in Ireland to Waterford Foods plc in a deal involving Virginia Milk Products, MacCormac Food Products, Express Foods Northern Ireland and 50% of Premier Tir Laighean.

In the year ending 31/3/92 GrandMet sold Express Dairy and Eden Vale to Northern Foods plc for £326m. Subsequent restructuring and rationalization by Northern Foods resulted in the closure of London head offices and Express Dairies at Morden and Cricklewood.

Sold its Express Foods cheese and foodservice interests to a management buyout team for £96m. Express has a turnover of about £270m and employs 1700 people. It currently has six cheese factories and two distribution depots.

GrandMet Food Service acquired the frozen products division of McGlynn Bakeries Inc., USA.

In Germany, Pillsbury GmbH acquired quality soup manufacturer Jurgen Langbein.

Brossard acquired the Italian baked goods manufacturer, Bistrial.

Acquired Kraft General Food's refrigerated dough business in Germany, which had sales of £12m in 1991. Knack & Back is the leading brand in Germany.

Sold Eatfresh UK to Buckingham Foods, Booker's sandwich-making subsidiary. Eatfresh, which produces sandwiches and chilled snacks, had sales of £11m in the year to 30/9/92.

IDV acquired a 33.8% stake in Buton for £19.7m. Buton owns the leading spirits brand in Italy, Vecchia Romagna. Subsequently Cinzano acquired the entire share capital of BIF for £36.8m. The result is that IDV now has an 85% shareholding in Buton, with the option to acquire the remaining 15%.

IDV exercised its option to dispose of its shareholding in Remy Cointreau.

Branded retailing and pubs division disposed of the Dominic Group of off-licences and Burger King's distribution operation.

1993 Sold the Oakland Fast Food burger bun and pizza dough business of Express Foods Group (International) Ltd for £2.2m to Northern Foods plc. Oakland has bakeries at Whitefield in Manchester and Crawley in Sussex and will provide synergy with Falconis/Kara Foods operations. Oakland also has an excellent distribution network.

Sold the Oakland Fast Food beefburger manufacturing operation to the Sims Food Group for £4m. Oakland's major customers are Burger King and Wimpy International. Sims inherits a new EC-approved plant in Hull and access to the European burger market.

Sold its chain of Chef & Brewer pubs to a rival brewer, Scottish & Newcastle, in a deal valued at £800m.

Pillsbury, GrandMet's North American operation has formed a joint venture with the South African company Foodcorp, involving its major subsidiary Table Top. The plan is to manufacture and sell food products in Southern Africa. Foodcorp is South Africa's third largest food company.

November 1993, IDV took a 21% share in Laurent-Perrier, an independent champagne house.

November 1993, Pillsbury in the USA, bought Roush Products, a baker with a turnover of about £34m ($50m).

Haldane Foods Ltd [2.3]

1991 Acquired Granose Foods, a vegetarian and health food company for an undisclosed sum.

Acquired Unisoy Milk 'n By Products of Stockport. Unisoy makes soya milk, yoghurt and ice cream.

1992 Acquired the vegetarian sausage manufacturer Buss Foods and the brands and goodwill of the Dietburger Co.

1993 Haldane sold its Saucemasters factory in Newport, Gwent to G. Costa.

David A. Hall Ltd [2.2]

1991 Acquired the McKellar Watt brand name and manufacturing rights for fresh meat products from Campbell's Freshbake.

H. Hargrave Ltd [2.3]

1991 Bought the Middlesbrough plant of Amberley Foods from Union International for an undisclosed sum. (Another Amberley Foods plant at at Tipton was acquired by Hazlewoods.)

Harrisons & Crosfield plc [2.2]

1992 Bought Unilever's BOCM Silcock Ltd's feeds business for £67m. New operation now called BOCM Pauls.

Sold Glentham International to McCormick (UK) plc for £15.5m. Glentham has been integrated with McCormick Stange's flavour business to form McCormick Glentham.

Pauls plc sold Soutra Foods (rusks and coatings) to Allied-Lyons. Soutra should complement Allied-Lyons' existing ingedients operation DCA Industries.

Haverhill Meat Products

(see NewMarket Foods Ltd [2.2])

Hazlewood Foods plc [2.1]

1991 Acquired a 50% holding in SA Plats Cuisine Hubsch for £3.1m.

Bought Sutherland Holdings plc, Sheffield for £35m which created a group with a turnover of about £650m. In 1990, Sutherlands made a loss of £0.5m but recovered in 1991, with a profit of £3m on sales of £58m.

Acquired the entire issued share capital of Loem-Wi BV for £0.2m.

Acquired the entire share capital of Hebro Gerechten BV for an initial payment of £1.4m, with a further £2.2m due in 1993.

Paid £2m for business and assets of Tipton-based Amberley Foods and George Hipkins (subsequently integrated into Baron Fresh Food Services).

1993 Bought Commerciale Choc'Ise SA and Pierre Iserentant SA for £0.8m in cash. Pierre Iserentant subsequently sold after year-end 31/3/93.

Bought West Country Frozen Foods (F.W. Farnsworth Ltd) at Plymouth from Northern Foods for £2.25m and integrated it into Hilliers Ltd.

Acquired the business and assets of Taborske Mrazirny in the Czech Republic.

Bought Waffle Technow BV for £1.3m.

Acquired the business and certain assets of Fri d'Or France SA.

Bought the remaining 24.25% of shares in its Portuguese subsidiary, Hectare Agro Industria SA, for £0.6m in cash.

Bought Welten Snacks BV for an initial consideration of £1.3m. Welten will be integrated into the Advang Group.

Sold Belgian chocolate decorations business Luijckx Beheer BV and its subsidiaries to Wessanen Nederland for £14.2m in cash.

H.J. Heinz Company Ltd [2.2]

1991 In August 1991 Heinz acquired Halls European Foods Ltd, a manufacturer and distributor of pizza products.

In October 1991, the company disposed of its interest in the Somycel group, inlcuding Darlington Mushroom Laboratories Ltd to Sylvan Foods Inc.

1992 Blue Prince Mushrooms was restructured with the closure of farms at Angmering and Bradford-on-Avon. Subsequently the frozen and processed product division was sold to a management buyout team. The new independent concern, Orchidwood Mushrooms, Beckley, East Sussex will continue to use the Blue Prince name under licence. Heinz retained the fresh mushroom operations.

Acquired the issued share capital of Sous Chef Ltd, of Deeside when the company got into financial difficulties. Sous Chef used novel technologies to produce foods for the catering industry.

1993 Custom Foods of Dundalk, Heinz's pizza-making subsidiary, has opened its new £15m factory and restructured its senior management.

Leading US chocolate manufacturer, Hershey Foods, takes over three Italian subsidiaries of Heinz. The companies concerned are Sperlari, Scaramelli and Dondi with a combined turnover of £60m.

Merger of Industrias De Alimentacao Ltda, Portugal with a fellow affiliate, resulted in H.J. Heinz Co. Ltd holding a 22.8% stake in the newly merged company.

Following a fall in net income of Heinz's American parent company, it has announced that £15m is to be spent on rationalizing and restructuring the UK operations.

Hillsdown Holdings plc [2.1]

1991 Sold petfood businesses, Sterling Wygate and John Morell to BP Nutrition.

Sold the edible oil division of Maple Leaf Foods (formerly Canada Packers) in which it had a 56% stake, to Can Amera Foods for C$105m.

HL Foods bought Foresight Canning Co. of Halstead, Essex. The company cans organically grown vegetables. It also bought ABF's canning and ready-meals business for £20.5m.

The Poupart Group (fresh fruit wholesalers) bought Cardiff-based wholesaler John L. Gilbart and the Pordage Group of companies in East Anglia.

1992 Bought J.P. Wood poultry business from Unigate for £37m. The deal included factories at Scunthorpe, Sandycroft (Clwyd) and Llangefni (Anglesey) plus feed mills at Flixborough and lay in Wrexham.

Jeco Conserven bought Holco, a Dutch mushroom grower and processor from Albert Fisher for £11.7m.

Tendercut Meat Ltd of Southampton was bought by the Towers Group. Tendercut produces a range of value-added meat products for Iceland Frozen Foods and other major retailers.

Associated Poultry sold Buxted Foods at Sutton Benger to Webbs Country Foods and Buxted Duckling to A.E. Button, part of Kerry Foods.

Formed a joint venture with Heinz, Magyar Foods, to take over a state-owned canning company near Budapest, Hungary called Kecskemeti Konzervgyar RT. This is Hillsdown's second venture in Hungary. Ross Breeders already have a division there.

Hillsdown combined all its fresh fruit and vegetable operations as Hillsdown Produce. The main companies represented in the group are Poupart, MBM Produce and Orchard World. The companies have retained their separate identities. Their combined turnover should exceed £140m.

1993 Bought part of the Beledia chilled foods group (Artal NorthV.) for £50m. It brought Nadler (Germany) Johma (Netherlands) and Kobenhavns Salat (Denmark) into the European operations.

Closed down Anglia Canners' Kings Lynn factory only 2 months after buying the company. Its operations will be transferred to Haywards Foods.

A proposal to merge Corporate Foods and Maple Leaf Foods was narrowly vetoed by minority shareholders. However Maple Leaf was successful in acquiring the remainder of McGavin Foods which has now been incorporated into Corporate Food's nationwide bakery network.

Acquisition of Vancouver Fancy Meats and Unox Meats establishes the group's position as leader in prepared meat operations in Canada.

North Devon Meat sold its Torrington abattoir to Goodman International. Goodman subsequently sold it on to St Merryn Meat.

FMC's abattoirs at Salisbury and Lowestoft were closed.

Lamberhurst and Basildon abattoirs sold to a new company Invicta Lamb Ltd, led by Dick Cawthorne, former head of FMC's North Devon Meat.

FMC's Bromsgrove abattoir was sold to a management team for £2.5m.

Northern Counties Meat Group (newly formed management buyout company) bought Hillsdown abattoirs in Sunderland, Bedale and Doncaster.

Aberdeen Meat Marketing [2.3] bought Inverurie and Edinburgh abattoirs.

Beck Food Group [2.3] bought the Perth abattoir and consumer-ready meat plant in Scotland, renaming it the Perth Fresh Meat Company.

Completed the sale of Cymru Chicken Ltd of Llangefni to Grampian Country Food Group [2.2] for an undisclosed sum.

Sold part of its Harvest Chicken operation to a management team lead by David Thacker for £6m. The new company is called Crown Chicken Ltd [2.3].

FMC sold its Plymouth Abattoir to Plymouth Quality Meat for an undisclosed sum.

Hillsdown's modern meat product plant at Pottington, Barnstaple has now joined its Harris Pork & Bacon Group as Harris (Barnstaple).

Premier Biscuits has gained its first manufacturing base on the Continent through Belgian company, Pirou Wafer. New wafer lines have been introduced under the Cadbury's brand.

Potato businesses of Mosspak and PMA were acquired and integrated into MBM.

Hughes Food Group plc

1991 Went into receivership with debts amounting to £18m. Only two of its 22 subsidiaries were insolvent. These were Eyemouth Cold Storage and Caledonian Fine Foods. Eight subsidiaries were put up for sale: S & A Foods; Peterhead Ice Co.; Idwal Fisher; Laurel Farm Chilled Foods; Polar Distribution Services; Eurocon Packaging; Peterhead Marine Electronics; Halas Pharmaceuticals.

S & A Foods was bought back from the receivers for £2.3m by the original owner Perween Warsi with funding arranged by 3i. Derby-based S & A employs about 100 people making ready-cooked Indian meals.

Laurel Farm, a Northern-based wholesaler and distributor, was bought from the receiver for an undisclosed sum. Laurel Farm provided distribution facilities for S & A Foods [2.3].

Hunter Saphir plc

(now part of Albert Fisher Group)

1991 Sold its Matthew Walker cake and Christmas pudding business to Northern Foods for £8.1m.

1992 Sold Butterkist, the group's popcorn business to Portfolio Foods for about £3m. It will be merged with Bard Bros. resulting in the loss of about 100 jobs. Butterkist was acquired by Hunter Saphir for £6.3m in 1987.

Sold its frozen ready meal business, Emile Tissot, to Cavaghan & Gray for £3.75m. It had previously sold the Torquay-based Emile Tissot Patisserie business in 1990 to a management team. The Torquay company trades as Dairyfresh Desserts Ltd.

1993 Bought by the Albert Fisher Group plc for £29.3m in January 1993. Nicholas Saphir, chairman of Hunter Saphir joined Albert Fisher as Chief Executive of its European Fresh Produce Group.

Hygrade Foods Ltd [2.2]

1992 Bought Ferndale Food Ltd from Union International for £0.9m.

1993 In February 1993, bought Borthwicks Consumer Products Ltd from Borthwicks for £275 000 in cash, who are currently implementing plans to withdraw entirely from the meat and meat product trade.

Hyman Foods Ltd [2.3]

1991 The company was bought by the giant American corporation Conagra Inc. Hyman makes a range of burgers, sausages and sandwich steaks under the Steak Express and Hickory Farm labels.

William Jackson & Sons plc

Sold a total of 19 Grandways Supermarkets to Kwik Save for £14.5m (seven in 1991 and a further 12 in 1992) and nine to the Argyll Group for £23m. This leaves Jacksons with about 30 of its smaller 'family' stores.

Jacobs Suchard Ltd

(see Kraft General Foods [2.2])

JLI Group plc [2.2]

1991 Acquired Langwood (Prepared Vegetables) from Heald's Foods for £0.87m. Langwood, which supplies prepared fresh vegetables to retailers and caterers, made a loss of £0.25m on sales of £4.8m in 1990.

Bought Playtime Foods Ltd, based in Leeds, a manufacturer of fresh and pre-packed popcorn sold mainly to cinemas, theme parks and other leisure outlets.

1992 Bought Golden Twin from Portfolio Foods for £2.1m. Golden Twin packs nuts for own labels.

It also acquired T.G. Smith (Fine Foods) Ltd for £0.8m. Together with Golden Twin and Playtime Foods it has formed the basis of Tee Gee Snacks Ltd.

The ingredients division acquired Cadec Ltd, the Birmingham-based nut supplier for £0.1m.

In September 1992, the entire issued share capital of Brookerpaks was bought for £2.4m, Brookerpaks is a leading importer and distributor of garlic and ginger.

1993 In October 1993, JLI bought Turban Foods from Field & Co. for £50 000 including intercompany debts of £1.7m. Turban processes and distributes nuts, dates and other dried fruits and has a turnover of £4.9m and operating profits of £200 000.

JLI paid up to £3m for Frozen Herbs Ltd of Suffolk.

Kellogg Company of Great Britain Ltd [2.2]

1993 Sold its packaging subsidiary, Cereal Packaging Ltd, to Low & Bonan plc for £33m.

Kerry Group plc [2.2]

1991 In May 1991, Meadow Meats, a subsidiary of Guinness Ireland with plants at Rathdowney and Waterford, was acquired by KMP (Exports) Ltd for a total of IR£10m. In 1990, Meadow had a turnover of £81m. All meat businesses were subsequently merged under the Meadow Meats identity.

Dairyland Products Inc., Minnesota a dried ingredients supplier to US food manufacturers, was acquired by Beatreme Foods Inc. for a total of £23.31m cash, paid in three stages. Its activities were subsequently merged with those of Beatreme.

During the second half of 1991, Tunney Meats plant at Monaghan was acquired.

In October 1991, Kerry Agribusiness acquired the Kantoher Co-operative.

Eastleigh Flavours Ltd in Southampton was acquired during 1991 and fully integrated into the Kerry Ingredients business.

1992 In July 1992, Buxted Duckling was acquired from Associated Poultry (part of Hillsdown Holdings). As part of this deal, Kerry Foods Catering (formerly McCallum Meats, Portsmouth) was disposed of.

In December 1992, the assets and business of Tuam Dairies in Ireland were acquired.

1993 In February 1993, Kerry acquired the flavours and seasoning business of Malcolm Food Specialities Inc., Ontario from Beatrice Foods, Canada. The purchase should assist expansion of the ingredients business in Canada.

Kraft General Foods Ltd [2.2]

1992 Kraft General Foods International gained control of Freia Marabou, Norway for £882m. An 18.6% stake in Freia, owned by Hershey Foods was sold to Kraft as part of the deal.

1993 Bought the Terry's Group from United Biscuits for £220m. Terry's offered excellent synergy with Kraft's Jacob Suchard operation. Terry's has factories at Bridgend and York employing about 2500 people.

Lion Mark Holdings Ltd [2.3]

1991 Bought Hereford Herbs, a supplier of fresh herbs.

McCormick (UK) plc [2.2]

1992 Bought Glentham International Ltd from Harrisons & Crosfield plc for £15.5m. The deal includes Glentham's UK plant in Northampton and its South African subsidiary.

McIntosh of Dyce [2.3]

1992 James McIntosh sold its McIntosh of Dyce operation (meat products) to the management. This leaves the parent company free to concentrate on its core slaughtering activities.

Marshall Food Group [2.2]

(formerly D.B. Marshall (Newbridge) Ltd)

1993 Bought Premier Poultry from Hillsdown Holdings plc. The deal included a processing unit at Sandycroft in Wales producing fresh, whole and portioned birds as well as broiler units, hatcheries and breeding farms.

Bernard Matthews plc [2.2]

1992 Acquired Savari Baromfiipari Rt., Hungary's leading poultry processing and marketing company.

1993 Unigate sold Turner Turkeys of Lincolnshire to Bernard Matthews for £18m in cash.

MD Foods plc [2.2]

1991 MD bought the former Scottish & Newcastle Brewery in Blackburn and refitted it to enable expansion of fruit juice production.

1992 Acquired the dairy interests of CRS for £65m gaining units in Barnsley, North London (Oakthorpe) and Bristol. The Barnsley operation was subsequently shut down and customers supplied from the Leeds operation.

MD extended its business in the North East of England with the purchase of Sunderland and District Creamery plc for £5.5m. Operations in Sunderland were subsequently closed down and customers supplied from the Newcastle dairy.

1993 Following the acquisition of Dairy Crest's Bamber Bridge dairy, MD closed its Accrington milk bottling plant, with the loss of 190 jobs, and transferred production to Bamber Bridge.

MD Foods UK, based at Redhill in Surrey, and responsible for marketing Lurpak butter and a range of Danish cheeses in the UK, completed its merger with MD Foods' sales and marketing activities based in Leeds.

Northern Foods bought the liquid milk business of MD Foods in the South West region for £7.2m. The deal included seven depots but not MD's Bristol Dairy which was closed.

Meadow Vale Foods Ltd [2.3]

1991 Meadow Vale acquired a controlling interest in Seafrost, a 9-year-old company with processing and freezing facilites in North Shields. Seafrost produces 20 different chicken and fish dishes.

Purchased a controlling interest in catering frozen food supplier, Arthur Barras, from Cocked Hat Food of Ripon, Yorkshire.

1992 Meadow Vale acquired two plants from Country Style Foods: a raw poultry processing plant at Mold in Wales and a cooked poultry meat plant in Wrexham.

Merrydown plc [2.3]

1992 Acquired Piermont from Taunton Cider.

Acquired Sorelle from Premier Teas.

Acquired PLJ and Schloer soft drinks brands from SmithKline Beecham in a deal valued at £8.5m.

1993 Merrydown moved from the unlisted securities market to a full stock market listing and changed its name from Merrydown Wine to Merrydown plc. It also combined Martlet Vinegars and West Country Honey Farms as one operating division: Martlet Natural Foods.

Milk Marketing Board [2.1]
(including Dairy Crest Ltd)

1992 Dairy Crest sold its 50% interest in Hoogwegt Food Ingredients BV to Hoogwegt Groep BV.

Dairy Crest closed creameries at Aston in Cheshire, Four Crosses in Powys and Torrington in Devon. It also closed bottling dairies at Kings Lynn and Luton and ceased cheesemaking at Maelor, Wrexham. Cheese pre-packing operations will be developed at the Maelor site.

Sold the creamery at Hawes in North Yorkshire to a management buyout team for £0.3m.

1993 Sold the Marshfield Dairy in South Wales to Unigate and the Bamber Bridge dairy business to MD Foods.

Major investments have been made in the sites at Crudgington in Shropshire, Cannington in Somerset and Davidstow in Cornwall.

Milk Marketing Board for Nothern Ireland [2.2]

1992 Sold its 50% holding in Ballyshrane Foods Ltd.

Moy Park Ltd [2.2]

1991 Formed a joint venture company, BMP SA, in France with Bourgoin SA to produce cooked, flavoured and enrobed poultry products for the European market. Moy Park's holding in the venture is 35%.

Nestlé UK Ltd [2.1]

1991 Findus closed its Grimsby coated fish factory, with the loss of 900 jobs. Frozen food production is now concentrated at the Longbenton factory, Newcastle-upon-Tyne.

1992 Sold Holgates to Peter Saunders, the Managing Director. The company employs about 75 people in a factory in Tywyn, mid-Wales, producing own-label confectionery. The company has now changed its name from Holgate Nutritional Products to Halo Foods.

Bought Manley Ratcliffe from Croda International. Manley's products will now be produced at the Rowntree Sun Pat site at Hadfield near Manchester.

Bought Clarke Foods (Lyons Maid ice cream) from the receivers, after Clarke's got into financial difficulties following a poor summer in 1991 and heavy investment in new plant and equipment. The deal was estimated at £30m.

A new company structure, Nestlé UK Ltd was established.

Following the Milk Marketing Board's announcement that it would be replaced by a voluntary co-operative, Milk Marque, Nestlé UK invited 8000 dairy farmers to sign contracts for milk supply from 1994.

Nestlé UK has invested £15m in a new, purpose-built factory for Polo Mints at the York site.

Nestlé SA took over Perrier.

1993 Nestlé Rowntree announced that it was seeking a buyer for its Glasgow site. Breakaway and Blue Ribband will now be made at Fawdon, Newcastle and Lion Bars will be made in France at the Dijon factory. If no buyer is found for the plant, then it will be phased out gradually and closed.

Newmarket Foods Ltd [2.2]

(formerly Haverhill Meat Products)

1990 The subsidiary J.S. (Market Drayton) Ltd (formerly Palethorpes Ltd) sold its business and assets to Northern Foods plc for £14m.

Nickerson Group Rothwell Ltd (The) [2.2]

1991 Following persistent losses and weak demand for ducks in the USA, the Group closed its US operation Concord Farms Inc.

Northern Foods plc [2.1]

1991 During the year ended 31/3/91 Northern Foods acquired:

The London Colney, St Albans contract distribution business from Hunter Saphir for £8.4m.

The share capital of Mollington Farmhouse Cheese Co. Ltd from Wild's Farm (Cheese Exports) Ltd for £1.3m.

The sausage, meat pie and pizza business from J.S. (Market Drayton) Ltd (formerly Palethorpes Ltd) for £14.0m.

The share capital of Square Investments Ltd, the holding company of Falconis Ltd, a manufacturer of speciality bread, from Mr P. Sacco for £21.4m.

The retail chilled cake business from Finnegans Ltd for £1.2m.

The share capital of Beck Smith Associates Ltd, an Irish distributor of canned goods from Mr J.B.D.E. Smith *et al.* for £1.5m.

The Bolton Cake Bakery ambient cake business from Warburtons (Soreen) Ltd for £4.4m. The bakery was subsequently renamed Park Cakes.

The Trafford Park meat pie business from Mattessons Wall's Ltd for £13.9m.

1992 In the year ending 31/3/92 Northern Foods bought the Express Dairy and Eden Vale from Grand Metropolitan for £326m. Subsequent restructuring and rationalization resulted in the closure of London head offices and Express Dairies at Morden and Cricklewood. Major investments were made at the Ruislip site to increase capacity and improve efficiency. Fresh dairy product manufacture was concentrated at factories in Shropshire and Cheshire. The Coldstream distribution business has been integrated into NFT.

Bodfari Foods, a supplier of milk to shops, bought for £20.7m, has been integrated into the Ashby Dairy operation.

Acquisition of Kara Foods and Grain d'Or from Hillsdown for £4.1m complemented the Falconis range of speciality bread products.

Further expansion of the speciality bakery interests of the group was seen in the purchase of a 24.9% holding in Fletchers Bakery, Sheffield and the acquisition of the Matthew Walker Christmas pudding business from Hunter Saphir for £8.1m.

Hildale Farms Ltd, a processor and distributor of milk and dairy products in W. Yorkshire was bought for £3.1m and integrated into the Northern Dairies business.

The dairy businesses of H.C. Baines and Thompson Dairies were acquired in April 1992 for £2.9m and £2.6m, respectively.

The West Country Frozen Foods factory at Plymouth was sold to Hazlewood Foods.

The Oakland burger bun and pizza dough business was bought from Express Foods Group (International) Ltd (GrandMet) for £2.2m. Oakland has bakeries at Whitefield in Manchester and Crawley in Sussex and will provide synergy with its Falconis and Kara Foods operations. Oakland also has an excellent distribution network.

1993 In June 1993, the Office of Fair Trading approved the purchase of part of the CWS liquid milk operation in Belfast, from Leckpatrick Dairies for £4.4m.

In July 1993, Northern Foods bought the West Country liquid milk business of MD Foods. The deal, worth about £7.2m, included seven depots but not MD's Bristol Dairy which was closed.

Pepsico Foods International

1992 Entered into a joint venture with General Mills Inc. called Snack Ventures Europe.

The Walkers and Smiths Crisps operations were fully integrated under the new name Walkers Smiths Snack Foods Ltd [2.2] at a new Head Office in Reading.

Perkins Foods plc [2.2]

1991 Bought G&K Tiefkuhlproduckte GmbH, a German distributor of frozen foods, from Hillsdown for about £1m.

Bought Anchor Seafoods Ltd, a supplier of frozen shellfish, based at Handcross in West Sussex for £5.1m. Its pre-tax profit for the year to April 1991 was £0.79m.

Paid about £16m for two companies: Cogel SA, Luxembourg, a manufacturer of frozen cannelloni and lasagne recipe dishes; Studleigh-Royd Ltd, a specialist in cooked chicken and turkey products, with factories in Halifax and Doncaster.

Sold poorly performing Sefton Meadow Seafoods to a management team.

1992 In March 1992, Group acquired Cedarbrim UK Ltd (trading as Rowleys Frozen Foods) for £0.6m in cash and shares.

Paid £0.4m for non-voting preferred ordinary shares in Brookfield Foods Ltd. Brookfield has a new Deeside factory and produces chilled beef and pork products. Perkins has an option to acquire the remainder of Brookfield's shares.

1993 Bought Cromer Crab Co. of Norfolk, in a share deal. The company offers synergy with Anchor Seafoods at Handcross in Sussex.

In December 1993, Perkins disposed of its Dutch canned mushroom activities.

Pigs Marketing Board (Northern Ireland) [2.2]

1992 In a complicated two-stage share 'swop', Uniport and Ulster Farmers Bacon Company (Newry) Ltd were merged to form an enlarged subsidiary, Bridgewater Country Foods Ltd.

Bridgewater Country Foods invested £250 000 (25% holding) in a new factory as part of joint venture with Ferne Foods Ltd to produce value-added egg and bacon products. Ferne Foods also produces breaded products.

Pinegain Ltd [2.3]

1992 Bought the fish products division of Wm Jackson & Sons which will trade as Junella Seafoods and service the wholesale sector.

Acquired Merson & Gerry which operates a 400-sq.m fish processing factory in Aberdeen.

1993 Bought Lewis Frozen Foods of Warrington.

Plumtree Farms Ltd [2.3]

1992 Acquired the assets of Chivers Farm Foods new cooked meats division. Ham products, frankfurters and other meat products will be sold to the catering trade.

Portfolio Foods plc [2.2]

Created from a £59m management buyout of the confectionery and snack food interests of Hazlewood Foods in 1990. The deal included Cravens of York, Sweetmate, Golden Twin, Red Mill Snacks (UK and Netherlands), Bard Bros, Dragees aus Weseke and Heidrich Confiserie in Germany.

1992 Acquired Barker & Dobson, Keiller's confectionery interests, Bensons and Milady confectionery businesses from the receivers of Alma Holdings for £3m.

Acquired Butterkist from Hunter Saphir for £2.9m. Prior to acquisition, Butterkist had been losing £700 000/year. Butterkist plant and equipment was transferred to Bard's operation in East London and the Dagenham factory closed.

Sold its Golden Twin own-label nut processing business to the JLI Group for £4m in April 1992.

Pride Foods Ltd [2.2]

1991 Gerber Foods acquired Prime Foods from the receiver. Gerber subsequently transferred its ready meals operation to Prime Food's purpose-built factory in Kettering, Northants. All the Prime Foods workforce were retained.

Quaker Oats Ltd [2.2]

1993 Sold Sutherlands Foods (turnover of about £12m), manufacturer of pastes and spreads to its competitor, C. Shippam Ltd, for an undisclosed sum. The sale is part of Quaker's new strategy of focusing on core petfood, breakfast cereal and health drinks businesses.

Ranks Hovis McDougall plc
(see also changes made by Tomkins plc)
Taken over by Tomkins plc in December 1992; subsequently de-listed from the stock exchange as a plc.
Prior to Tomkins [2.1] takeover:

1991 Acquired two Dutch flour milling companies – Meelfabriek A Kruyt BV and Moolach BV – and a Dutch bakery Vetrimo Beheer BV for a total of £19.1m.

Tenstar Products ceased to trade with the closure of its starch factory in December 1991.

1992 Also acquired a family-owned flour mill in Holland, Van der Zande Peelfabriek, for an undisclosed sum.

Bought the branded pasta business, James Marshall (Glasgow) Ltd, for £1.7m cash.

RHM bought Federal Bakeries from Dalgety for £28m. Federal Bakeries makes bread under the Family Loaf brand name at Avonmouth, Byfleet and High Wycombe.

John Rannoch Ltd [2.2]
1991 Withdrew from the egg processing business in January 1991.

Reckitt & Colman plc [2.1]
1991 Sold its Supersoft and Once brands to Schwartskopf which increases its brand share of the UK haircare market from 1.9 to 4.6%.

Sold its £6m shoe care business in the UK and the USA to Sara Lee Household and Personal Care.

1992 Disposed of its food interests in South Africa and its North American spice and seasonings businesses.

Samworth Brothers Ltd [2.2]
1992 Acquired Melton Mowbray pie-maker, Dickinson & Morris Ltd.

Scottish Milk Marketing Board
1991 Galloway Cheese Company based in Stranraer opened in October 1991. It is a joint venture between Scottish Pride, Unigate (St Ivel) and CWS. It is claimed to be the largest creamery in the UK providing 15 000 tonnes of cheddar/year as well as butter, whey, and skimmed milk. Scottish Pride hold a 60% stake.

1992 Scottish Farm acquired a 75% shareholding in Cuthbertson Foods Ltd.

Proposals that Scottish Pride should take over the fresh milk business of CWS were rejected by the MMC.

C. Shippam Ltd
1993 Acquired the Sutherlands Foods spreadables business from Quaker Oats Ltd for an undisclosed sum.

Sims Food Group plc [2.2]
1991 Acquired two small meat catering suppliers: O.G.F. (Butchers), operating in the London area and W. & W. Meat Supplies.

Sims' chicken abattoir at Gainsborough was closed.

1992 Bought the Hull-based burger manufacturing operation Oakland Fast Food (part of Express Foods (International) Ltd) from Grand Metropolitan for an initial £4.2m (up to a maximum of £10.25m). Oakland's major customers were Burger King and Wimpy International and their franchises, which give Sims a break into the European burger manufacturing business. Shouler, part of the Sims group, was already a supplier of meat to Oakland.

Bought Puritan Maid's sausage and burger manufacturing operation from Forte plc in July 1992 for £1.1m.

Canvin International bought the Pure Meat Co., a supplier of conservation grade meat mainly to ASDA, when it went into voluntary liquidation following the sudden death of the owner. The Pure Meat Co. had trading debts of £350 000, with Canvin International itself owed £42 500.

Acquired General Provisions and Buitelaar (Boston). Buitelaar (Boston) is a cow beef abattoir specializing in exports to France. The freehold plant and goodwill cost £1.25m.

1993 Difficult trading conditions led to the disposal of J. Redmond, Sims's food machinery business, and Country Feasts, a company which had never become profitable.

Sim's shares in the meat trading business T.S. & W. were sold.

Three Russell Hume Depots, supplying the catering trade were closed: Cambridge, Southampton and Motherwell.

The Bow abattoir in London was closed in June 1993.

SmithKline Beecham plc [2.2]

1992 Sold the rights to manufacture, distribute and sell PLJ and Schloer soft drinks brands to Merrydown plc in a deal valued at £8.5m.

Sutherland Holdings plc

1992 Acquired by Hazlewood Foods plc [2.1].

Sweet 'n Savoury Frozen Foods [2.3]

1993 Sweet 'n Savoury, based in Hartlepool, bought gateaux-making equipment from Allied Frozen Food's Silbury plant at Calne in Wiltshire.

Tate & Lyle plc [2.1]

1991 Paid £10m for a 30.3% stake in a Hungarian beet sugar factory, which claims to be the most modern in Hungary, and which supplies 15% of the nation's sugar requirements.

Acquired a controlling share in the Bundabery Sugar Co. of Australia.

1992 Acquired the outstanding minority in Alcantara, the leading Portuguese sugar refiner.

Thorntons plc [2.2]

1992 Following unsatisfactory results, the French business is to be restructured and about 25 shops sold.

Tomkins plc [2.1]

1992 Bought Ranks Hovis McDougall plc for £935m.

Following takeover Tomkins set aside £90m for the reorganization of RHM, cutting the workforce by 2000 within 6 months of taking over.

Flour bagging operations at RHM Foods, Nottingham were closed, as was the distribution depot at Chapelhall, Lanarkshire.

McDougall's Catering Foods at Boston and two of Chesswood's mushroom farms were closed.

British Bakeries plants at Cardiff, Exeter, Croydon and Dundee have been closed and capacity reduced at Wimbush in Birmingham.

In the US, three Red Wing operating companies were merged and surplus peanut butter manufacturing operations closed.

Top Hat Holdings Ltd [2.3]

1992 Took over Scotts Kitchen which produces hand-made cakes.

1993 Scotfruit bought the Rosemount fruit business of Howgate Foods of Dundee, leaving Howgate to focus on its dairy businesses.

Through Scotfresh Chilled Foods, Top Hat acquired the goodwill of Britannia Fresh Foods, a sandwich-maker based in Bathgate, Lothian. Production of Britannia's sandwiches will be transferred to Scotfresh's Cupar plant.

Trusthouse Forte plc

(see Forte plc)

Tulip International [2.2]

1990 In October 1990, Tulip International A/S, a new bacon processing group predicted to turn over £550m was formed from Tulip, Danepak, Jaka and Normeat. Head Office of the UK division will be at Danepak's Thetford, Norfolk factory. Jaka has two meat slicing plants in Liverpool and Normeat a plant in Wellingborough. Turnover of the UK group is likely to be £235m.

1992 In November 1992, Tulip International put in a bid for Plumrose (subsidiary of East Asiataic Co.). The bid was rejected as inadequate and Tulip withdrew from the bidding in January 1993.

Unigate plc [2.1]

1991 The loss-making J.P. Wood poultry business was sold to Hillsdown (Buxted) for a maximum of £36.7m, comprising an immediate payment of £18.5m, £8.2m over 3 years and up to £10m, depending on profits. The deal included the main factories at Scunthorpe, Sandycroft in Clwyd and Llangefni in Anglesey. J.P. Wood was acquired from Unilever in 1984. It processes 1.5 million birds a week. Plants at Great Longstone and Craven Arms had already been sold in 1989 and 1990.

Acquired Abbots Dairies in Kent.

St Ivel chilled products bought a frozen desserts and bakery facility at Paignton, Devon. The factory was previously owned by U.B. Ross Young's.

1992 The Arlington Motor Co. (Westbury) Ltd, Bullens Ltd and B & A Britton were also disposed of during the year ended 31/3/92.

Unigate Dairies closed its Eastbourne plant, and acquired Sussex Dairies. Sussex Dairies sells 160 000 pints/day and delivers to 60 000 doorsteps.

CWS milk businesses in London, the South East and Cornwall were purchased.

The small UK-based Giltspur International business was sold to P & O Exhibition services.

Sold its US Italian cheese businesses Frigo Food Inc. (which incorporated Frigo

Cheese and Gardenia Cheese) to Stella Foods (part of Artal NV, a Dutch company) for £45m.

Bought further shares in Nutricia at a cost of £18.1m, to bring the group's holding to 32.6%.

Sold Spronsen BV.

Acquired the J. Whittam business which operates a dedicated Sainsbury's distribution service in the London area.

1993 Subsequent to year-end on 31/3/93, the purchase of Cliffords Foods plc was agreed for a price of £50.4m.

Disposed of the Morton coatings and crumb business for £9.5m and the W.J. Oldacre feeds business for £20.9m, to Dalgety.

Sold Turner's Turkeys of Lincolnshire to Bernard Matthews for £18m in cash.

Unilever plc and NV [2.1]

1991 In 1991, a total of 27 businesses were acquired at a cost of £225m and 13 were sold with net proceeds of £286m (not all of these are noted below).

The group announced its intention to withdraw from its agribusiness activities with the exception of Plant Breeding International at Cambridge and its plantation interests.

The sale of 4P Packaging in Germany was completed.

Acquisitions in the food division included:
Benedicta in France (salad dressings); Rex Cheese in Brazil; Mora/Saltos, a snack foods business in Netherlands and Belgium; Fine Foods, a dry mix and bouillon business in Egypt; Molco, a leading frozen dough operation in Netherlands and Belgium.

Acquired a 49% interest in Jalostaja Oy from the Finnish confectionery, pharmaceuticals and food packaging company Huhtamaki. Jalostaja, which has a turnover of £80m, make ready meals, fish preserves and canned soup. (Full control of Jalostaja was subsequently achieved during 1992.)

Acquired another 50% of Margarine-fabrikken Alfa Solo in Denmark giving it total control. The business which makes dressings, mayonnaise and jam, has a turnover of about £45m.

A joint venture between Pepsico and Liptons Tea entered the ready-to-drink tea market in the USA with canned and bottled teas.

1992 In 1992, Unilever acquired 22 businesses for a total of £285m and sold 21, netting £203m (not all these are noted below).

Acquisitions included: edible oil businesses in Poland and the Czech Republic and the edible fats operations of Novenyolajipari es Mososzergyarto Vallalat in Hungary; Dos Alamos, a cheese manufacturer in Chile; Klondike ice cream business in the US and a number of other ice cream operations in Canada, Hungary and Thailand; Cica tomato products and canned vegetables in Brazil; Im. Van den Berg, a frozen food business in the Netherlands.

A joint venture with Komili in Turkey (edible oils) was set up.

Management at Birds Eye Wall's Treats Ice Cream factory in Leeds bought the factory for £22m.

Sold Marine Harvest to Marifarms of New Jersey for about £20m. The sale included operations in Scotland and Chile which had a combined turnover of £50m in 1991.

Marine Harvest accounts for 25% of farmed salmon produced in Scotland.

Sold BOCM Silcock and Paul & Vincent animal feeds businesses in the UK and Eire to Harrisons & Crosfield.

The group's 76% holding in AMI, an Italian maize company was sold, as well as its 60% stake in the Dutch grass breeding concern, Barenburg.

Quest International sold the DCL Yeast Company name and brands to the French yeast company, Lesaffre. Under the agreement, DCL baker's yeast will continue to be produced by Quest in Scotland, while Lesaffre will undertake sales and marketing for domestic and overseas customers.

1993 In January 1993 Unilever announced the merger of Mattessons Wall's and Van den Bergh & Jurgens to form Van den Bergh Foods.

Unilever expanded its baking product interests in Germany by acquiring Diamalt bakery goods from its joint owners.

Acquired Breyers, America's top selling ice cream brand for a price thought to be around £330m.

In November 1993, Unilever bought the edible oil business of Italian company Cirio-Bertorelli-De Rica. Bertorelli is a leading olive oil brand. The deal includes SME International, who are export agents for Bertorelli and Cirio.

Union International plc [2.1]

1992 Amberley Foods sold its fresh pork and bacon plant in Middlesbrough to H. Hargrave & Co. (wholesale pork). It sold the Tipton operation to Hazlewood Foods.

The group's chain of retail butchers trading under the Dewhurst, Alex Munro, West, Baxters, Matthews and Cobb facias were reduced from over 1000 to about 400 outlets generating about £20m. At least 1000 jobs were lost.

British Beef ceased trading at Scorrier, Fraserburgh and Market Harborough.

During 1992, Union International disposed of a number of non-core activities for a gross value of £80m. A number of unprofitable businesses in the UK, Europe and North America were closed.

Australian pastoral land and cattle herd were sold for A$44m, and Meatworks for A$38m.

A 30% share in Industria Envasadora Nacional C.A. (Venezuela) was sold for US$20m.

South African fruit and vegetable processing operations were sold for £12.6m.

1993 Following year-end 31/12/92, Union's investment in Ciabasa S.A., an Argentinian detergent manufacturer was sold for US$9.5m.

At an EGM in March 1993, approval was given for the disposal of Brazilian farming interests for US$45.5m and the purchase of Meatwork properties in Brazil for US$13.5m.

Sold Ferndale Foods to Hygrade Foods for £0.9m. Ferndale manufactures various value-added cooked meat and egg products including frozen ready-made omelettes.

United Biscuits (Holdings) PLC [2.1]

1991 The McVitie's Group acquired a majority holding (58%) in Oxford Biscuits, the largest biscuit maker in Scandinavia, whose annual sales are about £27m. It also bought an 84% stake in Gyori Keksz, a food manufacturer and processor in Hungary, and 49% of Fazer Biscuits in Finland.

KP Foods acquired a 49% share of the Portuguese snack food manufacturer Soc. Industrial de Aperitivos Ltda.

Keebler purchased the oriental food manufacturer, Chinese Food Processing Corp. (Golden Tiger).

Joint ventures with three snack food companies were agreed: Grupo Larios in Spain; Longa Vida in Portugal and Wardell Roberts in Eire.

The Terry's Group acquired a 74% share in Italian premium chocolate and confectionery manufacturer, Aura S.r.l.

Ross Young's sold Tudor Dairies to a buyout team led by Neil Smith for an undisclosed sum.

1992 KP Foods acquired Smith's Snackfood Company Ltd, the snack food division of Coca Cola Amatil in Australia for £195m. In 1991, Smith's had a turnover of £182.3m and a trading profit of £16.4m.

Keebler Company bought Bake-Line Products Inc. of Chicago USA for £47m in cash and assumed debt.

1993 KP Foods Group acquired Derwent Valley Food Group Ltd, manufacturer of adult snack foods, for £27m including assumed debt.

The Group announced the sale of the Terry's Group to Kraft General Foods International (part of Philip Morris) for £220m.

Warburtons Ltd [2.2]

1991 Disposed of 52% interest in its loss-making US associate Warburtons, Inc.

Made a profit of £2.9m on the sale of the Bolton Cake Bakery, part of Warburtons (Soreen) Ltd, to Northern Foods plc.

1993 Peter Hunts Bakery Foods began trading from a new £7m, 88 000-sq.ft factory at Kearsley.

Warner Lambert Confectionery Co. Ltd [2.2]

1992 Hall Brothers (Whitefield) was placed in voluntary liquidation in June 1992.

Waterford Foods plc [2.2]

1991 Bought United Co-Operative Dairies (UCD) for £37.5m. UCD joins Waterford's existing dairy subsidiary Heald's Dairy. Together the two concerns will have a combined turnover of £190m and process about 60 million gallons of milk per year.

1992 Bought the Express Food business in Ireland from Grand Metropolitan in a deal involving Virginia Milk Products, MacCormac Food Products, Express Foods Northern Ireland and 50% of Premier Tir Laighean. It also subsequently exercised its option to acquire the remaining 50% interest in Waterford Express Dairies Ltd (WED) from the Waterford Co-operative Society, giving it full control of both Premier and WED. The deal excluded Express's 80% interest in Carbery Milk Products, which is being acquired by Carbery Creameries.

The group made a loss of about £1.4m on the sale of its Eatwell meat pie business.

1993 Created Waterford Dairies in the UK, by combining Heald's, Rigby's and the United Co-Operative Dairies.

Webbs Country Foods Ltd [2.2]

1992 Bought a former Buxted poultry site at Sutton Benger near Chippenham in Wiltshire from Associated Poultry, part of Hillsdown Holdings.

Whitworth Holdings Ltd [2.2]

1992 Bought Quality Bakery Supplies, Whetstone, Leicester for £188 000 cash.

Added 16 new in-store bakery units to its retail division during 1992.

Acquired the master franchise for Delifrance, a French-style bakery and cafe. The first unit was opened at Peterborough, followed by two at Gatwick Airport.

Yorkshire Food Group plc [2.2]

1991 In March 1991, announced the purchase of Scotia Haven's dried fruits, nuts and pulses business from Berisford International for £12.8m.

1992 Bought American almond processor, Treehouse Farms, Inc. of Bakersfield, California from Berisford International for $11.6m in July 1992. This acquisition should increase the turnover of YFG's American operation from $16 to $30m.

1993 In July 1993, YFG bought an 80% share of the dried fruits business of the Amercian Del Monte Corporation for about £18m ($26.7m). The Del Monte Dried Fruits business is predicted to produce profits of $2.4m on sales of $58m.

Part Two

COMPANY INFORMATION

SECTION 2.1 LARGE COMPANIES

Companies with 5000 or more employees engaged in food manufacture:
Allied-Lyons plc
Associated British Foods plc
Booker plc
Cadbury Schweppes plc
Co-operative Wholesale Society Ltd
Dalgety plc
Albert Fisher Group plc
Grand Metropolitan plc
Hazlewood Foods plc
Hillsdown Holdings plc
Mars UK Ltd
Milk Marketing Board of England and Wales
Nestlé Holdings (UK) plc
Northern Foods plc
Reckitt & Colman plc
Tate & Lyle plc
Tomkins plc
Unigate plc
Unilever plc & NV
Union International plc
United Biscuits (Holdings) plc

ALLIED-LYONS PLC

Secretary and registered office

D.S. Mitchell, 24 Portland Place, London W1N 4BB; 071-323 9000

Directors

M.C.J. Jackaman (chairman Allied-Lyons), A.J. Hales (chief executive Allied-Lyons), R. Moss (vice-chairman Allied-Lyons and chairman Carlsberg-Tetley Brewing), D.W. Jarvis (chief executive Hiram Walker), P.F. Macfarlane (finance director Allied-Lyons), J.A.F. Trigg (chief executive Allied-Lyons Retailing), W.E. Mason*, M.J. Rivett-Carnac*, J.W. Robb*, N.J. Stapleton*
* non-executive

Allied-Lyons was formed in 1978 by the merger of J.J. Lyons & Co. and Allied Breweries. The group has interests in food, drink and hospitality worldwide, currently employing about 71 000 people across all five continents (July 1993). Under the difficult trading conditions prevailing during 1992, the company set four key strategic objectives: greater focus, international growth, increased market share, and greater operational efficiency. While pursuing these objectives, the beer sector also had to deal with the issues and legal requirements arising from the UK Government's Tied Estate Order. The Group now operates in four sectors:

Spirits and Wine (The Hiram Walker Group Ltd)

Hiram Walker has a range of well-known brands sold worldwide, including Ballantine's, Beefeater, Courvoisier, Teachers, Canadian Club and Kahlua. It has entered into a number of very successful joint ventures with Domecq (Spain and Mexico) and Suntory (Japan). In March 1993, Hiram Walker took equity interests in two major liquor companies in Venezuela (Carst) and Chile (Desa). It has also established a presence in Eastern Europe, including a whisky production unit in the Ukraine.

Retailing (Allied-Lyons Retailing Ltd)

This sector brings together 12 200 retail outlets worldwide including 4400 public houses, 1000 Victoria Wine Shops, 3500 Baskin-Robbins (USA ice-cream) parlours and 3300 Dunkin' Donuts Stores (USA, donuts and coffee). During the year 92/93, overall pub turnover was increased slightly with quality and 'concept' houses significantly out-performing traditional 'locals'. Victoria Wine had sales growth of 6% and gained both the Red and White Wine of the Year Award from Wine Magazine's 1992 International Wine Challenge. Two of the group's London pubs took first and second places in the Daily Mirror London Licensee of the Year award.

Brewing and wholesaling (Allied Breweries Ltd)

Six UK breweries and a Dutch brewery are combined with wholesaling and distribution operations. In December 1992 (following scrutiny by both the MCC and EC competition authorities), the merger took place of UK brewing interests with Carlsberg A/S of Denmark. Carlsberg-Tetley, now the third placed brewer in the UK, will focus attention on its three core brands – Carlsberg, Tetley and Castlemaine XXXX.

Food manufacturing (J. Lyons & Co. Ltd)

The food manufacturing sector has major interests in tea and coffee*, cakes and biscuits, ingredients, prepared mixes and coatings, with continued development of the seafood sector.

Development of the round Tetley teabag has helped the company gain market share around the world, maintaining brand leadership in the UK and Ireland, and extending its No. 1 position in Canada. Despite intense competition, overall bakery sales in Europe grew by 10%, with the new bakery in Madrid fully on-stream, and the acquisition of Picard in France, a producer of long-life toast and biscotte products. Acquisition of Soutra Foods from Pauls plc complements DCA's UK ingredients operations, and a joint venture with Interpegro Group in Poland (freezing and storage of fruits) will supply raw materials to Margetts Foods, the group's fruit ingredients company.

Interim results

Pre-tax profits for the 28 weeks ending 18/9/93 were up 15% to £285m (1992: £234m) on a turnover of £2.73bn (1992: £2.57bn).

***Recent developments**

Early in 1994 Allied-Lyons announced the sale of its instant coffee business to Kraft Jacob Suchard (formerly Kraft General Foods) and its roast and ground coffee business to Paulig, the Finnish food group.

Table 1 Principal subsidiaries and associates (7/6/93)

Spirits and Wine

Spirits and liqueurs

Allied Distillers Ltd (Scotland)
James Burrough Ltd (England)
Cockburn, Smithes & Cia Ltda (Portugal)
Courvoisier SA (France)
Harveys of Bristol Ltd (England)
Hiram Walker & Sons Ltd (Canada)
Hiram Walker Intl. Liqueurs (USA)
Maker's Mark Distillery (USA)

Wine

SCV de Chateau Latour (France) (sold in July 1993)
Calvet Group (France)
The Wine Alliance (USA)
Huadong Winery (China – 40% equity)

Retailing

Pubs

Alloa Pubs & Restaurants Ltd (Scotland)
Ansells Retail Ltd (England & Wales)
Ind Coope Retail Ltd (England)
Tetley Pub Company Ltd (England)

Franchising

Allied-Lyons Retailing (USA)
Allied-Lyons Retailing (International) (worldwide outside USA)

Off-licences

Victoria Wine Company Ltd (UK)

Brewing and wholesaling

Allied Breweries Overseas Trading Ltd (England)
Allied Breweries Nederland BV (Netherlands)

Food manufacturing

Bakery

Lyons Bakeries (UK) Ltd (UK)
Hooimeijer BV (Netherlands)

Table 1 ***Continued***

Hig Hubert Hagemann Internationale Gebackspezialitaten GmbH & Co. KG (Germany)
Panrico SA (Spain and Portugal)
Ets Picard SA (France)

Beverages
Lyons Irish Holdings plc (inc. in Ireland)
Lyons Tetley Ltd (UK)
Lyons Tetley International Ltd (inc. in England) (Europe and Australia)
Tetley Inc. (USA)

Ingredients
DCA Food Industries Inc. (USA)
Margetts Foods Ltd (UK)
Lyons Seafoods Ltd (UK)
Lyons Brooke Bond (Pvt) Ltd (Zimbabwe)
Lyons Brooke Bond (Zambia) Ltd (Zambia)

Table 2 Five-year review (6/3/93)

	1989	1990	1991	1992	1993
Turnover (£m)	4504	4731	5133	5360	5266
Pre-tax profit (£m)	474	525	396	439	505
Average nos employed	81 440	87 273	83 243	78 743	71 713

Table 3 Turnover (£m) – geographical area

	1991	1992	1993
United Kingdom	2994	3169	3263
Rest of Europe	753	815	737
USA and Canada	1167	1137	1169
Rest of world	219	239	97
Total	5133	5360	5266

Table 4 Divisional results

	Year	Turnover (£m)	Trading profit (£m)
Spirits and Wine	1992	1928	409
	1993	1957	421
Retailing	1992	1625	224
	1993	1604	217
Brewing and wholesaling	1992	1035	75
	1993	1103	67
Food manufacturing	1992	893	102
	1993	854	88

ASSOCIATED BRITISH FOODS PLC

Secretary and registered office

T.H.M. Shaw, Weston Centre, Bowater House, 68 Knightsbridge, London SW1X 7LR; 071-589 6363

Directors

Garry Weston (chairman), Harold W. Bailey, Trevor H.M. Shaw, Donald J. Tidey, W.G. Galen Weston*
* non-executive

Associated British Foods (ABF) commenced operation in the early 1930s as bakers and biscuit manufacturers and although the group has diversified into other areas – retailing, in particular – bread and cereal-based foods are still the group's most important products. The Weston family has always had a majority shareholding through Wittington Investments which holds 63% of the total issued share capital via George Weston Holdings, the immediate holding company. Group pre-tax profit for year ending 12/9/92 was £297m on a turnover of £3954m. (Comparable 1991 figures were £332m and £3510.) The fall in profits in 1992 is attributed to intense price cutting in a static wholesale bread market. The results include a profit of £139m and turnover of £699m for British Sugar acquired from Berisford in 1991.

Allied Bakeries Group

Principal trading names include Sunblest, Vitbe, Hi-Bran, Mighty White, Kingsmill, Baker's Oven and Speedibake. The wholesale bread market remained static with intense price competition reducing bakery margins. The need to cut costs led to further substantial plant and bakery rationalizations and a significant level of redundancies. Nevertheless, in the premium white bread market, Kingsmill brand showed strong growth and became the clear market leader. Retail bakeries continued to be converted to the Baker's Oven format and re-organised to meet the requirements of the Food Hygiene Regulations 1990. In total, 90 bakeries have been closed during this process with a further 38 scheduled for closure during 1993. Speedibake, which produces frozen and part-baked bakery products commissioned a third site at Bradford.

Cereal Industries Group

This incorporates milling, agricultural and grain trading divisions. Pressure on margins has been partly compensated by increased sales of specialist flours to food manufacturers. Westmill Foods (Tolly Boy brand) has newly commissioned an automated packing plant at Ely and a rice mill at Selby. The launch of Sunblest Muesli and retail flour have been particularly successful.

British Sugar plc

Despite severe weather in December 1991, and aggressive pricing in the UK sugar market, British Sugar showed profitable growth during the year. Acquisition of KW Agriculture strengthened the company's position in the feeds market by giving it direct access to the on-farm market.

Biscuits and Ryvita

In a static biscuit market, sales of Burton's branded biscuits grew by 25%. Ryvita increased turnover, despite the mature market for this product. During 1993, the closure of the Burton's Gold Medal Biscuit factory at Blackpool was announced.

Teas and coffees

In a stable tea commodity market, Twinings have maintained gross margins by management of product mix, new products (herbal teas) and extended TV advertising. 80% of Twinings' markets remain overseas, which have not been assisted by upheavals in the major currency markets. The new factory in Newcastle should greatly improve productivity.

Frozen foods and preserves

Allied Frozen Foods has ice cream factories (all BS 5750 accredited) at Ashford (Kent), Ivybridge and Bude (Devon) and a pizza factory at Devizes, Wiltshire, which holds an EC Meat Licence. In October 1992, ABF sold plant and equipment for gateaux-making at its Silbury frozen food operation at Calne to Sweet 'n Savoury Frozen Foods of Hartlepool. The Calne plant has since been upgraded and converted to ice cream production. Capital expenditure at Nelsons of Aintree should reduce costs and improve margins in the production of bulk jams.

Yeasts, fats and food ingredients

Rowallan Creamery, which produces margarines and shortenings for retailers' own-label products, was hit by higher raw material prices and produced poor results. Mauri Products improved sales of bulk cream yeasts, with increased bakery sales offsetting poor sales to the whisky industry. AB Ingredients and AB Technology (producing tin release, divider oil emulsions and distilled monoglycerides), performed well following heavy capital investment in new plant and product development programmes.

Retailing

In the grocery trade in Northern Ireland, ABF operates through Stewart's, Crazy Prices, Westside Discount Stores, Superdeal and Bloomfields. In the Republic of Ireland Quinnsworth and Crazy stores maintained their market share despite price cutting by competitors. ABF's two textile companies Penneys (Ireland) and Primark (UK) both produced good results despite difficult trading conditions.

Research and development

Weston Research Laboratories at Maidenhead, Berkshire provides analytical and product development services to the group.

Australia and New Zealand

The economy here has been similar to the UK with little growth and reduced margins. Nevertheless, George Weston Foods Group, in which ABF has a 78% interest continued to maximize efficiency, broaden its product range and seek opportunities in South-East Asia. The Group's main activities centre on: milling, bakery and ingredients; meat and dairy products; biscuit, cakes and pastries; adhesives, starches and chemicals.

Results for year ending 18/9/93

Turnover was £4.39bn and pre-tax profit was £338m. British Sugar's profit rose 17% to £162m.

Table 1 Divisional organization: principal trading subsidiaries

Allied Foods Group	*Cereal Industries Group*
Allied Foods Ltd	Allied Grain Group
Allied Frozen Foods	Allied Mills Ltd
Nelsons of Aintree	Chancelot Mill Ltd
Rowallan Creamery	Cranfield Brothers Ltd
Silbury Frozen Foods Ltd	Fishers Agricultural Holdings Ltd

Table 1 ***Continued***

Cereal Industries Group continued	*Other Companies*
James Neill Ltd	AB Ingredients Ltd
Mardorf, Peach & Co. Ltd	ABR Foods Ltd
Westmill Foods Ltd	AB Technology
Westmill Packing	Burton's Gold Medal Biscuits Ltd
Westmill Rice	Crazy Prices (N. Ireland)
Yorkshire Country Foods Ltd	C.W.I.I. Ltd (Jersey)
Allied Bakeries Group	Eastbow Securities Ltd
Allied Bakeries Ltd	Germain's (UK) Ltd
(32 bakeries throughout UK)	Germain's (Ireland) Ltd
Twining Crosfield Group	Seed Systems Inc. (Delaware, USA)
R. Twining and Company Ltd	Lax and Shaw Ltd
R. Twining & Co. Ltd (USA)	Mauri Products Ltd (50% holding)
Foods International S.A. (France)	Power Supermarkets Ltd (Ireland)
Grosvenor Marketing Ltd (USA)	Primark Ltd (Ireland)
Jacksons of Piccadilly Ltd	Primark Stores Ltd
Namosa Ltd (Belfast)	Provincial Merchants Ltd
George Weston Foods Group	The Ryvita Company Ltd
George Weston Foods Ltd (Australia)	Stewarts Supermarkets Ltd (N. Ireland)
N.B. Love Industries Pty Ltd (Australia)	Serpentine Securities Ltd
Allied Food Co. Ltd (New Zealand)	Walmsley Ltd
British Sugar plc	Weston Research Laboratories Ltd
Trident Foods	Wilsdon & Company Ltd
KW Agriculture Ltd	Hicore Ltd

Table 2 Five-year review. (Year-end was changed in 1991 from April to September. Figures quoted are those in Annual Report for year ending 12/9/92)

	1988	1989	1990	1991	1992
Turnover (£m)	2272	2496	2775	3510	3954
Pre-tax profit (£m)	210	237	284	332	297
Average nos employed	53 877	53 024	52 345	53 975	51 724

Table 3 Analysis of turnover and profits by main geographical areas (figures for year-end September 1991 and 1992)

Geographical area	Turnover (£m)		Profit (£m)	
	1991	1992	1991	1992
Europe, mainly UK and Ireland	3068	3576	206	238
Australia and New Zealand	424	362	36	31
North America	18	16	1	2

BOOKER PLC

Secretary and registered office

R.M.J. Andrews, Portland House, Stag Place, London SW1E 5AY; 071-828 9850; Fax 071-630 8029

Directors

Sir Michael Caine* (chairman, retiring June 1993), J.F. Taylor (chief executive, appointed chairman June 1993), C. Bowen (chief executive from June 1993), J.A. Haynes*, B.J. Skipper, D.J. Turner, R.C. Rockefeller*, E.C. Robinson, R.N. Quartano*, F.G. Hankins*, J.D. Nelson, Sir Antony Acland*
* non-executive

Booker is an international food and agricultural group, providing links in the food chain from genetics to distribution. It operates in three main areas: food distribution; agribusiness; fish and prepared foods. Successful integration and rationalization of Fitch Lovell operations during 1991 means that food distribution now accounts for 56% of pre-tax profits for the group (1992 figure). Further focusing on core businesses has resulted in the disposal of health food and fresh produce businesses.

Food distribution

Booker's food distribution division is the UK's largest food wholesaling and catering food service business. Blue Cap Distribution operates primary temperature-controlled distribution. Pullman Foods supplies frozen foods to the catering markets. Huttons of Hull supplies Cunard and other leading shipowners. Booker Belmont Wholesale Ltd includes Booker Cash & Carry with 159 depots and Booker Wholesale Foods supplying mainly independent retailers. The wholesale foods business was badly hit by the Sunday opening of the major multiple retailers. Booker Fitch Food Services offers a delivered food service to caterers and retailers. In 1991, Booker began a phased investment in Jeronimo Martins Holdings, which trades as Arminho and Recheio in Portugal, and currently holds 40%.

Prepared foods

Companies in this division are leading suppliers of bacon, meat products and other value-added fresh and frozen convenience foods (Stocks Lovell, Lovell & Christmas, Salaisons Le Vexin, W.A. Turner, Bluecrest Convenience Foods), dairy products (Loseley), stoneground flour (Allinsons) and sandwiches (Buckingham Foods).

Fish

Companies in the fish division specialize in frozen, chilled and fresh fish, fish products and fish trading. Rossfish (now integrated with Bluecrest Chilled Fish) is the leading fish merchanting business in the UK, and Scottish Seafoods is the largest UK scampi business. Arctic Seafare and Polarfrost Seafoods offers worldwide fish procurement capability and acquisition of Findus's Humberside site (six factories, 11 acres) will provide the fish division with a base from which to develop its export business in continental Europe.

Agribusiness

Booker's agribusinesses occupy leading positions in poultry breeding (Arbor Acres Farm, CWT Farms, Nicholas Turkey), salmon farming and live lobsters (McConnell Salmon, Atlantic Lobster), horticultural plant breeding (Daehnfeldt), farm and forestry management (Booker Farming, Tilhill), and mushroom growing (Middlebrook). Overseas operations include sugar estate management (Booker Tate), design and supply of cane and beet sugar industry equipment (Fletcher Smith) amd specialist marketing services for sugar industries (Booker Overseas Trading).

Other activities

Booker Entertainment (Agatha Christie television series); fluid handling equipment (Plenty), food sorting equipment (Sortex); Booker Prizes for Fiction and Russian Novel; Booker Countryside sponsorship of the Silver Lapwing Award.

Interim results

For the 24 weeks ending 12/6/93 turnover was £1508m (1992, £1514m) and pre-tax profit was £29.9m (1992, £36.8). During the half-year, the loss-making Middlebrook Mushroom plant at Gateforth in Yorkshire was closed and McConnell Salmon has returned to profit.

Recent developments

Early in 1994 Booker announced the disposal of Loseley Dairy Products in two parts. The ice cream business at Alton in Hampshire has been acquired by Tudor Dairies; its chilled foods and premium yoghurts businesses based in Guildford have been sold to Tim Henniker-Parker, formerly with Bridge Farm Dairies.

Table 1 Principal trading subsidiaries

Food distribution
- Blue Cap Distribution Services Ltd
- Booker Belmont Wholesale Ltd
  - Booker Cash & Carry
  - Booker Wholesale Foods
- Booker Fitch Food Services Ltd
- Huttons International Ltd
- Pullman Foods Ltd
- Recheio Distribuicao SA (40%, Portugal)

Fish and prepared foods
- Allinson Ltd
- Bluecrest Chilled Foods Ltd
- Bluecrest Convenience Foods Ltd
- Bluecrest Foods Ltd
- Buckingham Foods Ltd
- Freebooter Seafoods Ltd
- Kilron Seafoods Ltd (Scotland)
- Loseley Dairy Products Ltd
- Lovell & Christmas (Canada) Inc. (Canada)
- Lovell & Christmas (Ulster) Ltd (N. Ireland)
- Polarfrost Seafoods Ltd
- Rossfish Ltd
- Scotprime Seafoods Ltd (51%)
- Scottish Seafoods Ltd (Scotland)
- Societe de Salaisons Le Vexin SA (France)
- Stocks Lovell Ltd
- W.A. Turner Ltd

Agribusiness
- Arbor Acres Farm Inc. (USA)
  - Arbor Acres Farm (USA)
  - Nicholas Turkey Breeding Farms (USA)
  - CWT Farms International Inc. (USA)
  - Atlantic Lobster Inc. (USA)
  - Arbor Acres Farm BV (Netherlands)
  - Arbor Acres Japan Co. Ltd (50%, Japan)
  - Arbor Acres Thailand Co. Ltd (50%, Thailand)
- Booker Farming Ltd
- Booker Overseas Trading Ltd
- Booker Tate Ltd (50% Booker, 50% Tate & Lyle)
- Fletcher Smith Ltd (65%)
- L. Daehnfeldt (Denmark)
- Middlebrook Mushrooms Ltd
- McConnell Salmon Ltd
- Tlhill Economic Forestry Ltd

Other activities
- Agatha Christie Ltd (64%)
- Glidrose Publications Ltd (51%)
- Plenty Ltd
- Sortex Ltd
- Sortex-Scancore Inc. (USA)

Table 2 Five-year review (26/12/92)

	1988	1989	1990	1991	1992
Turnover (£m)	1839	2511	2926	3289	3395
Pre-tax profit (£m)	78.7	90.5	102.9	103.9	91.0
Average nos employed	17 166	18 778	22 762	25 574	22 743

Table 3 Analysis of turnover and profits by geographical area

	Turnover (£m)		Pre-tax profit (£m)	
	1991	1992	1991	1992
Operational analysis				
Food distribution	2573.7	2766.9	57.2	46.9
Agribusiness	196.8	184.2	20.7	20.3
Fish and prepared foods	318.4	355.0	18.4	20.5
Other activities	43.6	44.2	(2.1)	(1.3)
Geographical analysis				
United Kingdom	3027.4	3040.3	71.4	57.8
North America	201.3	150.1	27.5	29.0
Other overseas	60.4	204.4	5.0	4.2

CADBURY SCHWEPPES PLC

Secretary and registered office

M.A.C. Clark, 25 Berkeley Square, London W1X 6HT; 071-409 1313

Directors

Sir Graham Day* (chairman), T.O. Hutchison* (deputy chairman), N.D. Cadbury (group chief executive), D. Jinks (finance), I.F.H. Davison*, R.P. Hornby*, R.C. Stradwick (human resources), F.J. Swan (managing director, beverages stream), Mrs A.M. Vinton*, G.H. Waddell*, D.G. Wellings (managing director, confectionery stream), D.R. Williams (managing director, Coca-Cola & Scwheppes Beverages Ltd)
* non-executive

Beverages stream

The stream operates on a global basis through Cadbury Schweppes Beverages and Coca-Cola & Schweppes Beverages Ltd (CCSB), the latter being 49% owned by the Coca-Cola Co. Its well-known brands include Gini, Schweppes, Sunkist, Kia-ora, TriNaranjus, Apollinaris, Canada Dry, Solo, Mott's, Crush and Penafiel.

While volumes generally remained static in Europe, profit in France rose 70% following acquisition of the Oasis brand and Gini bottling rights in July 1990. In Austria and Germany, Apollinaris and Schweppes performed strongly with volume and profits up. The basis of the business was broadened with the acquisition of Aguas Minerales, the leading mineral water business in Mexico for £184m. Mexico has the second largest soft drinks market in the world. In Spanish markets the recession deepened, beverage volumes declined and price competition was fierce. In 1993, the presence of Schweppes and Canada Dry in Hungary, Poland, Slovakia and the Czech republic will be strengthened as the result of a licensing agreement with Pepsi-Cola bottlers there.

Confectionery stream

As with beverages, Cadbury's brands are known worldwide and include Trebor, Bassett's, Pascall's, Sharps Toffees, Porlain, Crystal and Huesco.

Despite difficult economic conditions and slow market growth, trading profit in 1992 increased by 2.9%. Restructuring of Cadbury Ltd during 1991/92 and rationalization of its management has led to significant job losses over this period. Its position in continental Europe was enhanced by the acquisition of a 7% interest in Piasten, a leading German manufacturer of boxed chocolate assortments. Chocolat Porlain in France developed strongly in 1992, with a new factory at Blois fully operational. In the UK, the Group maintained its position as market leader in both chocolate and sugar confectionery with Cadbury Ltd and Trebor Bassett Ltd operating as separate manufacturing and marketing businesses. Cadbury's brand development policy which involves licensing its use in related products was extended in 1992 to other well-known names, including Birds Eye Walls, St Ivel, Premier Brands, Manor Bakeries and Hedges and Butler.

Recent developments

The Group has acquired a further 50% of the equity of its Egyptian associate, The International Confectionery Company, so that it has a controlling shareholding of 80%. A further 11% of shares in Cadbury India Ltd were purchased, and following an increase in its shareholding to almost 26%, talks are underway with America's third largest soft drinks company Dr Pepper/Seven-Up. It is also investing £20m in a joint venture in China to build a factory near Beijing.

Interim results

Sales for the 24 weeks to 19/6/93 were up 16% to £1709m (£1469.9 for 1992). Pre-tax profit rose 32% to £166.2m.

Table 1 Principal associated undertakings

Cadbury Nigeria Ltd	International Confectionery Company (Cadbury Egypt)
Cadbury India Ltd	Apollinaris & Schweppes GmbH & Co.
Schweppes (Central Africa) Ltd	Amalgamated Beverages Industries (Pty) Ltd
Crystal Candy Private Ltd	

Table 2 Principal subsidiary undertakings (export, franchise, finance and holding companies not included)

United Kingdom
Cadbury Ltd
Coca-Cola & Schweppes Beverages Ltd (51%)
ITnet Ltd
Sodastream Ltd
Reading Scientific Services Ltd
Trebor Bassett Ltd

Europe
Cadbury Ireland Ltd (Ireland)
Schweppes France SA (France)
Chocolat Poulain SA (France)
Chocolates Huesco SA (Spain)
Schweppes SA (Spain)
Citricos y Refrescantes SA (Spain)
Schweppes Belgium SA (Belgium)
BV de Faam (Netherlands)
Cadbury Schweppes Bebidas Portugal SA (Portugal)
Piasten Schokoladenfabrik Hofmann GmbH & CoKG (Germany)

Americas
Cadbury Beverages Inc. (USA)
Cadbury Beverages Canada Inc. (Canada)
Trebor Canada Inc. (Canada)
Cadbury Aguas Minerales, SA de CV (Mexico)

Other Overseas
Australia
  Cadbury Schweppes Australia Ltd
  Cadbury Schweppes Pty. Ltd
  Cottee's Ltd
  Beatrice Australia Ltd
New Zealand
  Cadbury Confectionery Ltd
South Africa
  Cadbury Schweppes (South Africa) Ltd
  Bromor Foods (Pty.) Ltd
Ghana
  Cadbury Ghana Ltd
Kenya
  Cadbury Kenya Ltd
Zambia
  Cadbury Schweppes (Zambia) Ltd
Malaysia
  Cadbury Confectionery Malaysia SB
  Trebor (Malaysia) SB
Indonesia
  P.T. Trebor Indonesia
Japan
  Trebor Japan Ltd

Table 3 Five-year review (2/1/93)

	1988	1989	1990	1991	1992
Turnover (£m)	2360.4	2776.7	3146.1	3232.3	3372.4
Pre-tax profit (£m)	282.3	258.7	276.6	314.7	332.7
Average nos employed	28 874	34 982	35 653	35 372	36 579

Table 4 Analysis of sales and operating profit by stream

	Sales (£m)		Operating profit (£m)	
	1991	1992	1991	1992
Confectionery				
United Kingdom	767.0	785.9	84.1	83.0
Europe	207.1	268.7	21.7	25.5
Americas	30.9	31.7	9.2	9.2
Pacific	284.7	273.2	43.3	45.2
Africa and others	97.6	109.7	13.3	12.9
Beverages				
United Kingdom	739.0	760.3	76.1	89.6
Europe	448.7	432.2	57.9	29.9
Americas	407.4	482.1	33.3	62.9
Pacific	206.5	175.1	20.6	10.7
Africa and others	43.4	54.4	11.6	15.7

CO-OPERATIVE WHOLESALE SOCIETY LTD

Secretary and registered office

G.J. Melmouth, New Century House, Manchester M60 4ES; 061-834 1212

Directors

G.L. Fyfe (chairman), A. Arlow (vice-chairman, retiring 31/3/93), D. Skinner (chief executive), W.A. Prescott (controller CWS finance and property), D.R. Robinson (controller CWS corporate affairs), J.B. Owen (controller CWS retail), C.W. Weir (controller CWS production), T.J. Thomas (managing director Co-operative Bank plc)

The Co-operative Wholesale Society, originally founded in 1863, still supports a significant proportion of the trading requirements of retail co-operative societies through its factories, farms, production units and distribution depots. It is also a major retailer in its own right in Scotland, Northern Ireland and parts of England, with 700 retail food outlets (450 Late Shop C-stores, 223 supermarkets and 36 superstores). There is also substantial involvement in banking, insurance, funeral direction, travel and property sectors. Over the period reviewed, a number of societies were transferred to CWS: Cumbrian, Greater Nottingham, South Midlands, Enfield and St Albans and Kilsyth.

Recessionary pressures, including intense competition from discount stores in the last 3 years have affected total turnover of CWS Production, which declined from £975.4m (1991) to £877.3m for 1992. A major cause of reduced turnover was the disposal by certain societies of their milk businesses which reduced the amount of raw milk sold by the Milk Group. Trading profits were hit by the lower margins earned by the Food Manufacturing Group. Nevertheless, it continued to widen its customer base so that 20% of the fast moving goods' turnover is now to major UK multiples, contract and overseas customers. Own-label sales both under the Co-op label and those of other major retailers are handled through F.E. Barber in Manchester. F.E Barber also manufactures under several other brand names, e.g. Shire and Barber.

Further investment was made in the Crewe Tea and Coffee Centre during 1991, with the successful launch of round teabags early in 1992 and a widened range of coffee products which boosted sales by 20%. Harlow biscuit factory suffered low demand, while the breakfast cereal centre on Deeside increased trade mainly through export orders. Edible oils are now sold to 50% non Co-op customers and further improvements to the Lowestoft factory have greatly extended the range of pasta, cook-in sauces and ready meals produced there. Soft drink sales suffered a decline.

In the Milk Group, the milk businesses of the Leicestershire and Greater Nottingham societies were fully assimilated, and proposals were agreed to sell the Scottish liquid milk business to Scottish Farm Dairy Foods, a subsidiary of the Scottish MMB (see section 2.2). These proposals were subsequently rejected by the MMC in December 1992. Agreements to sell liquid milk businesses in London and Cornwall to Unigate, and in Northern Ireland to Leckpatrick, were finalized in July 1993. CWS's manufacturing creameries suffered from further reductions in milk quotas, with butter and milk powder production severely restricted at Llangadog, but with Llandyrnog benefitting from a recovery in the cheese market.

During 1993/4, the CWS Milk Group will be merged with Associated Co-operative Creameries Ltd, its own wholly owned subsidiary. The enlarged group will have a turnover of £600m and trade as Associated Co-operative Creameries.

A possible merger with Norco to alleviate its financial difficulties was turned down by CWS in July 1993 on the grounds that Norco's best assets had already been sold.

Recent developments

In February 1994, the financial press reported that CWS planned to reduce its £300m of debt by selling its food manufacturing operations to Hobsons (a buyout team) for about £100m.

Table 1 Principal subsidiaries and associates

Principal subsidiaries
Co-operative Insurance Society Ltd
CIS Mortgage Maker Ltd
Neale House Holdings Ltd
Hornby Road Investments Ltd
CIS Unit Managers Ltd
CIS (Shudehill) Ltd

Co-operative Bank plc
Unity Trust Bank plc
Co-operative Bank Financial Advisers Ltd
Co-operative Investment Management Ltd

Trading
Belfast Co-operative Society Ltd (N. Ireland)
Capra Ltd
Glenhaven Funeral Services Ltd
Shield Packaging Ltd
Millgate Insurance Brokers Ltd
York Street Developments Ltd
Associated Co-operative Creameries Ltd
James Windsor & Son (Mansfield) Ltd

Associates
Industrial and Provident Societies
National Co-operative Chemists Ltd
Shoefayre Ltd
Kent Co-operative Society Ltd
Gilsland Spa Ltd
Companies
Galloway Cheese Co. Ltd (Scotland)
Co-operative Pensions Funds Unit Trust Managers Ltd
Whatco Holdings Ltd
Crownco Holdings Ltd
Bryco Developments Ltd
Waltham Cross Developments Ltd
Shudehill Developments Ltd

Table 2 Five-year review (9/1/93)

	1988	1989	1990	1991	1992
Turnover (£m)	2558.8	2678.0	3026.9	3141.3	3253.4
Surplus (deficit) before taxation (£m)	41.6	26.4	24.4	7.5	(0.4)
Employees	23 800	23 600	28 300	30 300	36 400

Table 3 Divisional analysis by turnover and trading profits

	Turnover (£m)		Trading profits (losses) (£m)	
	1991	1992	1991	1992
Retail and services				
Retail	593.4	946.5	8.7	13.6
Travel, funeral, optical	138.2	192.1	7.7	9.5
Services (food and non-food)	1875.7	1691.1	1.9	1.3
Production and property				
Food manufacturing	322.2	304.7	15.6	11.9
Milk	449.1	408.0	5.5	7.2
Agriculture	22.4	24.0	4.0	4.2
Engineering	171.4	136.8	(0.9)	(0.5)
Quality and consumer care	10.3	3.8	0.1	(0.1)
Finance and property	38.1	34.9	10.0	6.7
North-Eastern Region	365.4	396.0	5.6	4.8
Associated Co-operative Creameries	150.7	210.9	4.3	4.7
Unallocated items	–	–	(12.1)	(12.8)
Deduct inter-group trade	(995.6)	(1095.4)	–	–
Total	3141.3	3253.4	50.4	50.5

DALGETY PLC

Secretary and registered office

B.E. Gandy, 100 George Street, London W1H 5RH; 071-486 0200

Directors

M.E. Warren (chairman), R.J. Clothier (chief executive), R.N. Harris (commercial director), E.C. Humphreys, P. Kirk, J.R. Martyn (finance director), J. Rowell, P.G. Birch*, B.F. Baldock*, The Rt. Hon. Peter Walker*
* non-executive

Dalgety is a major food group with leading positions in snackfoods, petfoods, grocery products, food ingredients, food distribution and agribusinesses. It also carries out research and development at its Food Technology Centre in Cambridge.

Agribusiness

Dalgety Agriculture Ltd operates in the areas of animal feed manufacture, the supply of seed grain and pulse production, the distribution of fertilizers and other agrochemicals. The Pig Improvement Co. (UK and USA) supplies genetically improved pigs to businesses all over the world, specializing particularly in pigs which are free from porcine stress syndrome. In the USA, PIC has pioneered a scheme to use wetland plants to treat animal wastes.

In September 1993, Dalgety bought the W.J. Oldacre feeds business from Unigate for £20.9m.

Consumer foods

This division is divided into snackfoods, fresh produce and grocery product groups. It comprises Homepride, Dalgety Produce and Golden Wonder, which together generate 50% of the group's profits (compared with 20% in 1989). The acquisition of Sooner Snacks has lifted Golden Wonder's share by 20% giving it five of the top 20 best-selling UK snacks. In Holland, Golden Wonder's Nibb is the leading brand. Golden Wonder 'pots' are leaders in the instant hot snacks market and have recently been launched in Holland. Dalgety Produce improved its results despite extremely adverse trading conditions with oversupply and weakened consumer demand.

Homepride is brand leader in baking mixes, cooking sauces, and domestic flours. At the Maryport factory, £1.5m has been spent on the glass jar line for curry and French-style sauces. During 1992, flour packing operations were concentrated at Gainsborough and baking mixes at Thurcroft, Yorkshire. The Hammonds Sauce business was sold to Albert Fisher for £12.24m.

Petfoods

The petfood division, which now comprises Spillers Foods and Spillers Deutschland, weathered the recession well, with overseas sales now accounting for 20% of turnover. Kattomeat, the premium catfood, has been re-launched as Arthur's – the white cat used to promote it, and a new dry dog food – Complete Choice – introduced. In November 1993, Dalgety bought Paragon Petcare from BP Nutrition.

Food ingredients

Milling and baking suffered from over-supply and following a difficult year Federal Bakeries were sold to RHM for £22m. Spillers Milling concentrated on technological innovation, value-added products and speciality flours (e.g. Elephant chapatti flour) and introduced Cosmos, a new bread improver. A further contract to manage Lesotho Flour Mills in S. Africa was secured.

Lucas Ingredients introduced a textured gluten called Wheatpro and Masterspice, a range of herbs and spices processed by combining extrusion and pasteurization. Renovation of the Gainsborough plant offers Lucas opportunities to develop novel crumb products and extruded vegetable proteins; the acquisition of Morton Foods from Unigate in September 1993, for £9.5m, provided synergy with Lucas's crumbs and coatings. France Ingredients and ABC Paniermehl in Germany improved results by expanding into new markets and broadening product ranges. In the US, Modern Maid Food Products was sold to allow the group to focus its effort on UK and European ingredient markets. Spillers Premier Products increased sales of soy flour and exports of bread improvers.

Food distribution

Martin-Brower's largest customer, MacDonald's, benefitted from economic recovery in the US and Canada, and together with a new business through National Accounts with Chili's and TGI Friday's case deliveries went up by 9%. A new distribution centre in Atlanta Georgia, servicing 1100 restaurants was opened.

Table 1 Divisional organization: principal operating subsidiaries

Foods	*Agribusiness*
Dalgety Foods Ltd	Dalgety Agriculture Ltd
Homepride Foods	Pig Improvement Co. Ltd
Dalgety Produce	Pig Improvement Co. Inc. (USA)
Dalgety Spillers Foods Ltd	
Golden Wonder	*Food ingredients*
Spillers Foods	Lucas Ingredients Ltd
Food distribution	Spillers Milling Ltd
The Martin-Brower Co. Inc. (USA)	

Table 2 Five-year review (30/6/93)

	1989	1990	1991	1992	1993
Turnover (£m)	4757	4634	3659	3982	4470
Pre-tax profit (£m)	110.4	118.1	110.9	111.6	112.2
Employees	22 503	22 700	16 753	16 073	15 417

Table 3 Divisional analysis by turnover and operating profits

	Turnover (£m)		Operating profits (£m)	
	1992	1993	1992	1993
Divisional analysis				
Foods	589	594	59.9	65.7
Food ingredients	310	278	22.8	17.4
Food distribution	2119	2545	12.0	15.2
Agribusiness	1021	1110	28.1	32.1
Intercompany t/o	(57)	(57)	–	–
Total	3982	4470	122.8	130.4
Geographical analysis				
UK	1690	1693		
Other Europe	105	170		
USA	1942	2330		
Canada	245	277		
Total	3982	4470		

ALBERT FISHER GROUP PLC

Secretary and registered office

Mary Sweeney, 'C' Sefton Park, Bells Hill, Stoke Poges, Bucks SL2 4HS; (0753) 524380

Directors

S. Walls (chairman), A. Whittaker (deputy chairman), T. Howden (chief executive Europe), L. Pippin (chief executive North America), Ian Quinlan (finance), R. Edwards, R. Portergill, H. Ashton*, N. Freeman*, A. Wambold*
* non-executive

Over the past 10 years, Albert Fisher (AF) has grown from a loose collection of small businesses to an international food group with a turnover of £1.2bn. The group's interests are based in North America, the UK and continental Europe.

In North America the company is currently the third-largest purchaser of fruit and vegetables, which it distributes fresh and pre-cut to the food service industry. During 1992, prices and demand in the US remained weak and led to a fall in operating profits of £8.1m when compared with the 1991 figure. In July 1993 the group announced the acquisition of Reddi-Made Foods, Tampa, Florida for £5.3m.

Albert Fisher Europe operates through three divisions: fresh produce, food processing, and seafood.

Fresh produce

Operating profits fell from £15.2m in 1991 to £7.5m in 1992, brought about by weak markets and over-supply in some areas. Nevertheless, two companies were acquired during the year – Fruco Fruchthandel GmbH (Germany) and Jacques Onona (UK). The Stokes Bomford potato packing operation performed poorly and the Worcester plant was closed in July 1992.

In February 1993, Hunter Saphir was purchased for £29.3m, with Hunter's Chairman, Nicholas Saphir heading the European operation of the enlarged group. The takeover increased AF's European fruit and vegetable sales by £138m to a total of £1.4bn. Immediately following acquisition, British Pepper & Spice and Euroma Holding BV were sold to the Australian Burns Philp Group for £25m.

Food processing

Restructuring has left the division with five businesses: Fisher Quality Foods (canned goods, pickles, sauces and salad dressings), Fisher Frozen Foods (frozen vegetables), Mondi Foods BV (frozen fruits and concentrates), Broadheath Foods (chilled and frozen pastries and desserts) and Delby's NV (food distribution in Belgium). In the process, the bakery mix business of Evedale Foods was sold to Dawn Foods (UK), Hammonds Sauces were acquired from Dalgety and Unimerchants was sold to a management team. The division has factories at Stratford, Bradford, Glasgow, Littleborough, Falkirk and Barrowford.

In Holland, Uyttewaal (pickles) increased sales and in Spain, Albert Fisher Larios acquired a majority interest in Conservas Fernandez (canned fruit and vegetables). The European mushroom processor Holco was sold.

Seafoods

In September 1992 four companies were brought together to form the new division: Peter Vassallo (UK; chilled seafoods and modified atmosphere packaging of fish), Sea Products International (SPI) (UK; sourcing and supply of warm-water prawns to the food service industry), Roem van Yerseke (Holland; cockles and mussels processor), Morubel (Belgium; processor of warm-water prawns and smoked salmon). The companies, which work closely together, had a combined turnover of £115m and operating profits of £7.7m for 1992. The most recent development is the acquisition of the entire share capital of Forman Fish in Peterhead, a primary fish processor.

Results for 31/8/93

Turnover was £1284.2m and pre-tax profit was £31.5m. These compare with restated results for 1992 (FRS No. 3) of £1197.6m and £25.9m. Figures in the tables below contain the original figures from the 1992 Annual Report.

Table 1 Principal subsidiaries (wholly owned by Albert Fisher Group plc except where indicated*)

Registered in England
Broadheath Foods Ltd
Evedale Foods Ltd
Fisher Frozen Foods Ltd
Jacques Onona & Co. Ltd
S.P.I. plc.
Charles Sidney Holdings
Stratford-upon-Avon Foods Ltd
Peter Vassallo Ltd*
Wentworth Import & Export Ltd
A Wood & Sons (Detling) Ltd
Saphir Produce Ltd*
Portway International Ltd*

Registered in Scotland
Rowats Foods Ltd

Incorporated in Holland
Aartsenfruit Breda BV*
Citronas BV*
Mondi Foods BV*
Roem van Yerseke*
Conservenfabriek L.P. Uyttewaal BV*
Van Soest BV*

Registered in Belgium
Rokerij la Couronne*
Delby's NV*
Fruitasol bvba*
Norubel NV*

Registered in France
Agricommerce SA*

Registered in Netherlands Antilles
Albert Fisher Finance NV

Registered in Germany
Fruco Fruchthandel GmbH*
R. Horstmann GmbH*

Registered in Canada
I & S Produce Ltd*
Albert Fisher Canada Ltd*

Registered in USA
American Produce & Vegetable Co.*
Carnival Fruit Company*
G & G Produce Co. Inc.*
Lee Ray-Tarantino Co. Inc.*
Mitt Parker Company, Incorporated*
Movsovitz & Sons of Florida, Inc.*
Red's Market, Inc*
Royal Foods Company, Inc.*
Southern Produce, Inc.*
Sunny Sally Acquisition Corp.*
P. Tavilla Co. (Miami), Inc.*
Tavilla Marketing, Inc.*
Velvet Tomatoes Companies, Inc.*
Reddi-Made Foods Co. Inc.

Table 2 Five-year review

	1988	1989	1990	1991	1992
Turnover (£m)	536.1	830.6	1037.1	1095.5	1176
Pre-tax profit (£m)	33.4	44.9	74.4	89.0	52.1
Employees	3789	5100	6335	6629	6688

Table 3 Financial record by division and geographical area (figures not restated)

	Turnover (£m)		Profits (losses) (£m)	
	1991	1992	1991	1992
Division				
European Food and Distribution	724.4	834.5	53.9	43.6
N. American Food Services	371.1	341.5	18.9	10.8
Cash and Interest			14.2	7.7
Exceptional items			2.0	(10.0)
Total	1095.5	1176.0	89.0	52.1
Geographical analysis				
United Kingdom	322.1	369.8		
Germany	151.2	209.2		
Holland	118.1	112.4		
Other European	129.5	132.2		
USA	318.3	297.7		
Canada	56.3	54.7		

GRAND METROPOLITAN PLC

Secretary and registered office

Roger H. Myddleton, 20 St. James's Square, London SW1Y 4RR; 071-321 6000

Directors

Sir Allen Sheppard (chairman and chief executive), R. Giordano (deputy chairman), I. Martin (group managing director), D.P. Nash (group finance director), J.B. McGrath, G.J. Bull, D.E. Tagg, Sir Colin Marshall*, D.A.G. Simon*, Sir John Harvey-Jones*, Professor Dr. Gertrud Hohler*
* non-executive

Grand Metropolitan (GrandMet) is an international group specializing in branded consumer businesses in food, drinks, retailing and pubs. It currently employs 102 405 people worldwide, a figure down by 20 000 on the 1991 figure as a result of the group's divestment of non-core businesses and the refocusing of existing ones. In 1992 operating profits were made up of 56% from drinks, 20.6% from food and 23.4% from branded retailing and pubs.

Food

GrandMet Foods–Europe is based in Paris; Grandmet Foods–Americas, in Minneapolis USA. Well-known branded goods include Pillsbury, Haagen-Dazs, Green Giant, Peters Savoury Products, Brossard, Jus-rol, Knack & Back. Recent well-publicized disposals centred on the Express Food and Dairy businesses. Somerwest, part of the Peters Savoury Products pork and bacon business was sold to Dalehead Foods. In 1993, Peters acquired part of the Gillard operation (see section 2.3). Eatfresh sandwiches and chilled snacks were sold to Buckingham Foods (part of Booker plc). The Oakland Fast Food burger business was sold to the Sims Food Group, and the burger bun and pizza base part of the business to Northern Foods.

Drinks

International Distillers and Vintners (IDV) had excellent results in difficult trading conditions. IDV owns 11 of the world's top 100 spirit brands and its volume sales have increased through acquisition, brand growth and geographic expansion. Following acquisition of a further 75% share it has now added Cinzano to a list already including Gilbey's Gin, Malibu, Piat d'Or, Metaxa, Aqua Libra, Smirnoff, J & B whisky, Bailey's Irish Cream and many other leading brands.

Branded retailing and pubs

Burger King has more than 6600 owned or franchised restaurants and is the second-largest restaurant chain in the world. It has recently sold its distribution subsidiary and created Restaurant Services Inc., a non-profit making venture with franchises, which operates as the US system's sole purchasing agent and saves millions of dollars through economies of scale.

Chef & Brewer, with 1600 pubs and restaurants, was sold to rival Scottish and Newcastle Breweries in a deal valued at £800m in September 1993, leaving GrandMet with its 6580 Inntrepreneur pubs, jointly owned with Courage (UHC Fosters Brewing Group). The value of the properties owned by Inntrepreneur Estates has fallen in line with the overall fall in the UK property market.

Table 1 Principal group companies

Food
ALPO Petfoods, Inc. (USA)
Brossard Surgeles, SA (France)
Conservas Chistu, SA (Spain)
The Haagen-Dazs Company, Inc. (USA)
The Pillsbury Company (USA)
Pillsbury Canada Ltd (Canada)
Pillsbury GmbH (Germany)
Pilstral SA (France)

Drinks
AED SA (Spain)
RGA Bailey & Company (Ireland)
Carillon Importers Ltd (USA)
Cinzano SpA (Italy)
Croft & Ca Lda (Portugal)
Croft Jerez SA (Spain)
Gilbey Canada Inc. (Canada)
Heublein Inc. (USA)
International Distillers & Vintners (England)
Justerini & Brooks Ltd (England)
S & E & A Metaxa SA (Greece)
Sovedi France SA (France)
The Paddington Corporation

Branded Retailing & Pubs
Burger King Corporation (USA)
Grand Metropolitan Estates Ltd (England)
Inntrepreneur Estates Ltd (England; 50%)
The Chef & Brewer Group Ltd (England) (sold Sept. 93)
Pearle Inc. (USA)

Corporate
Grand Metropolitan Finance plc (England)

Table 2 Five-year review (30/9/92)

	1988	1989	1990	1991	1992
Turnover (£m)	6029	9298	9394	8748	7913
Pre-tax profit (£m)	906	1222	971	448	925
Employees	89 753	143 403	138 149	122 178	102 405

Table 3 Financial records by division and geographical area

	Turnover (£m)		Operating profits (£m)	
	1991	1992	1991	1992
Division				
Food	2618	2647	257	186
Drinks	2471	2858	451	505
Retailing	1560	1540	187	211
Discontinued operations	2099	868	132	47
Total	8748	7913	1027	949
Geographical area				
UK and Ireland	2940	1872	354	319
Continental Europe	862	1292	104	117
USA	4433	4184	504	445
Rest of North America	216	185	20	19
Africa and Middle East	145	155	10	7
Rest of world	152	225	35	42
Total	8748	7913	1027	949

HAZLEWOOD FOODS PLC

Secretary and registered office

K.M. Higginson, Rowditch, Derby DE1 1NB; (0332) 295295; Fax (0332) 292300

Directors

P.E. Barr (executive chairman), C.J. Ball (chief executive UK), A.P.P.M. van Gurp (chief executive Europe), J.M. Simons (finance), F.H. Lee*, G.F. Risley (sales and marketing), H. Hooykaas*, R.C. Lund*, R.D. Thomas*, D.C. Truran*
* non-executive

Hazlewood Foods plc, originally a small sauce manufacturer, has become an international food group through an acquisition policy which began at the end of the 1970s with the purchase of the onion grower, Polderman. It has a strong commitment to quality, innovation and technical excellence. The merger with Sutherland Holdings in 1991, brought eight UK fresh food factories to the group and additional turnover of £98m. In the UK 35% of group sales are of chilled, frozen, value-added convenience foods to the major supermarkets. In continental Europe it has a strong position in frozen snack foods and sees potential growth in own labels which are less developed than in the UK. It also has bakery, fresh produce, fish and shellfish interests.

Convenience foods Group

Products include pizza, quiche, ethnic foods, soups, airline meals, salads, sandwiches, pies and desserts. Investment in this part of the group has included a £3.5m extension to the Wrexham ready meals plant, a new £4m factory nearby and £1.7m worth of improvements to the St Martin's salads plant at Waltham Abbey. West Country Frozen Foods at Plymouth was bought from Northern Foods and its operations merged with Hilliers. Hazlewood Desserts now has an additional factory at Stroud, and Baron Fresh Food Services Ltd offers full UK distribution from 14 depots following the integration of Hipkins.

Ready meals Group

Products include snacks, fish products, ice cream and desserts. The group expanded into chilled meals while maintaining its core frozen food business. Sales increased significantly at Rowan Foods (ready meals). Pann Crisp, a dedicated chilled fish operation launched 12 new products during the year. Ice cream markets remained competitive and Creamery Fare operations were rationalized on to one site in London.

Produce Group

A glut in Europe in 1992 resulted in price deflation, however Van Heyningen Bros Ltd remained profitable (tomatoes, peppers, herbs and cress). The Portuguese subsidiary Hectare provided a quality beef tomato over the whole winter period.

Meats Group

Sales expansion in delicatessen and charcuterie were led by Roman Delicatessen, J. & J. Transfield and Hazlewood Trading (importer of meat products). Beef and pork butcheries in Scotland and Lincolnshire enjoyed continued success (Sandyford Foods, Prestwick and Gainsborough Pork & Bacon Co.).

Bottling Group

Hazelwood Preserves, Selby, is now the main centre for pickle and sauce production, the Derby factory closing in 1991. Product development has focused on Hazlewood's range of cook-in and stir-fry sauces. Sales of mineral water by Campsie Spring increased by 3%.

Bakery Group

Beaverlac (Christmas cake and puddings) and Michael Scott Cakes showed year-end growth despite the adverse effect of sterling devaluation on raw material prices.

Non-foods

Paper products, disposable nappies and feminine hygiene items are manufactured at sites in Bolton (F.H. Lee and Charles Turner) and Eire (Maharry Holdings). They made up 37% of group turnover and 42% of operating profits in 1993.

European Operations

These comprise ready meals, fish, shellfish, meats, produce and bakery. Representation in the Netherlands is particularly strong with major manufacturing sites at Bergen-op-Zoom (Fri d'Or) and Tilburg (Fri d'Or, Ad van Geloven, Hebro and Loem-wi). During the year Fri d'Or acquired a company in the Czech Republic (frozen fruits and potato products) and Welton Snacks BV. Welton will add Indonesian rice snacks to Ad van Geloven's existing range of satays and spring rolls. Belgian Luijckx Chocolate decoration company was peripheral to Hazlewood's core business and was sold to Wessanen Nederland for $14.2m. The group continues to benefit from high-quality fish processing skills (Orca-vis, Diepvries Monnickdam, Sterk, Hansen and Vishandel Gebr. van der Veen).

Interim results

For the 6 months ending 2/10/93 Hazlewood made a pre-tax profit of £25.6m on sales of £403.9m. Sales in the UK and Ireland accounted for £258m and those to the rest of Europe £146m.

Table 1 Principal subsidiaries

Hazlewood International BV – wholly owned by parent company – operates through Trading Board of Directors who report to supervisory Board of Directors

Grocery and non-food Division

Bottling
- Hazlewood & Co. (Selby) Ltd
- Samuel Moore Foods Ltd
- Manor Vinegar Brewery Co. Ltd
- Pandora Pickle Co.
- Campsie Spring Scotland Ltd
- Agrozumos SA (Spain)

Bakery
- Beaverlac (Fine Foods) Ltd
- Michael Scott Cakes Ltd
- Wafel Janssen BV
- Belgpraline NV

Non-food
- F.H. Lee Ltd
- Charles Turner & Co. Ltd
- F.A. Wyatt & Co. Ltd
- Drammock Ltd

Fresh and Produce Division

Meat (UK)
- Faux Brand Foods Ltd
- Roman (Delicatessen Products) Ltd
- J & J Tranfield Ltd
- Hazlewood Trading Ltd
- Baron Meats Ltd
- Gainsborough Pork & Bacon Co.
- Sovereign Foods Ltd
- Sandyford Foods Ltd

Meat (continental)
- Feldhues Group (Germany & Eire)
- Henri van de Bilt BV (Netherlands)

Table 1 ***Continued***

Frozen food division	
Fish	Convenience foods
Orca Vis BV (Netherlands)	Forrester Foods Ltd
Gebr. Sterk BV (Netherlands)	E.T. Sutherland & Son Ltd
Diepvries Monnickendam BV (Netherlands)	Derbyshire Chilled Foods Ltd
Gebr. van der Veen BV/Hansen Vis BV (Netherlands)	Hilliers Ltd
	Saint Martin Food Products Ltd
	Food Enterprises Ltd
Cockles	Baron Fesh Food Services Ltd
International Shellfish BV/Van der Endt de Jager BV (88%, Netherlands)	Hazlewood Desserts Ltd
Wirral Seafoods Ltd	
L. Mieras & Co. BV (Netherlands)	
Lenger Seafoods (Netherlands)	
Prawns	Produce (UK)
Heiploeg Beheer BV (Netherlands)	Van Heyningen Brothers Ltd
Ready meals	Produce (continental)
Hazlewood Frozen Foods Ltd	Evers Specials BV
Chic-o-Roll Co. Ltd	
Creamery Fare Ice Cream	
Fri d'Or BV (Netherlands)	
Fri d'Or Tilburg (Netherlands)	
Fri d'Or Spol Sro (Czech Rep.)	
Ad van Geloven Snacks BV (Netherlands)	

Table 2 Five-year review (31/3/93)

	1989	1990	1991	1992	1993
Turnover (£m)	453.7	576.7	541.3	632.4	761.8
Pre-tax profit (£m)	45.3	46.0	60.5	51.3	55.0
Employees	7156	7583	6597	7898	9749

Table 3 Financial record by division and geographical area

	Turnover (£m)		Profits (losses) (£m)	
	1992	1993	1992	1993
Division				
Grocery and non-food	156.6	169.9	15.5	15.9
Frozen foods	277.0	285.1	24.1	27.1
Fresh foods	198.8	306.8	19.0	21.9
Confectionery and Snacks (licence fee, disposal)			1.5	1.9
Total	632.4	761.8	60.2	66.8
Geographical area				
UK and Republic of Ireland	353.4	477.1	33.9	38.9
Belgium	0.8	0.4	0.04	(0.2)
France	15.8	12.5	0.4	(0.02)
Netherlands	241.1	250.6	23.2	25.9
Portugal	0.4	0.4	0.5	0.5
Spain	11.6	9.9	0.5	0.1
Germany	9.2	8.9	1.7	1.6
Czech republic	–	2.0	–	0.02
Total	761.8	632.4	60.2	66.8

HILLSDOWN HOLDINGS PLC

Secretary and registered office

K.M. Buchanan, Hillsdown House, 32 Hampstead High Street, London NW3 1QD; 071-794 0677; Fax 071-435 1355

Directors

Sir John Nott (executive chairman), J.A. Jackson (deputy chairman), D.A. Newton (chief executive), M.N. Buswell, B.J. Constandse, K.W. Lillington, R.J. Mackie (finance), M.J. Teacher (non-food activities), Sir Roy Griffiths*, Sir Harry Solomon*
* non-executive

Since its formation in 1975, Hillsdown has grown at a phenomenal rate. Under the Chairmanship of Sir Harry Solomon (retired April 1993) the company acquired some of the best known brands in the food industry as well as operating successful furniture, housebuilding and property companies. In 1992, its three food divisions accounted for 87% of group turnover. Difficult trading conditions in the last two years have forced Hillsdown to dispose of less profitable businesses and rationalize the remaining ones. Half-year profits to June 1993 showed a drop in pre-tax profits of 17% to £62.3m.

Food processing and distribution Division

Activities include Premier Brands and HL Foods in the UK, Maple Leaf Foods in Canada, Hillsdown International BV in Holland and the Towers Group (meat importing and trading). In September 1992, Hillsdown bought part of the Beledia chill food group (Artal N.V.), bringing Nadler and Hamal (Germany), Johma (Netherlands) and Kobenhavns Salat (Denmark) into the European operations. Premier Brands has been restructured so that Premier Beverages (worth £180m in sales) now encompasses Typhoo, Glengettie and Ridgways teas, Bournvita, London Herb & Spice Co's herbal teas and Cadbury's Chocolate Break and Marvel. Premier Ambient Foods (sales of £300m) has been created from the preserves, desserts and pickle activities of Chivers–Hartleys and the fruit, vegetable and meat canning facilities of HL Foods. Similarly it has streamlined its chilled foods (Telfers, Pinneys and Smedleys) and combined its fresh produce operations into Hillsdown Produce (Poupart, MBM and Orchard World).

Pinney's Scottish smoked salmon performed well, exporting widely in North America and France. In Canada, Maple Leaf Foods bought McGavin Foods and sold its edible oil division.

Fresh meat and bacon Division

Profits in its meat and bacon division slumped from £32m to £12.8m as abattoirs (FMC and Harris Pork & Bacon Group) were affected by changes in EC subsidy payments to farmers and overcapacity in the UK slaughtering trade. Following these problems Hillsdown undertook a rapid disposal programme and by November 1993 had closed or sold all its abattoirs (about 30). Against market trends Fairfax Meadow Farm (formed from Pyke Biggs Fairfax and Meadow Farm in 1991) improved profitability and is now the leading fresh meat supplier to hotels, restaurants and airlines in the UK and continental Europe. In Canada, Hoffman was integrated into the Fearman's pork processing plant.

Poultry and eggs Division

This division covers poultry breeding, rearing and processing, egg production, added-value poultry products and animal feed supplements. Daylay provides a total range of egg products and is the leading supplier to major retailers in the UK. It maintained profitability despite depressed egg prices.

Imports of cheap poultry from France in 1991 put Buxted and its other poultry business under pressure. Following the purchase of the J.P. Wood business from Unigate, the group restuctured its poultry processing operations, selling Harvest Poultry (Weybread & Kenninghall sites) to a management buyout team for £6m (now Crown Chickens), Buxted's Sutton Benger plant to Webb's Country Foods and Buxted Ducklings facilities to Kerry Foods (A.E. Button). Through closures and sales, capacity has been reduced from over 3 million to under 2 million birds processed per week. In Canada, Maple Leaf Food's poultry processing operations have a significant market share, selling whole-bird, portion and value-added products to retailers and caterers. In the Netherlands, Hencu and G. Jansen Bodegraven Beheer BV produce poultry for Benelux and Germany. Moorlands Poultry and Foods business are major turkey producers and value-added chicken and turkey product manufacturers. Ross Breeders continues to grow rapidly and increase world market share, winning a Queen's Award for Export Achievement.

Table 1 Principal UK subsidiaries

Belwood Foods Ltd
Associated Poultry Ltd
  Buxted Chicken Ltd
  Devon Crest Poultry Ltd
  Harvest Poultry Ltd (sold 4/93)
  Premier Poultry Ltd
  Ross Poultry Ltd
Carleton Furniture Group Ltd
Channel Foods Ltd
Christie-Tyler plc
Colloids Ltd (51% holding)
Daylay Foods Ltd
Fairview New Homes plc
Farm Kitchen Foods Ltd
Firstan Ltd
Formwood Group Ltd
FMC plc
  C & T Harris (Calne) Ltd
  FMC (Meat) Ltd (sold 1993)
Vic Hallam plc
Henry Telfer Ltd
HL Foods Ltd
  Haywards Foods Ltd
  Smedleys Foods Ltd
H.I.T. Investments plc
MBM Produce Ltd
Moorland Foods Ltd
Moorland Poultry Ltd
North Devon Meat Ltd (sold 1993)
Pinneys of Scotland Ltd
Poupart Ltd
Premier Brands Ltd
Fairfax Meadow Farm Ltd
Ross Breeders Ltd
Rugby Securities Ltd
  (84% holding)
St Andrews Land Ltd
Strong & Fisher (Holdings) plc
Towers & Co. Ltd
Walker & Homers plc
J.J. Yates & Co. Ltd (75% holding)

Table 2 Five-year review (31/12/92)

	1988	1989	1990	1991	1992
Turnover (£m)	3548.8	3688.2	4215.2	4656.6	4373.0
Pre-tax profit (£m)	144.7	189.0	191.2	186.8	154.1
Employees	42 832	40 558	47 504	48 652	44 196

Table 3 Financial record by division and geographical area

Divisional analysis	Turnover (£m)		Profits (losses) (£m)	
	1991	1992	1991	1992
Activity				
Food processing and distribution	2189.5	2121.0	137.3	123.4
Poultry and eggs	823.9	868.0	26.3	27.3
Fresh meat and bacon	1041.4	880.8	32.0	12.8
Total food	4054.8	3869.8	195.6	163.5
Non-food	601.8	503.2		
Total	4656.6	4373.0	231.7	184.9
Geographical origin				
United Kingdom	2503.2	2580.9	131.3	102.2
Canada	1512.7	1141.2	70.4	54.6
Europe	403.5	478.9	21.2	24.4
USA and other countries	237.2	172.0	8.8	3.7
Total	4656.6	4373.0	231.7	184.9

MARS UK LTD

Secretary and registered office

S. Guthrie Brown, Dundee Road, Slough, Berkshire SL1 4LG; (0753) 550055; Fax (0753) 531532

Directors

S. Bullimore, M.R. Davies, J.B. Dale, R.D. Raper

The ultimate holding company is Mars, Inc. of Delaware, USA. It has over 50 autonomous units operating from 80 locations throughout the world. Group sales are made up as follows: snack foods 43% (mainly confectionery products); pet care products 44%; main meal, electronics and drinks 13%. Sales by region are: Europe 56%; Americas 36%; Australasia, Japan and West Pacific 8%.

The immediate parent company in the UK is Food Manufacturers (G.B. Company) Ltd, a holding company not only for Mars UK but also Hoops Ltd and Mars (GB) Ltd.

Confectionery

The first Mars Bar was produced in a small rented factory in Slough by Mars Confections Ltd in 1932. Since then the company has grown to become the UK's leading manufacturer of chocolate confectionery. It also manufactures a number of well-known sugar confectionery brands. With a turnover of £200m, it produces 2 million Mars Bars a day, along with Twix, Bounty, Maltesers, Galaxy, Milky Way, Topic, Snickers, M&M's, Tracker, Opal Fruits, Revels, Skittles, Tunes, Lockets. The Mars brand is now also applied to ice cream and milk. About 2300 people are employed by Mars Confectionery in the UK. Its sister company Effem Management Services Ltd (Effemex) is responsible for sales and marketing in over 100 countries.

Petcare

The petfood division (part of Mars (GB) Ltd) encompasses Pedigree Petfoods, the overall market leader in petfoods, Thomas's Petcraft and the Waltham Centre for pet nutrition. Brands include Bounce, Cesar, Chappie, Pal, Pedigree Chum, Brekkies, Frolic, Whiskas, Kitekat, Thomas Smackos and Biscrok.

Grocery

Master Foods based at King's Lynn manufactures a range of pasta products and sauces under the Dolmio brand. Other grocery brands handled by the group are Tyne (formerly Tyne Brand) canned and frozen meat products, Uncle Ben's rice, and Yeoman dehydrated potato products.

Other group companies

Four Square, a Division of Mars (GB) Ltd manufactures and markets machines and beverages for Klix, the best-selling in-cup drinks vending system in many countries in Europe and the Far East. Flavia is a new vending system developed by Four Square for freshly brewed ground coffee and leaf tea.

The need for reliable vending systems, originally for Mars Confectionery bars, led to the formation of Mars Electronics International which is now a leading manufacturer of electronic payment systems for food products, amusements, telecommunications and transport.

Table 1 Five-year review – Food Manufacturers (G.B. Company) Ltd (26/12/92)

	1988	1989	1990	1991	1992
Turnover (£m)	1185.1	1260.8	1377.5	1442.7	1430.1
Pre-tax profit (£m)	76.53	76.43	77.05	48.52	4.59
Employees	7036	6736	6834	6492	6009

Table 2 Five-year review – Mars UK Ltd (26/12/92)

	1988	1989	1990	1991	1992
Turnover (£m)	538.7	571.6	654.3	704.1	726.4
Pre-tax profit (£m)	30.73	34.00	41.51	39.21	50.52
Employees	2848	2787	2857	2852	2613

Note: No segmental information is given in the annual returns because the group claim that disclosure of turnover and operating results by class of business would be prejudicial to the interests of the group

MILK MARKETING BOARD

Secretary and registered office

P.W. Beswick, Milk Marketing Board, Thames Ditton, Surrey KT7 0EL; 081-398 4101

Directors

R.S. Stevens (chairman), B.B. Kent (vice-chairman), A.R. Dare (chief executive), C. Melchers (finance), A. Iveson (M.D. milk marketing), Dr J. Craven (M.D. Genus), F. Harding (technical director), R. Turner (director NMR). (In addition there are appointed special board members as well as regional representatives.)

DAIRY CREST LTD

Secretary and registered office

R.J. Newton, Dairy Crest House, Portsmouth Road, Surbiton, Surrey KT6 5QL; 081-910 4000.

Directors

G.R. John (chairman), W.J. Houliston (chief executive), J.W.D. Hall (M.D. Dairy Crest Foods Ltd), D.R. Lewis (finance), C. Fidler, P.A. Jacobs, B.B. Kent, N.G. Saphir.

Set up in 1933, the Milk Marketing Board (MMB) has had monopoly status in the marketing of milk throughout England and Wales. In April 1992, the Board signalled its intention to end this monopoly in its proposal entitled 'New Ways of Buying Milk'. Under the scheme its wholly owned subsidiary Dairy Crest Ltd would be transferred to dairy farmers, who would be allocated shares on the basis of the quantities of milk produced. Dairy Crest's flotation is planned for 1994. It is likely that Dairy Crest would be protected from takeover for a limited period.

Milk Marque, the new co-operative launched in January 1993 should become a practical reality in 1994. Major manufacturers will be at liberty to buy milk direct from farmers or set up their own co-operatives and Northern Foods and Nestlé already have plans to do this.

Milk marketing

In 1992/3 a total of 28 459 wholesale producers sold 11 029 million litres of milk to the MMB, a drop of 1% on the previous year's figures. During the year milk quality improved again, with butterfat levels at 4.1%, protein at 3.29% and total bacterial counts of 13 000 organisms/ml. The average price of milk sold over all manufacturing markets was 20.6 ppl. During the year price differentials between doorstep deliveries and shops widened, with shop milk now about 9 p/pint cheaper than doorstep milk. Low-fat milk sales continued to expand so that they now amount to almost 50% of the market for liquid milk.

National Milk Records (NMR) and Genus

During the year NMR and Genus (a breeding, animal health and management consultancy) laid the foundations for becoming independent commercial ventures offering services to the dairy industry.

Dairy Crest Ltd

Dairy Crest (DC) is a wholly-owned subsidiary of the MMB. Results for year ended 31/3/93 showed an increase in operating profits of 5%, a reduction in gearing from 49 to 33% and an increase in return of capital employed to 22.4%. These have been achieved mainly through large-scale job losses and plant closures necessitated by reduced milk availability and the need to reduce costs prior to stock market flotation.

Dairy Crest continues to be the leading producer of English Cheddar and regional cheeses, with cheesemaking capacity now concentrated at the two largest creameries, Davidstow and Aspatria. It has recently introduced Dovedale, a soft blue cheese, and a smoked Sturminster Newton Cheddar. In the blended spreads market, Clover is claimed to be No. 1 in Great Britain. As a leading member of the English Butter Marketing Co., Dairy Crest manufactures and packs 50% of Country Life butter, as well as Dairy Crest branded butters and own-label butter. It supplies milk for doorstep delivery and to supermarkets. Its fleet of 900 insulated tankers collects a substantial proportion of the milk produced in England and Wales. Through its Technical Development Centre at Crudgington it is expanding its dairy-based ingredients business. In 1992, it also entered a joint venture with Yoplait SA to produce Petit Filous, YOP, Ile Flottante and Fruit on the Bottom fromage frais.

Subsidiary

Philpot Dairy Products Ltd (handles exports of DC products, 100% owned by DC).

Associated undertakings

- English Butter Marketing Co. Ltd (50%)
- Mendip Foods Ltd (45%)
- Yoplait Dairy Crest Ltd (49%)

Table 1 Milk Marketing Board – Two-year review (31/3/93)

	1992	1993
Sales of milk (£m)		
Liquid	1435.3	1430.2
Manufacturing	935.6	1080.1
Market development	3.2	5.1
Technical Division	0.5	0.7
Total milk marketing turnover	2392.5	2516.1
National Milk Records	9.7	10.3
Bull & Cow Evaluation Unit	0.1	0.2
Genus		
Breeding	32.3	37.3
Management and Animal Health	9.0	9.5
Total turnover	2443.6	2573.4
Employees	2517	2396

Table 2 Dairy Crest Ltd – Five-year review (31/3/93)

	1988	1989	1990	1991	1992
Turnover (£m)	896.5	1254.0	1177.0	1156.8	1155.8
Pre-tax profit (£m)	50.0	39.1	31.9	24.6	28.2
Employees	8163	12 755	11 937	10 450	9287

Table 3 Turnover by geographical destination

	1992	1993
United Kingdom	940.8	938.3
Other EC countries	52.1	47.2
Rest of world	64.5	56.9
Diverted milk – all UK	99.4	113.4
Total turnover (£m)	1156.8	1155.8

NESTLÉ HOLDINGS (UK) PLC

Secretary and registered office

R.P.J. Sonneborn, Nestlé UK Ltd, St George's House, Croydon, Surrey CR9 1NR; 081-686 3333

Directors

P.H. Blackburn, F.R. Edwards, J.R. Strachan, R. Masip, D.F. Harris, G. Millar

The ultimate holding company is Societe des Produits Nestlé SA in Vevey in Switzerland, which was formed by the merger in 1905 of Farine Lactée Henri Nestlé and the Anglo-Swiss Condensed Milk Co. Following this merger the company had 18 factories, five of them in the UK. By 1992 there were 428 factories in 69 countries employing 218 005 people. Sales rose by 8% in 1992, reaching figures of SFr 54.4bn. Perrier, consolidated in July 1992, contributed SFr 934m to total sales. Sales volumes increased slightly in Europe overall, although markets were stagnant in Italy while Switzerland and the UK saw modest declines. With the exceptions of Brazil and Canada, sales trends were highly favourable in North America and the Far East.

Acquisitions during the year totalled SFr 4758m and included Perrier, Cokoladovny (Czech Republic), certain of Quintie's activites in Hungary, Hirz (Switzerland), Clarke Foods (UK) and Sopharga (France). Disposals netted SFr 457m.

Nestlé UK Ltd Organization

This has 26 factories in the UK employing about 20 000 people and operates through four divisions. In 1991 a major reorganization of the UK business took place and as a result the principal trading activities of the group have been carried on by Nestlé UK Ltd.

In July 1992 Nestlé acquired the Manley Ratcliffe honey business from Croda International and in November 1992, the trade and assets of Clarke Foods (UK) Ltd (ice creams) from the receivers.

Grocery

This division manufactures and sells instant coffee, milks, creams, coffee whiteners, juices, drinks and petfoods. It also handles the sales of products from Cereal Partners, the joint venture between General Mills and Nestlé (Shredded Wheat, Clusters, etc). Famous brands include Nescafé, Libby's Orange C, Carnation, Nesquik, Friskies, Go-Cat and Go-Dog. Nescafé and Gold Blend are manufactured at the Hayes factory.

Nestlé Rowntree

Nestlé acquired Rowntree Mackintosh in 1988, integrating it fully into its UK business structure in 1992. Confectionery, including chocolate blocks, seasonal goods and candies are the main products. Famous brands include Polos, Smarties, Kit-Kat, After Eight Mints, etc. The main administrative functions of the division are at the former Rowntree H.Q. in York. Chocolate and confectionery research and development are undertaken at Yorkreco.

Nestlé Food

This includes Findus frozen foods, Chambourcy chilled desserts, Rowntree desserts, Herta chilled meats, Buitoni pasta, Crosse & Blackwell canned goods, Branston Pickle, Sarson's and Dufrais Vinegars, Waistline salad dressings, Gales Honey and Sun-Pat peanut butter.

Nestlé Food Services

Food services include ambient and frozen catering products from the Nestlé ranges and the Multisnack vending operation with 2000 machines; Healthcare – dietetic products like Slender, Nesfit, and Build Up; Nestlé Food Imports – products imported for special UK markets from Nestlé companies throughout the world; Libby Purchasing Co. – coordinates all the purchasing requirements of Libby's canned fruit and meat for European markets; Food Ingredients Specialities – supplies and advises on the use of ingredients, especially flavourings, imported from Nestlé companies overseas.

Operating Companies of the Nestlé Group

Too numerous to be included in this guide.

Table 1 Nestlé UK Ltd – Four-year review (31/12/92)

	1989	1990	1991	1992
Turnover (£m)	1541.8	1558.4	1630.0	1677.1
Pre-tax profit (loss) (£m)	31.7	7.6	(8.6)	22.6
Employees	–	21 165	21 003	19 890

Table 2 Nestlé SA – Five-year review (31/12/92)

	1988	1989	1990	1991	1992
Turnover (SFr.m)	39 502	48 036	46 369	50 486	54 500
Consolidated net profit (SFr.m)	2058	2412	2272	2470	2698
Employees	197 000	197 000	n/a	201 139	218 005

Table 3 Divisional analysis by geographical area and product division

	Turnover (SFr.m)*		Trading profit (SFr.m)	
	1991	1992	1991	1992
Geographical area				
Europe	24 350	26 632	2276	2359
North and South America	18 098	19 214	1861	2158
Rest of world	8038	8654	949	1120
Product Division				
Beverages	11 917	13 521	2136	2415
Dairy and dietetics	13 924	14 890	1192	1394
Prepared dishes and cooking aids	14 956	15 718	782	806
Chocolate and confectionery	8077	8598	685	685
Pharmaceuticals	1612	1773	291	364

* Swiss Francs/£ = 2.53 in 1991 and 2.21 in 1992

NORTHERN FOODS PLC

Secretary and registered office

J.N. Wild, Beverely House, St Stephen's Square, Hull HU1 3XG; (0482) 25432; Fax (0482) 226136

Directors

C.R. Haskins (chairman), M. Clark (finance), M.A.J. Morgan (personnel), W.B. Howard (deputy chairman)*, R.P. Eddy*, J.M. Fry*
* non-executive

Northern Foods plc (NF) claims to be the UK's most broadly based fresh food manufacturer employing 30 219 people. Turnover is now derived entirely in the UK where its main business is the development and production of high-quality chilled foods under the own-labels of major retailers as well as the Ski, Eden Vale, Pork Farms, Bowyers and Hollands brands.

In the last 2 years NF has acquired businesses which complement existing operations and offer scope for improving sales and reducing costs, particularly by devolving responsibility to manufacturing units and closing costly head offices. Group sales leapt by 40% to over £2bn in 1993 following the purchase of Express Dairy and Eden Vale. Operating profits also increased sharply (31.5%) to £180.8m. The Dairy and Convenience Foods Division both increased their profits by about 50% in the year to 30/6/93 through organic growth and the acquisition of Express and Eden Vale. Grocery companies' sales grew by 5% and profits by 13%. Meat Products had a very difficult year with sales up 2% and profits down 16%. Pork Farm's van sales to small independent retailers were hit by the recession and Sunday opening by the multiples.

With the end to the MMB in 1994, NF has declared its intention to buy as much milk as possible directly from producers and has been active in launching the Northern Milk Partnership.

Dairy (sales £962m; employees 8340)

NF holds 24% of the market for liquid milk in the UK, delivering to 3 million homes under the Dale Farm and Express brands. Ashby Dairy Co. holds a strong position in the supply of own-label milk to supermarkets in non-returnable containers (cartons and plastic bottles). In Northern Ireland, Dale Farm Dairies have important shares in both liquid milk and ice cream markets and have recently acquired the CWS milk business in Belfast. The Eden Vale Food Ingredient's creameries in Yorkshire and Somerset supply bulk milk products to NF group companies and other food manufacturers. There is a liquid egg facility near Sheffield.

Convenience foods (sales £553m; employees 10 502)

Major product areas include recipe dishes for major retailers (Lenton, Fenland and Pennine Foods); pies, quiches and pizzas (Evesham Foods, Riverside and Gunstones Bakeries); sandwiches and flour confectionery (Gunstones); cakes and Christmas puddings (Park Cakes and Matthew Walker); speciality breads (Falconis, Kara Foods, Grain d'Or, Oakland Burger Bun Co. and Fletchers Bakeries (25% holding); fresh dairy products (Eden Vale, Ski, Munch Bunch from factories in Cuddington and Minsterley and Rawmarsh Foods). L.O. Jeffs is a small fresh produce business. NFT, the group's chilled distribution business has its primary depot at Alfreton, near the M1 motorway.

Meat products (sales £315m; employees 7055)

This division includes Nottingham-based Pork Farms, Hollands, Fleur de Lys (North West and West Midlands catering meat pies); Bowyers Wiltshire (sausages and meat pies) and a modern meat pie bakery at Manchester making own-label products for

Tesco. Wall's, Richmond's and Lawson's branded goods are also produced there under contract. Palethorpes and the Pizza Factory produce savoury pies, pastries, pizza, etc. for supermarkets. Mayhew Chicken (Aberdare and West Sussex) produces fresh and processed chicken for retailers, caterers, and manufacturers.

Grocery (sales £197m; employes 4320)

Fox's dominate the quality biscuit assortment market. Elkes makes Farley's rusks for Boots and a range of 'value' lines under retailer own-brands. Smith and Garratts flour mills supply NF companies.

In Eire, Batchelors produces canned vegetables and fuit juices; Green Isle has expanded its frozen foods and pizza businesses.

Interim results to September 1993

Pre-tax profit rose 5.4% to £72.1m for the half year, and turnover was up 5% to £1bn.

Table 1 Principal operations

Dairy	Northern Dairies Dale Farm Dairies Ashby Dairy Co. Express Dairy Eden Vale Food Ingredients Dale Farm Dairies (NI)	
Convenience foods	Pennine Foods Fenland Foods Lenton Foods Riverside Bakery Evesham Foods Gunstones Bakery Matthew Walker	Falconis Kara Foods Park Cakes Eden Vale Rawmarsh Foods L.O. Jeffs NFT Distribution
Meat products	Pork Farms Walter Holland & Son Bowyers Dorset Chilled Foods Trafford Park Bakery Palethorpes	The Pizza Factory Trentham Savoury Foods (closing 1993) Mayhew Chicken Binghams Cooked Meats
Grocery	Fox's Biscuits Elkes Biscuits Smith Flour Mills	Batchelors (Dublin) Green Isle (Co. Kildare)

Table 2 Five-year review (31/3/93)

	1989	1990	1991	1992	1993
Turnover (£m)	1041.3	1094.4	1187.0	1442.2	2026.1
Pre-tax profit (£m)	73.2	81.9	106.7	126.2	153.2
Employees	21 864	21 795	22 924	27 002	30 219

Table 3 Financial record by division

	Turnover (£m)		Profits (losses) (£m)	
	1992	1993	1992	1993
Dairy	601.9	1032.1	62.5	92.0
Convenience foods	396.0	637.3	29.4	44.4
Meat products	309.1	315.1	24.1	20.2
Grocery	199.7	211.5	21.5	24.2

RECKITT & COLMAN PLC

Secretary and registered office

P.D. Saltmarsh, One Burlington Lane, London W4 2RW; 081-2994 6464; Fax 081-2994 8920

Directors

Sir Michael Colman (chairman), V.L. Sankey (chief executive), C.C.B. Brown, A.J. Dalby*, J.C.L.de Mel, I.G. Dobbie, R.M.M. Foster, G.J. Hearne*, Professor J-C Larreche*, P.J. Maydon, M.F. Turrell, M.R. Valentine*
*non-executive

The merger of Reckitt with the operations of Colman's in 1953 resulted in the formation of one of Britain's largest companies with many well-known brands in the UK. It employs 20,800 people worldwide, with 2800 in the UK. Following the sale of its industrial cleaning divisions in 1988 along with the wine operations, the group has concentrated on three major areas: household and toiletries, food, and pharmaceuticals. In 1992 the group's total sales amounted to £1.9bn, of which one-fifth was derived from the UK. Food sales were £353.42m in 1992, a drop of 8% on the 1991 figures.

Food

The Group's food interests are focused in two main regions – North America and the UK. Products include mustards, sauces, condiments, savoury mixes and other flavour enhancement products sold under the Colman's brand. Recent additions to the list include cook-in sauces for chicken and curries. While Robinson's soft drinks, supported by a vigorous marketing campaign, gained market share during 1992, interim results for 1993 show that in the UK soft drinks markets sales growth was difficult to achieve.

Babyfoods are also marketed under the Robinson's brand.

In the USA, French's mustard and RedHot cayenne pepper sauce maintained market share despite strong competition.

During the year the group withdrew from its North American spice and seasoning businesses and disposed of its food activities in South Africa. These divestments contributed to a fall in sales of 8.1% and operating profits of 1.3% when compared with the 1991 figures.

Household and toiletries

These account for the largest percentage of group sales (£1.26bn in 1992) and the widest international coverage. Well-known brands include Harpic, Dettox, Woolite, Brasso, Mr Sheen, Immac and Steradent.

Pharamceuticals

Contributing £177m to group sales in 1992, this division covers over-the-counter and prescription products. Main markets are in the UK, Australasia, Asia and South Africa. Well-known brands include Dettol, Disprin, Lemsip and Senokot.

Principal subsidiary undertakings

- Reckitt & Colman Products Ltd (UK)
- R & C Products Pty Ltd (Australia)
- Representacoes Reckitt & Colman Brasil Limitada (Brazil)
- Reckitt & Colman France SA (France)
- Reckitt & Colman South Africa (Pty) Ltd (South Africa)
- Reckitt & Colman Inc. (USA)

Table 1 Five-year review (2/1/93)

	1988	1989	1990	1991	1992
Turnover (£m)	1394.0	1565.8	1763.6	1986.9	1904.3
Pre-tax profit (£m)	191.3	217.4	235.2	233.97	181.73
Employees	21 700	22 000	23 800	23 000	20 800

Table 2 Financial record by division and geographical area

	Turnover (£m)		Operating profits (£m)	
	1991	1992	1991	1992
Geographical area				
United Kingdom	364.72	356.47	68.60	70.12
Euope (excluding UK)	509.15	519.86	79.14	71.63
North America	464.70	469.93	41.84	39.55
Australasia and Asia	211.20	208.09	42.27	40.48
Africa	114.27	92.60	21.21	28.23
Latin America	190.54	207.40	27.66	31.69
Discontinued operations	132.31	49.95	11.85	(9.59)
Total	1986.89	1904.30	292.57	272.11
Division				
Household and toiletries	1247.79	1261.94	186.45	185.59
Food	384.64	353.42	45.50	44.90
Pharmaceuticals	1687.16	176.73	39.99	41.56
Other activities	53.99	62.26	8.78	9.65
Discontinued operations	132.31	49.95	11.85	(9.59)
Total	1986.89	1904.30	292.57	272.11

TATE & LYLE PLC

Secretary and registered office

N.J. Nightingale, Sugar Quay, Lower Thames Street, London EC3R 6DQ; 071-629 6525 (No fax)

Directors

N.M. Shaw (chairman), S.R. Brown (chief executive), P.S. Lewis (finance), M.D. McEwen, P.M.E., Callebaut*, Lord Jellicoe*, R.M. Powers*, Lady Prior*, H.S. Tate*, J.F. Taylor*, L.R. Wilson*, Sir Brian Hayes*, Lord Walker*
*non-executive

The Tate & Lyle (T & L) group of companies originated from the amalgamation of the sugar and syrup refining businesses of Henry Tate and Abram Lyle in 1921. Since then, T & L has grown to become one of the world's leading sugar, cereal, sweetener and starch groups. It has subsidiaries and associates in 30 countries. A major position in North America was secured through the $1.5bn acquisition of Staley in June 1988 followed by the $305m acquisition of Amstar in September 1988 and the outstanding shares in Redpath, the group's Canadian subsidiary, for C$286m in May 1989. In Europe, £44m was spent on Campo Ebro, one of Spain's major cereal sweetener and starch manufacturers.

Following pre-tax profits of £230.8m in 1991, the group suffered its first profit setback in 14 years. In 1992, the US sweetener businesses A.E. Staley and Western Sugar suffered reduced profits brought about by bad weather and low sugar prices following a good beet crop. Staley's high-fructose corn syrups (HFCS) were affected by over-supply and low demand. In Europe, cereal sweetener and starch operations improved profitability and in Australia, Bundaberg Sugar Co. performed well in its first full year with T & L. As well as significant investments in process plant worldwide, T & L has increased European investment in a number of ways:

- acquisition of nearly all the outstanding minority at Alcantara Portugal;
- investment of 5% (£8.4m) in Gardini et Associes, an international holding company formed to respond to consumer markets in Eastern Europe;
- investments by CST (Cereal Science & Technology Group) in Italy and in joint ventures with Archer Daniels Midland Company (ADM) (USA) and Agrana (Austria) in Hungary and with ADM in the Czech Republic.

Cane and beet sugar

Main operating companies include Tate & Lyle Sugars (UK), Alcantara (Portugal), Bundaberg Sugar (Australia), Domino Sugar (USA), Western Sugar (USA), Redpath Sugars (Canada). There are also holdings in Kaba (Hungary), Azucarera (Spain) and Zimbabwe Sugar Refineries. At 989 300 tonnes, T & L's Thames Refinery achieved its highest ever annual output in 1992 following substantial capital investment (£17m) in the plant. Productivity in tonnes/employee has increased by 50% since 1987.

Cereal sweeteners and starches

Over-capacity in the industry reduced Staley's output of HFCS, but its new fat replacer 'Stellar', and 'Krystar', a crystalline fructose launched in 1987, have been gaining acceptance and selling well. There has also been expansion of Ethylex production to cope with increased demand for hydroxyethylated starches from the paper and textile industry. CST (Cereal Science & Technology Group) increased profits by 6% and in former East Germany, investment in the Prignitze-Starke potato starch plant at Dallmin brought productivity improvements. CST also invested in Hungrana (maize-based starches, Hungary), Sedamyl (wheat starch, glucose and alcohol, Italy) and Stovamyl (maize starch and glucose, Czech Republic).

Sugar by-products

Providing 8.5% of group profits, this division is involved in molasses trading, animal feeds, feed mill engineering, speciality chemicals and international sugar trading.

Other businesses

Speciality Sweeteners produces the low-calorie sweetener Sucralose under the brand name 'Splenda'. Booker Tate is a sugar management business owned jointly by Booker plc and Tate & Lyle (50/50). There are also group companies involved in shipbuilding, insurance, cane crushing equipment and rum production.

Results for year ending 25/9/93

A successful cost reduction programme plus devaluation of the green pound resulted in record profits for the European sugar business. Pre-tax profits rose from £189.5m to £222.5m on sales up from £3.37bn to £3.82bn. In the UK, pre-interest profit for sugar improved from £50.3m to £67.7m with sales increasing from £678.4m to £724.6m.

Table 1 UK-based subsidiaries and main operating units

UK-based subsidiaries	*Main operating units of Tate & Lyle Industries Ltd*
The Molasses Trading Company Ltd	Charles E. Ford
Richards (Shipbuilders) Ltd	Four-F Nutrition
Tate & Lyle Holdings Ltd	Kentships
Tate & Lyle Industries Ltd	Micronized Food Products
Tate & Lyle International Finance plc	Milltech
Tate & Lyle Investments Ltd	Rumenco
Tunnel Refineries Ltd (66.7% of share capital)	Speciality Sweeteners
United Molasses (Ireland) Ltd (50% of share capital)	Tate & Lyle International
	Tate & Lyle Sugars
	United Molasses

Table 2 Five-year review (26/9/92)

	1988	1989	1990	1991	1992
Turnover (£m)	2281.3	3465.5	3445.2	3262.5	3366.3
Pre-tax profit (£m)	118.9	196.6	214.3	230.8	189.5
Employees	14040	18606	16670	16159	17004

Table 3 Financial record by division and geographical area

	Turnover (£m)		Profit before interest (£m)	
	1991	1992	1991	1992
Cane and beet sugar				
North America	684.1	633.4	59.4	39.6
Europe, Africa, Pacific	797.4	932.9	65.1	71.2
Cereal sweeteners and starches				
North America	677.1	625.0	96.8	65.3
Europe	458.8	475.3	51.1	54.0
Sugar by-products	370.0	374.1	17.5	20.1
Other businesses and activities	275.2	325.6	(9.4)	(12.8)
Total	3262.5	3366.3	280.5	237.4

TOMKINS PLC

Secretary and registered office

R. N. Marchant, Tameway Tower, Bridge Street, Walsall, West Midlands WS1 1JZ; (0922) 723372; Fax (0922) 644495

Directors

M.R.N. Moore (chairman)*, G.F. Hutchings (chief executive), I.A. Duncan (finance), A.J. Reading, J.D.S. Stark, R.M. Muddimer
*non-executive

Ranks Hovis McDougall

R.M. Muddimer (chairman), G. Eaton, B. Harris (finance) P. Baker, P. Wilkinson, B. Cook, H. Mann.

In December 1992, Ranks Hovis McDougall (RHM) was taken over by Tomkins plc, an industrial management group which specializes in revitalizing low-risk technology companies. Other Tomkins group companies have interests in:

- fluid controls: water heating, ventilating and air conditioning valves and plumbing fittings;
- services to industry: conveyor and materials handling systems, fittings and fasteners, automotive components and protective footwear;
- professional, garden and leisure products, finely engineered products including power mowers, bicycles and handguns, handcuffs and buckles;
- industrial products, plastic and fibre glass mouldings, doors and windows, control instrumentation, disc brakes and clutches.

Ranks Hovis McDougall

RHM, a major food group, manufactures and markets a wide range of branded and own-label products and ingredients for the consumer, caterer and food manufacturer in the UK, continental Europe and the USA. RHM Retail operates 300 shops and catering outlets. Well-known brands include Robertson's jams and Golden Shred, Bisto, Atora, Mr Kipling and Cadbury's Cakes, McDougalls, Sharwood's, Rombouts, Hovis, Just Juice, Capri Sun and Mother's Pride. In the last 2 years profits had slumped as over-capacity in the UK milling and baking industries affected RHM's British Bakeries. Manor Bakeries (Mr Kipling) also showed falling profits as did the US businesses.

Following the takeover, Tomkins set aside £90m for the reorganization of RHM, cutting the headcount by 2000 within its first 6 months of ownership. Emphasis was placed on improving operating efficiency and cost control.

At RHM Foods Ltd, flour bagging operations at Nottingham and the distribution depot at Chapelhall, Lanarkshire were closed. Production problems with Rombouts coffee were addressed and new products introduced at RHM Ingredients. The Boston factory of McDougalls Catering Foods was closed and two of Chesswood's mushroom farms.

J.A. Sharwood, a major supplier of oriental foods, and Tiffany Sharwoods Frozen Foods both performed well, increasing profits and introducing innovative products. Avana Bakeries at Cardiff and Rogerstone have been extended and the management reorganized. British Bakeries plants at Taffs Well (Cardiff), Exeter, Dundee and Croydon have been closed and capacity reduced at the Wimbush bakery in Birmingham. For Mother's Pride and Hovis, the rationalization has been accompanied by a considerable increase in advertising spend.

Ranks Hovis UK and Ranks Meel in the Netherlands operate 14 flour mills and are one of Europe's largest flour milling groups. Here again, rigorous cost reduction programmes were implemented by Tomkins to cope with the harsh trading conditions.

In the US, the merger of three Red Wing operating companies and closure of surplus peanut butter capacity has reduced the labour force and improved plant utilization and operational efficiency. A poor strawberry crop depressed results for Carriage House Fruit.

In the 5 months to Tomkin's year-end in May 1993, RHM contributed £30.1m to operating profits on a turnover of £662.8m. The workforce of the group had been reduced to 24 500.

Table 1 Ranks Hovis McDougall – principal trading subsidiaries

Abercroft Cakes Ltd	RHM Grocery International Ltd
Avana Bakeries Ltd	RHM Ingredients Ltd
British Bakeries Ltd	RHM Research & Engineering
R.F. Brookes Ltd	RHM Retail Ltd
Chesswood Produce Ltd	J.A. Sharwood Ltd
Holgran Ltd	Supreme Salt Company Ltd
Manor Bakeries Ltd	Tiffany Sharwood's Frozen Foods
McDougalls Catering Foods Ltd	RHM Foods (Ireland) Ltd (Eire)
Overseal Foods Ltd	Ranks Meel BV (Holland)
Pasta Foods Ltd	Vetrimo Beheer BV (70%; Holland)
Rank Hovis Ltd	The Red Wing Co. Inc (USA)
RGB Coffee Ltd	Carriage House Fruit (USA)
RHM Computing Ltd	Sofrapain (Societe Francaise de Panification et de Patisserie-Lyon) SA (France)
RHM Foods Ltd	

Table 2 Five-year review (figures for RHM up to 5/9/92, prior to the Tomkins' takeover)

	1988	1989	1990	1991	1992
Turnover (£m)	1668.6	1786.1	1771.4	1531.4	1559.6
Pre-tax profit (£m)	156.6	176.5	133.2	150.2	92.6
Employees	38 195	39 189	36 155	30 340	28 551

Table 3 Financial record by geographical area

	Turnover (£m)		Profits (losses) (£m)	
	1991	1992	1991	1992
United Kingdom	1271.0	1278.6	129.4	85.7
Continental Europe	73.6	81.1	2.6	4.1
USA	186.5	199.5	13.7	3.2
Other countries	0.3	0.4	–	–
Net interest	–	–	6.7	1.8
Supplemental interest	–	–	(2.2)	(2.2)
Total	1531.4	1559.6	150.2	92.6

UNIGATE PLC

Secretary and registered office

P.N. Heriz-Jones, Unigate House, Wood Lane, London W12 7RP; 081-749 8888

Directors

Sir Brian Kellett (chairman), R. Buckland (chief executive), J.S. Kerridge*, C.A. Lawrence, E.H. Sharp*, A.W. P. Stenham*, G.C. Summerfield, J.G. Worby (finance)
* non-executive

In 1990, the Unigate Group had international interests in some widely disparate businesses ranging from dairies through poultry breeding to car auctions. Following poor performances by the group over the last 3 years, loss making and non-core operations have been sold or run down and acquisitions have focused on strengthening mainstream businesses in dairy products, fresh foods and food and non-food distribution.

Unigate Dairies

These contribute about 30% of operating profits and supply bottled milk and fresh foods to 1.6m households in the UK. Sussex Dairies and CWS milk businesses in London, the South East and Cornwall were acquired in 1992/3. Immediately following the year-end (31/3/93) Unigate bought Clifford's Foods plc for £50.4m. This strengthened and extended its milk delivery business, so that Unigate now has 16% of the liquid milk market in England and Wales. St Ivel Westway in London produces milk for supermarkets in cartons and plastic bottles under the Shape brand.

Fresh foods

St Ivel, Fermanagh Creameries (N. Ireland) and Wexford Creamery (Eire) form the basis of this division. Operating profits fell 15% from £19.9m (1992) to £16.8m in 1993. St Ivel's Shape yoghurts achieved volume growth and Gold low fat spread maintained its market share. New products included St Ivel baby and toddler yoghurts. Milk shortages adversely affected butter, cheese and milk powder – although a move into the premium cheese market helped margins. Major investment at the St Ivel Wootton Bassett factory allowed production of fromage frais which was previously sourced from France.

Farm foods

Comprising Malton Bacon, Turner Turkeys and Oldacre animal feeds, the contribution to operating profits in 1993 was 8% (£7.5m). Malton's factory improvements led to a 27% increase in sales of bacon and ham. Oldacre benefitted from the sale of the loss-making J.P. Wood poultry business and the cessation of grain trading activity. Turner's Turkeys sales were static.

Wincanton

The distribution group (food and non-food) provides 16.5% of profits, which have risen by 20% in the last year. Two recent acquisitions have strengthened the group: Glass Glover for £54m and J. Whittam (dedicated Sainsbury's distribution). Disposals include Sprinsen and Ebrex Holding, both fresh produce distributors in the Netherlands.

US restaurants

The group owns two US restaurant chains – Black-Eyed Pea (Prufrock) and Taco Bueno (Casa Bonito). Plans in April 1993 to raise capital by floating the business were postponed in May 1993 following storms which adversely affected sales. Nevertheless, 1993 operating profits were up 7.5% on the 1992 figure.

Other businesses

In line with its policy of shedding non-core businesses, Unigate began a controlled run down of its car contract hire. It also sold part of Giltspur International exhibitions and its US Italian cheese manufacturers Frigo and Gardenia Foods. This leaves the group with vehicles sales and service (Arlington, NCA, Wincanton Rental and Contracts), modular display equipment (Marler Hayley) and exhibitions (Giltspur Inc.).

Associated undertakings

Subsidiaries of Unigate plc also hold 32.6% of Nutricia, Belgium, 50% of Stovold and 20% of Galloway Cheese Co.

Recent developments and interim results

Turner's Turkeys continued to suffer from contracting margins and was sold to Bernard Matthews in December 1993 for £18m in cash. Oldacre animal feeds, and Morton Foods (acquired with the Clifford's takeover) were sold to Dalgety for £20.9m and £9.5m, respectively. Half-year results to 30/9/93 showed sales up from £726.2m to £820.6m and pre-tax profit up from £42m to £50.3m. Talks on a joint cheese manufacturing venture between Unigate and Golden Vale are to be discussed further in January 1994.

Table 1 Principal subsidiary undertakings

Unigate (UK) Ltd	Unigate Distribution Services Ltd
St Ivel Ltd	Wincanton Ltd
Unigate Dairies Ltd	Wincanton Vehicle Rentals Ltd
Malton Bacon Factory Ltd	Marler Hayley Exposystems Ltd
St Ivel Farm Foods Ltd	Glass Glover Group plc
Turners Turkeys*	NCA (Auctions) Ltd
W.J. Oldacre Ltd†	Ebrex holdings BV (sold July '93)
Fermanagh Creameries Ltd (N. Ireland)	Giltspur Inc (USA)
Wexford Creamery Ltd (Eire)	Prufrock Restaurants Inc (USA)
	Casa Bonito Inc (USA)

* Sold to Bernard Matthews in December 1993; † sold to Dalgety in September 1993

Table 2 Five-year review (31/3/93)

	1989	1990	1991	1992	1993
Turnover (£m)	2202	2241	2142	2021	1925
Pre-tax profit (£m)	166.8	91.1	(47.1)	65.7	104.1
Employees	35 195	31 316	29 618	27 698	27 257

Table 3 Financial record by division and geographical area

	Turnover (£m)		Operating profits (losses) (£m)	
	1992	1993	1992	1993
United Kingdom	1751	1825	88.9	92.7
USA	200	221	17.1	18.1
Other countries	29	31	1.4	0.6
Discontinued operations				
UK	96	6	(14.4)	0.5
USA	157	83	4.5	–
Other countries	17	11	(1.4)	(0.3)
Total	2021	1925	77.6	92.9
UK food	1072	1081	55.2	54.7
US restaurants	124	142	13.3	14.3
Wincanton	171	205	12.8	15.4
Giltspur	92	94	4.9	4.6
Vehicles sales and service	248	266	3.7	4.4
Other activities	44	37	(1.0)	(0.7)
Discontinued operations				
J.P. Wood	83	–	(13.7)	–
Other	187	100	2.4	0.2
Total	2021	1925	77.6	92.9

UNILEVER PLC AND NV

Joint Secretaries and registered office (for the PLC)

J.W.B. Westerburgen and S.G. Williams, Port Sunlight, Wirral, Merseyside L62 4ZA

Directors

F.A. Maljers (chairman NV), M.S. Perry (chairman plc), M. Tabaksblat (vice chairman NV), J. Peelen (US foods executive), O.O.H. Müller (N. Europe foods executive), C. Miller Smith (S. Europe foods executive), N.W.A. FitzGerald (detergents), A. Burgmans (personal products), H. Eggerstedt (finance), A.S. Ganguly (research and engineering), A.C. Butler (personnel), C.M. Jemmett (Latin America and Central Asia), A. Kemner (East Asia & Pacific), R.O. Brown (Africa and Middle East). There is also a team of 11 advisory directors

The Unilever Group is one of the world's largest businesses, employing 287 000 people worldwide. There are two parent companies – Unilever NV and Unilever PLC – which have the same Board of Directors and operate essentially as a single entity. Of the two parent companies, Unilever NV is the larger with a turnover (1992) of £17.2bn compared with £7.5bn for Unilever PLC. Reporting to the Board are specialist divisions or departments, product management groups and regional managements and it is through these that Unilever's global activities are coordinated. In the case of foods – which make up 51% of both turnover and operating profits – there is a Foods Executive consisting of three directors (see above). About 60% of group turnover and operating profits are generated in Europe, the remaining 40% being split equally between North America and the rest of the world.

During 1992, 22 companies were acquired and 21 disposed of. Recent disposals of the 4P Packaging group and a major part of the agribusiness operations means that four core product groups – foods, detergents, personal products and speciality chemicals – now account for 96% of sales. Despite the recession, the group maintained its commitment to research and development, spending £461m in 1992 (£426m in 1991) and to graduate recruitment, in the belief that competitive advantage lies in the quality of the workforce.

Main UK food companies

- Birds Eye Wall's Ltd – frozen foods and ice cream: BEW now has factories at Gloucester (ice cream, frozen cakes and desserts), Lowestoft (frozen vegetables and meat products), Grimsby and Hull (fish products, ready meals). Well-known brands include Birds Eye Menu Master, Country Club, Healthy Options, Gino Ginelli and Fish Cuisine; Wall's Too Good to be True, Magnum, Romantica and many others. During the last 2 years it has faced serious competition from supermarket own-label frozen foods which offer quality products at lower prices.
- Van den Bergh Foods Ltd – margarines, low-fat spreads, cheese, cooking aids, meat products, oils and shortenings. In January 1993 Unilever announced the merging of Mattessons Wall's with Van den Bergh & Jurgens to form Van den Bergh Foods Ltd, a company with a turnover of about £600m. Well-known brands include Wall's, Mattessons and Richmond's meat products, Flora, Delight, Elmlea, Boursin, Country Crock and Crisp'n Dry.
- Brooke Bond Foods Ltd – tea and tea products, dried soups and savoury products, canned vegetables, ambient microwaveable meals, meat extracts and stock cubes. Annual sales are over £400m and there are about 3600 people in the workforce. Well-known brands include Oxo, Batchelors, PG Tips, Ragu, Red Mountain and the Sir Thomas Lipton range of prestige teas. In March 1993, BBF sold the Fray Bentos canned meat products brand, worth about £35m, to Campbell's Food UK.
- John West Foods Ltd – canned meats, fish and fruits; microwaveable long-life ready

meals; jars of spices. The company has about 145 approved canneries in 35 countries and claims a 32% share of the £283m canned fish market. Future plans include the development of more value added canned fish products, e.g. tuna in sauces.

Half-year results to June 1993

Pre-tax profits rose 7% to £984m and turnover by 6% to £12.7bn. Comparable 1992 figures were £919m and £11.9bn.

Recent developments

In September 1993, Unilever acquired America's top selling ice cream brand, Breyer's for an estimated £330m. The edible oil business of Italina Cirio-Bertorelli-De Rica was acquired in November 1993, and in January 1994 a family-owned deep frozen and pre-baked bread and dough business in France, with a turnover of £20m was purchased.

In February 1994, Unilever announced a major worldwide restructuring programme costing £490m which will result in 7500 job losses, about 1000 of them in the UK. The sale of John West Foods was also disclosed.

Table 1 Principal group companies of Unilever PLC in the UK (31/12/92)

Foods
Birds Eye Wall's Ltd
Brooke Bond Foods Ltd
Lipton Ltd
Lipton Tea Company Ltd
Loders Croklaan Ltd
* Mattessons Wall's Ltd
* Van den Bergh & Jurgens Ltd
John West Foods Ltd

Personal products
Calvin Klein Cosmetics (UK) Ltd
Cheseborough-Pond's Ltd
Elida Gibbs Ltd
Elizabeth Arden Ltd
Rimmel International Ltd

Detergents
Lever Brothers Ltd
Lever Industrial Ltd

Other operations
H. Leverton Ltd
Plant Breeding International Cambridge Ltd
UAC Ltd
UAC International Ltd
UML Ltd
Unilever Export Ltd
Unilever U.K. Central Resources Ltd

Speciality chemicals
Joseph Crosfield & Sons Ltd
Laing National Ltd
National Starch and Chemical Ltd
Quest International Ltd
Unichema Chemicals Ltd
Unipath Ltd
Vinamul Ltd

* Merged January 1993

Table 2 Five-year review (PLC & NV at 31/12/92)

	1988	1989	1990	1991	1992
Turnover (£m)	17 625	20 962	22 734	23 163	24 700
Pre-tax profit (£m)	1454	1703	1782	1792	2029
Employees	291 000	300 000	301 000	298 000	287 000

Table 3 Financial record by division and geographical area (31/12/92)

	Turnover (£m)		Operating profit (£m)	
	1991	1992	1991	1992
Division				
Foods	11 923	12 731	1057	1088
Detergents	5218	5877	374	429
Personal products	2823	3200	242	312
Speciality chemicals	1879	1962	229	229
Other operations	1320	930	96	64
Total	23 163	24 700	1998	2122
Geographical area				
Europe	13 764	14 516	1242	1240
North America	4769	4846	336	385
Rest of world	4630	5338	420	497
Total	23 163	24 700	1998	2122

UNION INTERNATIONAL PLC

Secretary and registered office

J.R. Cuthbert, 13/16 West Smithfield, London EC1A 9JN; 071-248 1212

Directors

E.H. Vestey (chairman), T.J. Robinson (chief executive), P.A. Taylor (finance), The Lord Vestey, The Hon. M.W. Vestey, T.R.G. Vestey

Union International is one of Britain's oldest meat businesses involved in ranching, slaughtering, meat processing, wholesaling and retailing, shipping, cold storage and distribution. The ultimate holding company is Western United Investment Company.

In the last 3 years the group has experienced severe financial difficulties culminating in a pre-tax loss of £102.4m in 1991. Following a refinancing package agreed with banks in November 1992, the group has improved its management performance, reduced its debt, disposed of or closed non-core or loss-making businesses and returned to profit (£32.5m pre-tax profit for 1992).

Foods Division

Amberley Foods (with meat processing operations at Tipton and Middlesbrough) was sold. The Middlesbrough site was acquired by H. Hargrave (Wholesale Pork) Ltd (Section 2.3) and the Tipton business sold to Hazlewood Foods (Section 2.1). Ferndale Foods (formerly Downsfood) which manufactures cooked meats and egg products was sold to Hygrade Foods for £0.9m.

Dewhurst – retail butchers

Hard hit by consumers turning to major supermarkets for their meat, the Dewhurst chain has shrunk from 1200 outlets in 1989 to 400 in 1992. The number of suppliers has been reduced from 1200 to 200, central quality control introduced and a number of new retail formats trialled. These measures have returned Dewhurst to profit for the first time in several years.

British Beef (BB)

Abattoir operations in the UK have been affected by over-capacity and the implementation of new EC regulations which set higher standards and demand expensive veterinary supervision. In 1992, BB ceased trading at Scorrier, Fraserburgh and Market Harborough. Investment in the group's remaining abattoirs to meet EC standards, combined with restructuring on a site-by-site basis should ensure they continue to provide an acceptable return on investment.

Weddel Swift – meat wholesalers

There are 25 depots throughout the UK which service 25% of all UK high street butchers. A deport upgrading programme is currently in hand.

Weddel Swift Trading Desk

This manages Union's GATT licences and operates the primary meat exporting and importing activities of the group. While maintaining GATT activity at a fairly constant level, it has recently expanded its trading base, through agency representations, exports to Germany, former Eastern Bloc countries and the west coast of Africa, and developed additional areas for the trade in meat products. These non-GATT activities are now a major contributor to profits.

UK Property

Union's UK property porfolio had to be significantly written down to current market values, but sales during 1992, including the Dewhurst retail outlets netted £30m.

Australia

During 1992 pastoral operations, meatworks and a substantial amount of property were sold leaving Southern Country Foods Pty Ltd (canned meats), a fish catching and processing company, a retail division with 40 outlets and property interests owning 2000 ha of land with development permits.

New Zealand

Combined effects of new NZ labour legislation and the implementation of a quality management programme have raised trading profits from NZ$3.5m in 1991 to NZ$24m in 1992.

Venezuela

Ranching increased profits by 18% in 1992.

Brazil

Ranching, standing crop agriculture and food processing interests performed satisfactorily and there are plans to purchase Meatworks properties.

International Division

Union services airlines, hotels and catering outlets in the Far East, with turnover and profits increasing sharply in the last 2 years. Canned product distribution in the USA and Canada also remains profitable.

Table 1 Principal subsidiaries (31/12/92)

Registered in Great Britain	*Registered overseas*
Anglo Cold Storage & Produce Co. Ltd	W. Angliss Australia Investments Pty. Ltd (Australia)
British Beef Co. Ltd	Agropecuria CFM Ltda (Brazil)
Vestey Estates Ltd	S.A. Frigorifico Anglo (Brazil)
J.H. Dewhurst Ltd	Weddel Ltd (Canada)
Weddel Swift Ltd	The Hong Kong Refrigerating Co. Ltd (Hong Kong)
Union Traffic Ltd	The Malayan Refrigerating Sdn Bhd (Malaysia)
Wheatear Ltd (reg'd in the Isle of Man)	Weddel New Zealand Ltd (New Zealand)
	The Malayan Refrigerating Co. (Pte) Ltd (Singapore)
	The Tupman Thurlow Co. Inc. (USA)
	La Companania Inglesa (I.C.I.) C.A. (Venezuela)

Table 2 Five-year review (31/12/92)

	1988	1989	1990	1991	1992
Turnover (£m)	1233.3	1316.3	1386.7	1300.9	1091.2
Pre-tax profit (£m)	18.91	23.03	9.80	(102.4)	32.5
Employees	23 161	23 812	22 479	21 301	17 903

Table 3 Financial record by division and geographical area (by origin)

	Turnover (£m)		Profits (losses) after interest and exceptional items (£m)	
	1991	1992	1991	1992
Division				
Farming	25.4	53.6	29.5	29.2
Food manufacturing	457.7	515.5	(4.4)	14.5
Food wholesaling	562.1	397.5	(11.9)	8.9
Food retailing	225.7	88.9	(18.9)	2.9
Property holding	10.8	17.1	(32.9)	8.0
Other	19.2	18.6	(25.2)	0.09
Total	1300.9	1091.2	(63.8)	63.5
Geographic area				
United Kingdom	559.6	440.2	(87.3)	12.0
Europe (excluding UK)	106.6	44.5	(3.2)	1.3
USA and Canada	167.5	28.4	(0.4)	1.0
South America and West Indies	80.6	133.1	28.0	16.7
Australia and New Zealand	271.4	373.8	1.6	29.2
Far East	64.6	71.3	(4.3)	3.4
Africa	10.5	–	1.7	–
Interest			(38.5)	(31.1)
Total	1300.9	1091.2	(102.4)	32.5

UNITED BISCUITS (HOLDINGS) PLC

Secretary and registered office

D.R.J. Stewart, 12 Hope Street, Edinburgh EH2 4DD

Directors

Sir Robert Clarke (chairman), E.L. Nicoli (group chief executive), J.A. Warren (finance), B.D. Chadbourne (President, Keebler Co.), D.L.G. Hearn (MD, KP Foods), Lord Prior*, N.M. Shaw*, Lady Howe*, Sir Charles Fraser*, G. van Schaik*, C.M. Short*, Dr. T. Spickschen*, R.S. Napier*, T.H. Wyman*
* non-executive

The UB Group is a leading producer of snackfoods and biscuits with operations in 28 countries, employing about 40 000 people. It supplies retail and catering markets with branded and own-label biscuits, savoury snack, and chilled and frozen foods. Following the sale of its confectionery division in March 1993 (Terry's Group), it now operates through four divisions: KP Foods, Ross Young's, Keebler Company and McVitie's.

In difficult trading conditions, the group increased sales to just over £3bn in 1992 (£2.97bn in 1991) but had pre-tax profits 18% down, mainly attributable to a poor performance by Keebler in the USA.

McVitie's Group

This is the UK's No. 1 biscuit manufacturer, No. 2 in Europe. It employs 7500 people at ten UK factories. As well as its branded biscuits and cakes it also produces 42% of the own-label biscuit market. It has subsidiaries or associated undertakings in France, Spain, Netherlands, Hungary, Denmark, Norway, Finland, Japan, and the Philippines. In 1992, UK operations had sales of £594m and operating profits of £92.7m. Well-known brands include Jaffa Cakes, Digestives, Rich Tea, Hob-nobs, Penguin and Boasters.

Keebler Company

This is the second-largest manufacturer of cookies and crackers in the USA. It employs 10 400 people and has a turnover of around £1.05bn. In 1992 profits slumped by 60% to £26.2m. Nevertheless, there were successful launches of deluxe chocolate chip cookies and Sweet Spots. O'Boisies and re-launched pretzels gained share in the salty snacks market. In January 1993, Bake-Line Products Inc. was acquired, which is a major manufacturer of own-label cookies.

KP Foods Group

This is the leader in the UK snack market with a 40.2% share in 1992. Its brands include Frisps, Roysters, Hula Hoops, Skips, McCoys, Brannigans and KP Nuts. It also manufactures many own-labels. In Europe it markets products under the Croky label. Its Westimex business in Belgium had a particularly good year. In November 1992 it acquired the Coca Cola Amatil business in Australia, and re-named it The Smith's Snackfood Company and in February 93 it bought Derwent Valley, owner of the Phileas Fogg range of adult snacks. Derwent Valley had a turnover of about £24m and pre-tax profits of £2m in 1992. In 1993, the majority of the group's sales will come from outside the UK.

Ross Young's

Chilled and frozen food operations benefitted from the rationalization and cost-reduction programme imposed in 1991. In a static frozen food market, Ross Young's increased its market share to 18.2% with notable successes through San Marco pizzas, Oriental Express ready meals and value-added fish products for the Young's brand. Major investments were made at the Young's chilled fish factory at Grimsby and a new bakery desserts line at Okehampton making McVitie's cheesecakes and Cafe Continental pastries.

Table 1 Principal subsidiaries

UB Investments plc (Scotland)	Koninklijke Verkade NV (Netherlands)
United Biscuits (UK) Ltd (Scotland)	NV Westimex (Belgium) SA (Belgium)
UB (Ross Young's) Ltd	KP Foods France SA (France)
Keebler Company (USA)	Productos Ortiz SA (Spain)
Shaffer, Clarke & Co. Inc. (USA)	KP Larios (SA) (75%) (Spain)
U.B.F.C. Inc. (USA)	Industria Confezioni Alimentari Spa (60%) (Italy)
Oxford Biscuits Holding A/S (57.9%) (Denmark)	United Biscuits (China) Ltd (90%) (China)
Györi Keksz Kft (84%) (Hungary)	

Table 2 Associated undertakings

Aguia SA (45%, Brazil)	KP Ireland Ltd (50%, Ireland)
Meiji McVitie Ltd (50%, Japan)	Sociedade Industrial de Aperitivos Ltda (49%, Portugal)
CFC McVitie Inc (40%, Philippines)	Fazer Keksit Oy (49%, Finland)
Consolidated Biscuits (Trinidad) Ltd (28%, Trinidad)	

Table 3 Five-year review (2/1/93)

	1988	1989	1990	1991	1992
Turnover (£m)	2380.3	2723.7	2723.6	2979.1	3118.2
Pre-tax profit (£m)	177.1	229.3	181.0	197.2	162.3
Employees	47 004	46 958	40 565	40 226	38 698

Table 4 Financial record by division and geographical area (origin) (2/1/93)

	Turnover (£m)		Operating profits (£m)	
	1991	1992	1991	1992
Division				
McVitie's Group	787.3	830.6	106.7	111.8
Keebler Company	983.0	1050.6	65.6	26.2
KP Foods Group	504.3	530.2	43.5	44.8
Ross Young's	542.9	519.9	27.5	28.5
Terry's Group	153.4	176.5	14.3	13.4
Development operations	11.0	15.3	(1.7)	(1.5)
Inter-company sales	(2.8)	(4.9)		
Overheads/central svs			(12.5)	(14.1)
Exceptional items		(5.6)		
Total	2979.1	3118.2	243.4	203.5
Geographic area				
UK and Ireland	1627.4	1630.3	156.1	155.3
Continental Europe	336.1	399.3	20.0	25.4
USA	1006.7	1076.7	66.6	21.5
Rest of world	8.9	11.9	0.7	1.3
Total	2979.1	3118.2	243.4	203.5

SECTION 2.2 MEDIUM-SIZED COMPANIES

Companies with 500–5000 employees engaged in food manufacture.
Acatos & Hutcheson plc
Geo. Adams & Sons (Holdings) Ltd
Anglo Beef Processors Ltd
Avonmore Foods plc
F.W. Baker Ltd
A.G. Barr plc
Bass plc
Bensons Crisps plc
Berisford International plc
Borthwicks plc
Brake Bros. plc
B. Brooks (Norwich) Ltd
BSN (UK) Ltd
Campbell's UK Ltd
Cargill plc
Carr's Milling Industries plc
Cavaghan & Gray Ltd
CPC (United Kingdom) Ltd
Dalehead Foods Holdings Ltd
Dalepak Foods plc
Ess-Food (UK) Ltd
Express Foods Ltd
Faccenda Holdings Ltd
Favour Parker Ltd
James Finlay plc
Fyffes plc
Geest plc
Golden Vale plc
Grampian Country Food Group Ltd
Greggs plc
David A. Hall Ltd
Harrisons & Crosfield plc
H.J. Heinz Company Ltd
Hygrade Foods Ltd
William Jackson & Son plc
JLI Group plc
Kellogg Company of Great Britain Ltd
Kennerty Farm Dairies Ltd
Kerrygold Co. Ltd (The)
Kerry Group plc
Kitchens of Sara Lee – UK Ltd
Kraft General Foods Ltd
Lloyd Maunder Ltd
Marshall Food Group Ltd
Bernard Matthews plc
McCain Foods (GB) Ltd
McCormick ((UK) plc
MD Foods plc
Milk Marketing Board for N.I.
Moy Park Ltd
NewMarket Foods Ltd

Nickerson Group Rothwell Ltd (The)
G.W. Padley (Holdings) Ltd
Perkins Foods plc
Pigs Marketing Board (NI)
Portfolio Foods Ltd
William Price & Sons Ltd
Pride Foods Ltd
Quaker Oats Ltd
Raines Dairy Foods Ltd
John Rannoch Ltd
Christian Salvesen plc
Samworth Brothers Ltd
Scottish Milk Marketing Board
C. Shippam Ltd
Sims Food Group plc
SmithKline Beecham plc
Swizzels Matlow Ltd
Thorntons plc
Tulip International (UK) Bacon Division Ltd
Thomas Tunnock Ltd
Walkers Smiths Snack Foods Ltd
Warburtons Ltd
Warner Lambert (UK) Ltd
Waterford Foods plc
Webbs Country Foods Ltd
Weetabix Ltd
Whitworths Holdings Ltd
Wrigley Company Ltd, (The)
Yorkshire Food Group plc

ACATOS & HUTCHESON PLC

Secretary and registered office

P.G. Watson, Orchard Place, London E14 0JH; 071-987 2066; Fax 071-538 1796

Directors

I.S. Hutcheson (chairman), K.M. Ashken*, I.C. Caunt, G.A. Johnson, R.W. Seabrook*, Sir Ruthven Wade*, J. Weir
*non-executive

The group manufactures boxed cooking fats and bottled oils under the Pura label, and own-label margarines and cooking oils for several supermarkets. It also supplies bulk fats and oils to the bakery trade.

A fall in pre-tax profits from £12.25m in 1987 to £3.1m in 1990 led to the sale or closure of non-core businesses in the UK and Spain. In 1992, all the group's edible oils and fats businesses were combined into one company, Pura Food Products Ltd. The accompanying management reorganization, integration of procurement activity and reduced manning levels have reduced costs and improved quality at all three refineries (Orchard Place, Bootle and Trafford Park).

Half-year pre-tax profits for March 1993 were £5.14m compared with £4.23 (restated) for 1992 showing that the group has benefitted from major investments at the Orchard Place margarine factory, from lower interest rates and the devaluation of sterling. Nevertheless, prices in the retail sector remain under pressure and there is increased competition for the Peerless Food Products division in the supply of bulk fats and oils.

P & W Printers' performance was adversely affected by a substantial bad debt following the collapse of Clarke Foods (ice cream), but the Birkenhead haulage, storage and processing operations and the animal feed operations performed satisfactorily. The food packaging company P & W Plastics Ltd was sold and a settlement reached on an outstanding loan from the joint venture with Macisa in Spain. Following disposal of all the group's Spanish interests, substantially all profits are now earned in the UK.

Principal subsidiaries

Pura Food Products Ltd (formerly Central Edible Oils)
Pura Foods Ltd
P.I. Transtore Ltd
Vitafoods Ltd
P & W Printers Ltd
P.L. Belgium S.A. (99.6%, Belgium)
Wallworths Ltd
Agrocomex S.A. (90%, Spain)
Peter Smith (Bakers & Caterers) Ltd (42.5%)

Table 1 Five-year review (27/9/93)

	1989	1990	1991	1992	1993
Turnover (£m)	260.7	248.2	228.5	200.9	222.0
Pre-tax profit (£m)	4.5	3.1	6.7	7.6	10.2
Employees	1086	1093	1106	895	–

Table 2 Geographical analysis of group turnover

	1991	1992
United Kingdom	214.4	185.1
Continental Europe	13.3	15.6
Other	0.8	0.2

GEO. ADAMS & SONS (HOLDINGS) LTD

Secretary and registered office

J.N. Adams, Crescent House, Fulney Lane, Spalding, Lincs. PE12 6EZ; (0775) 766161

Directors

G.C. Adams (chairman and joint managing director), G.C.M. Adams, J.N. Adams (joint managing director)

The company was founded by the father of the group's present chairman, George Adams, who opened a butcher's shop in Spalding in 1910. Originally, the company concentrated upon developing a chain of retail butchers shops and a slaughtering facility but in the 1960s and 1970s the group began wholesaling, and acquired factories at Ruskington and Frampton.

Its EC-approved abattoir has a weekly throughput of 10 000 pigs, 5% of which comes from its own two pigs farms. It also handles about 100 cattle and 1000 lambs/week (1992 figures). 18% of output from the plant goes to manufacturing units at Spalding, Ruskington and Frampton; 16% of the fresh meat sales are now into export markets.

In 1990 a new pie and prepared foods factory was opened at the Crescent House site in Spalding. The Ruskington plant also make pies and pastries as well as sausages (500 tonnes/year) for Asda, Safeway and the Co-op. Refurbishments have been made at all the group's sites to ensure they meet legally required hygiene standards and as a result they now operate the British Meat Manufacturers' Association (BMMA) self-regulation scheme.

Difficult trading conditions in fresh meats means that the group has focused on value-added products and in 1992 its 'Pork Nutters' won Retail Pork Product of the Year. The frozen food operation suffered a fall in sales in the last 2 years with retained profits of £17 324 in 1991 falling to £9311 in 1992.

Some 60% of the group's manufactured products are sold under own-labels for the major supermarkets. However, it is planning to extend the range of products under the Adams brand. It also operates a van sales force from depots at Spalding, Ruskington, St. Neots, Leeds, Norwich, Braintree, Wolverhampton and Reading.

Table 1 Five-year review (31/3/92)

	1988	1989	1990	1991	1992
Turnover (£m)	15.4	19.5	24.3	29.3	31.6
Pre-tax profit (£m)	0.58	0.83	0.3	1.5	1.6
Employees	687	678	754	918	962

ANGLO BEEF PROCESSORS LTD

Secretary and registered office

J. McLaughlin, Battlefield Road, Harlescott, Shrewsbury, Shropshire SY1 4HA

Directors

A.G. Rees, P.M. Goodman, L.J. Goodman, J. O'Donnell, R.M. Cracknell

Anglo Beef Processors Ltd (ABP) is the UK subsidiary of the major Irish meat company Goodman International. Goodman has a turnover of around £900m and processes 2.5 million animals/year through Anglo Irish Beef Processors. Its ultimate parent company in Eire is Mourne Holdings Ltd.

Following the collapse of Middle Eastern markets for meat in 1991, particularly the Iraqi market, Goodman faced severe financial difficulties. In addition, a 'World in Action' TV progamme alleged that the company was dealing with intervention beef and the associated EC subsidies in a dishonest manner. Inevitably the company denied the allegations, but the Irish Courts appointed an examiner to investigate the affairs of Goodman International and the result was a debt restructure agreement, in which Larry Goodman passed control of 60% of his companies to a consortium of banks leaving him with 40%.

ABP was set up in 1991 to handle all the UK business of Goodman and includes the former operations of Euroscot Meat (now ABP Scotland), City Meat Wholesalers in Nuneaton, Warwickshire and Ellesmere Shropshire (now ABP Ellesmere and ABP Nuneaton).

ABP specializes in traditional beef. It is currently involved in the Partnership in Livestock group, through which farmers supply ABP Ellesmere with premium cattle destined for Sainsbury's 320 stores and Savacentres, under the Traditional Beef label. Farmers in the group qualify for a £6/head bonus if they meet certain stringent specifications and are assured of quick payment.

Financial information

The only figures available appear to be for the 11 months ending 31/3/92 for which there was a pre-tax profit of £0.17m on a turnover of £148.5m. The company had 1557 employees at that date.

AVONMORE FOODS PLC

Secretary and registered office

T.B. Graham, Avonmore House, Partick Street, Kilkenny, Ireland; (056) 22907; Fax (056) 21909

Directors

J. Duggan (chairman), J.F. Logan (vice-chairman), P.A. O'Neill+ (group managing director), R.P. Bennett, P. Delaney, H. Drea, P.J. Flaherty, E.Holland, T. Mather, G.J. Meagher+, W.G. Murphy+, R.J. O'Connell, P. O'Hanlon, R.E. O'Mahoney+, E.S. Power+, E. Stallard, G.E. Stanley, M. Walsh
+executive director

Despite the recession, Avonmore has achieved outstanding growth, more than doubling turnover between 1989 and 1992, making a number of successful strategic acquisitions, and taking further steps in its declared aim of becoming an international food group with turnover in excess of 1 billion IR£. Spread of the group's activities include food processing in Ireland, UK, Europe and the USA, with sales and marketing throughout Europe, USA, Mexico, Japan, Korea and Australia.

Discussions on a possible merger with Waterford Foods plc begun in May 1991 were discontinued in September 1991.

Operating divisions are dairy products, meat and agricultural trading. In 1992, acquisitions expanded both meat (notably Barretts & Baird, West Midlands, UK; United Meat Packers and Master Pork in Ireland; and Harzland Fleisch) and liquid milk (Whitecroft, Bennetts,Salisbury and Tom Parker' s dairies). In July 1993, the business and assets of the Midlands and Western divisions of Dairy Crest were also added to Avonmore for IR£21.62m. These had a turnover of £87m and pre-tax profits of £4.6m at the time of acquisition. Avonmore is now the fifth largest supplier of liquid milk in the UK.

Joint ventures were undertaken with Kempico in Belgium (51%, Golden Foods NV, cheese manufacture and marketing) and Hungary (75%, Avonmore Paszto Kft, milk products).

Table 1 Five-year review (2/1/93)

	1988	1989	1990	1991	1992
Turnover (IR£m)	295.4	403.1	499.7	575.2	831.4
Pre-tax profit (IR£m)	9.5	13.2	13.3	16.3	25.0
Employees	1646	1892	2771	3293	4228

	Turnover (IR£m)		Operating profit (IR£m)	
	1991	1992	1991	1992
Division				
Dairy products	364.9	421.5	15.1	18.5
Agri-trading	100.9	108.0	6.4	7.0
Meat products	109.4	302.0	5.2	11.0
Total	575.2	831.5	26.7	36.5
Geographic area				
Ireland	395.4	536.0	19.3	25.9
Europe (excluding Eire)	98.7	196.2	5.7	8.1
USA/other	81.1	99.3	1.7	2.5
Total	575.2	831.5	26.7	36.5

F.W. BAKER LTD

Secretary and registered office
J. Baker, Crick, Northampton NN6 7TZ; (0788) 823711; Fax (0788) 822960

Directors
J.W. Harrison (chairman), R.F. Baker, J. Baker, E. Baker, G.W. Baker

The main activities of the group are farming, meat wholesaling, pet food manufacture and transportation. Members of the group are Midland Meat Packers (EC-approved abattoir, meat wholesalers, beefburger production); Midland Pet Food Canners Ltd (pet food manufacturers using meat by-products and cereal from group farms); Baker Group Distribution (transport services); Ashby St Ledgers Farms Ltd (farming). The group farms 4000 acres, slaughters 4000 cattle and produces 600 tons of beef per week.

In a difficult year for the meat trade, group turnover was down only slightly. The pet food canning operation continued to make a significant contribtion to profits. Its Butcher's Tripe Mix brand has been the subject of heavy promotional activity recently with several new flavours added to the range.

Table 1 Five-year review (31/7/92)

	1988	1989	1990	1991	1992
Turnover (£m)	71.7	89.5	103.6	122.4	122.0
Pre-tax profit (£m)	1.28	3.1	3.4	5.0	2.8
Employees	581	604	643	807	859

Table 2 Geographical distribution of turnover (£m)

	1991	1992
United Kingdom	102.1	99.0
Continental Europe	20.3	23.0

A.G. BARR PLC

Secretary and registered office

G.M. MacLaren, 1306 Gallowgate, Glasgow G31 4DS; 041-554 1899; Fax 041-554 5768

Directors

W.R.G. Barr (chairman), W.G. Barr*, M.R.C. Barr, A.A. Bibby, J.E. Dawson
*non-executive

This Scottish-based soft drinks company claims to be the third largest soft drinks manufacturer in the UK with an 8% share of the soft drinks market (1992 figures). Its flagship brand Irn-Bru was first produced in 1901 and still has a 39% share of the carbonated soft drinks market in Scotland. Other well-known brands include Tizer, with a 2.3% brand share of 'other flavour carbonates' and St Clements high juice squash with about 25% of the high-juice sector. The diet carbonates sector is growing at three times the rate of the total carbonates market and all Barr brands are now available in the diet format.

The group's results for 1992 were adversely affected by a 16% price increase in sugar following the UK's departure from the exchange rate mechanism, fluctuating weather conditions, and extensive discounting at the commodity end of the soft drinks market to unrealistic levels. Barr's branded products continued to perform well showing a 4% increase in sales in 1991/2 while sales of retailer's own-labels declined. Nevertheless, the group continued to invest in its Scottish operations with a new in-line bottle blower for 1/4 litre PET bottles being installed at its Parkhead site in Glasgow. All Barr's production units now hold BS 5750 accreditation.

Reduction in demand for returnable glass bottles means that the Sheffield production site and depots at Leeds and Mansfield will close during 1993.

Interim Results – half year to 1/5/93

Pre-tax profits fell by 31% to £2.1m on sales 1% lower at £42m. Sales of Barr's branded products Tizer and Irn-Bru rose by 6% but sales of own-labels continued their decline.

Operating subsidiaries (all incorporated in England)

Barr Leasing Ltd
Hampshire Mineral Water Co. Ltd
Iceborg Ltd
Stotherts Ltd

Table 1 Five-year review (31/10/92)

	1988	1989	1990	1991	1992
Turnover (£m)	73.1	88.5	89.8	90.0	94.7
Pre-tax profit (£m)	3.5	4.3	3.6	3.3	6.0
Employees	1391	1509	1441	1382	1228

Note: the figure for turnover includes exports of £2.1 m which are not considered significant enough to warrant allocation to the markets involved

BASS PLC

Secretary and registered office

F.S. Wigley, 20 North Audley Street, London W1Y 1WE; 071-409 1919; Fax 071-409 8503

Directors

I.M.G. Prosser (chairman and chief exectutive), Lord Bancroft*, P. Bowman (finance), O.C. Darby, K.H.M. Dixon*, B.D. Langton, A.E.R. Manners (personnel), Sir Peter Middleton*, G.J. Mulcahy*, M.S. Perry*, A.D. Portno, J.R.D. Swan
*non-executive

Though best known for its brewing activities, Bass also owns the global Holiday Inn hotel chain and 50% of Britannia Soft Drinks, itself a major shareholder in Britvic Soft Drinks Ltd (see below). Major divisions within the group are Bass Brewers, Britvic Soft Drinks, Bass Taverns, Bass Leisure, and Holiday Inn Worldwide.

Britvic Soft Drinks

This company is held by Britannia Soft Drinks (90%) and Pepsico (10%). Britannia itself is held by Bass (50%), Whitbread (25%) and Allied-Lyons (25%). Its well-known brands include Tango, Britvic juices, Citrus Spring, Seven-Up, Corona, Quosh, Idris, R. Whites and Shandy Bass. Pepsi is also made and sold under a UK franchise. During 1992, sales volumes fell by 3% in a flat soft drinks market. Turnover increased by 0.6% to £442m while operating profit increased by 5.7% to £45m.

In 1991, it opened a new £5m break centre in Glasgow, which specializes in breaking down ingredients into mixed quantities of product for delivery to customers and depots. Factories at Chelmsford and Leeds (Swinnow Lane) produce soft drinks in returnable glass bottles for the 'on premises' market. Production sites at Leeds (Pudsey), Rugby, Wigan and Beckton (East London) produce drinks in cans and PET containers for the take-home market. Pudsey and Chelmsford also produce post-mix syrups for draught soft drinks in pubs and catering outlets. Britvic also has a network of 21 distribution depots.

Table 1 Five-year review (30/9/92)

	1989	1990	1991	1992	1993
Turnover (£m)	4036	4461	4383	4307	4450
Pre-tax profit (£m)	465	535	430	501	508
Employees, group	90 138	98 345	90 104	84 095	81 105
Employees, soft drinks			3998	3766	3526

Table 2 Financial record by division and geographical area (origin)

	Turnover (£m)		Operating profit (£m)	
	1991	1992	1991	1992
Division				
Drinks: Brewing	1056	1166	185	210
Drinks: Soft drinks	405	407	42	45
Retail: Pubs	1173	1091	231	174
Retail: Leisure	993	950	62	65
Hotels	570	510	103	116
Other	186	183	(1)	2
Total	4383	4307	622	612
Geographical area				
United Kingdom	3767	3807	519	496
Rest of Europe	174	134	15	15
United States	325	301	80	85
Other regions	117	65	8	16
Total	4383	4307	622	612

BENSONS CRISPS PLC

Secretary and registered office

A.M. Fiddian, Marquis Street, Kirkham, Preston PR4 2HY; (0772) 683383; Fax (0772) 671180

Directors

M.W. Jones (chairman and managing director), A.M. Fiddian (finance), G. Beech (production), D.B. Campion*
*non-executive

Bensons is the fourth largest manufacturer of crisp and snack foods in the UK with 6.7% of the crisp and 3.3% of the snack food markets. Its Retail Van Sales delivers direct to a range of licensed, CTN and independent outlets in England and Wales. Brands include Hedgehog, Crunch Crew, Lunchbox, Mammoth Bites and La Mexicana Tortilla Chips.

During 1991/2 the company has been adversely affected by a number of factors which have contributed to falls in turnover and profits. These include increased consumer purchases of low margin multipacks through the major retailers, increased prices for imported potato pellets and edible oils following sterling devaluation, poor quality old potatoes in 1991 which had to be replaced by more expensive alternatives and competition for independent retailers caused by Sunday opening of the main supermarket chains.

In 1991, investment in multi-pack equipment at the South Wales plant helped to increase trade with supermarkets and in April 1993 a successful rights issue raised £5m to be invested in a new factory and warehousing at Kirkham. Savings of about £1m are anticipated when all three Lancashire operations are centralized.

The acquisition of A.K. Snacks, Gerald Harvey Ltd and four other small businesses, led to some short-term problems during their integration, but subsequently led to a turnover of £10.2m for Retails Van Sales and an increase in its national licensed and catering customer base.

The group also has a subsidiary, Sight and Sound Animations, which produces models and displays. In 1991, it acquired Loxton Laminates, a small subcontractor.

Interim results

Interim results to 29/5/93 were disappointing, showing an increased loss of £386 000 (1992 loss of £191 000).

Table 1 Five-year review (30/11/92)

	1988	1989	1990	1991	1992
Turnover (£m)	14.7	18.2	22.0	26.3	33.2
Pre-tax profit (£m)	1.0	1.3	1.4	1.0	0.5
Employees	445	586	589	753	762

Table 2 Financial record by division and geographical area

	Turnover (£m)		Profits (losses) (£m)	
	1991	1992	1991	1992
Division				
Food Products	25.9	32.7	0.985	0.490
Models and displays	0.4	0.5	0.017	0.010
Total	26.3	33.2	1.000	0.500
United Kingdom	25.9	33.1		
EC	0.3	0.1		
Rest of world	0.1	–		
Total	26.3	33.2		

BERISFORD INTERNATIONAL PLC

Secretary and registered office

P.F.O'Connor, Berisford Wing, One Prescot Street, London E1 8AY; 071-481 9144

Directors

John Sclater (chairman), Frank Knight (deputy chairman), Alan Bowkett (chief executive), Peter Butler (chief financial officer), Barry O'Connor (company secretary), Jay Pritzker (USA)*, Robert Pritzker (USA)*, Dr N Brian Smith*
*non-executive

Following disastrous losses during 1989/1990, Berisford set about disposing of company assets to reduce its debts. At one stage the group had £1.2bn of debts funded by 68 banks in three separate banking pools. In 2 years this has been reduced to £15m funded by just four banks.

The sale of British Sugar plc on 2/1/91 to Associated British Foods plc netted £880m. Other disposals in the food and agribusiness group included the sale of Scotia Haven Foods to the Yorkshire Food Group for £11.1m, the sale of H.E. Daniel's flavours business to CPL Group, Bishop's Stortford. It also disposed of its almond processing business in California to a subsidiary of the Yorkshire Food Group, Treehouse Holdings Inc. in a deal involving cash plus preferred stock redeemable for cash over the next 5 years. Disposal of its sugar trading interest, J.H. Rayner, to C. Czarnikow Sugar Ltd left Berisford with a 15% holding in Czarnikow.

Berisford has also restructured its financial obligations to Rayner Coffee International, which has now ceased trading and sold its entire interest in its biotechnology company Enzymatix Ltd. Its residual cocoa interest (Bristar Trading) provided a satisfactory return.

In the year to October 1993, Berisford, effectively now a shell company, made a loss of £5m. Debt has now been eliminated and the company has £3.5m in cash.

Table 1 Principal subsidiaries, associates and investments as at 30/9/92

Food and Agribusiness
W-L Research Inc. (USA) (100%)
T & T Industries Inc. (USA) (100%)
Germains, Inc. (USA) (100%)
Bristar Trading Ltd (England) (49%)
Hunter Saphir plc (England) (20%)
Other operations
Ketlon (UK) Ltd (England) (98.6%)
Euro Trailer Rental Ltd (England) (95%)
Nacolah Holding Corp. Inc. (USA) (23.9%)
Pallas Invest SA (Luxembourg) (9.2%)
Property
Berisford Property Group Ltd (England) (100%)
Berisford Property Developments Ltd (England) (100%)
Berisford Property Investments Ltd (England) (100%)

Table 2 Five-year review (30/9/92)

	1988	1989	1990	1991	1992
Turnover (£m)	2645	1927	1949	694	130
Pre-tax profit (loss) (£m)	101.6	107.3	(96.1)	(20.5)	(22.3)
Average nos employed	9108	6717	n/a	2113	909

BORTHWICKS PLC

Secretary and registered office

B.E. Buck, Bank House, Primett Road, Stevenage, Hertfordshire SG1 3UQ; (0438) 743900; Fax (0438) 746406

Directors

J.M. Thomson (non-executive chairman), P. Brackenridge (chief executive), J. Beaumont (finance), D.C.P. Carey, G.R.H. Clemons

In 1986, Borthwicks began a radical transformation of the group into a natural flavours and food ingredients business. It disposed first of its meat and pastoral operations in New Zealand and Australia and more recently of its abattoirs in the UK. During 1993, refocusing has been completed by the sale of Burton Son & Saunder's bakery business and the Consumer Products division which made prepared meals. The remaining abattoir at Ringer in Sussex was sold to Anglo Dutch Meats (Charing) for £285 000. Despite a pre-tax loss of £2m (March 1993) resulting from disposal costs of £2.4m, Borthwicks is now smaller with a sounder financial base and aspirations to join the world's top ten flavour companies. In the latter half of 1993, Globe Extracts and Barnett & Foster were combined to form Borthwicks Flavours Ltd.

Natural flavours

Barnett & Foster: 1992 figures show record sales in the UK and overseas, with sales to food manufacturers exceeding those to soft drink manufacturers. Despite the recession and poor summer weather which depressed sales of flavours to soft drink manufacturers, 1993 operating profits were maintained at 1992 levels. Cooke, Tweedale and Lindsay was fully integrated into the Wellingborough plant, and the redundant facility at Manchester sold. Sales from the European Regional Centre in Holland have increased by 43%.

Major International: this provides natural stock products mainly for the catering industry. It is a joint venture company with Major Products Inc. of the USA. Catering exports to Europe were satisfactorily initiated. Catering sales in the UK remained depressed.

Global Extracts, USA: acquired in 1989, and subjected to considerable management and product rationalization, it now has a reputation for supplying a range of high-quality, higher margin flavours. 1993 results show a profit increase of 16% for Globe.

Speciality fats

Broadlands Foods: this is a speciality fats business based in Birmingham, which supplies manufacturing and retail markets.

Interim results for 6 months to 25/9/93

Sales rose from £12.88m (1992) to £14.51m and pre-tax profit from £0.83m (1992) to £1.13m.

Table 1 Five-year review (27/3/93)

	1988	1990*	1991	1992	1993
Turnover (£m)	184.3	122.8	51.3	44.4	43.0
Pre-tax profit (loss) (£m)	1.0	(1.2)	0.4	1.3	(1.99)
Employees	1434	760	827	579	566

* 18 months September 1988–March 1990

Table 2 Financial record by division and geographical area (destination)

	Turnover (£m)		Profits (losses) (£m)	
	1992	1993	1992	1993
Divisional analysis				
Continuing operations	25.3	25.6	2.1	2.0
Discontinued operations	19.1	17.4	(0.3)	(0.7)
Total	44.4	43.0	1.8	1.3
Continuing operations				
United Kingdom	14.6	15.3		
Europe	1.2	1.6		
North America	6.8	6.5		
Rest of world	2.7	2.2		
Discontinued operations				
United Kingdom	17.1	15.6		
Rest of world	2.0	1.8		
Total	44.4	43.0		

BRAKE BROS PLC

Secretary and registered office

J.C. Huber, Enterprise House, Godinton Road, Ashford, Kent TN23 1EU; (0233) 637370; Fax (0233) 648359

Directors

W.T. Brake (chairman), F.R. Brake (managing director), M.J. Champion (finance), B.J. Martin*
*non-executive

The principal activity of Brake Bros is supplying frozen foods to the catering trade in the UK. It also has plans to acquire and integrate a number of distributors of chilled and frozen foods in France building up a business comparable with that in the UK. It currently owns Anjou Surgelation SA in Western France and Cogehalles SA covering the Paris region.

Despite the recession, the business has performed well with an increase in pre-tax profits of 3.7% in 1992. The acquisition and integration of Everfresh Frozen Food Ltd and Midfish Ltd have yet to make an impact on profits, and development of the Fine Food Division (mainly the business of London Larder Ltd) has proved more expensive than envisaged.

There has been considerable capital investment (£17.9m in 1992) on cold stores at Highbridge (Somerset), Bishop's Stortford (Hertfordshire), and Hemsworth (Yorkshire), and satellite depots at Bodmin (Cornwall) and Aberdeen (Scotland), a specialized delivered meals factory at Kettering (Northants) and conversion of the Lenham (Kent) factory to pudding production.

Recent developments

In May 1993, Brake Bros bought Country Choice Foods for £10m in shares. Country Choice supplies frozen bakery products to bakeries, fish and chip shops and snack bars. In the year to February 1993, it made profits of £1.55m on a turnover of £40m. It employs 250 staff and operates from eight depots.

Table 1 Five-year review (31/12/92)

	1988	1989	1990	1991	1992
Turnover (£m)	124.8	145.3	195.4	223.0	280.5
Pre-tax profit (£m)	9.6	11.8	14.9	15.4	16.0
Employees	1408	1584	1979	2261	2718

Table 2 Geographical analysis of turnover (£m)

	1991	1992
United Kingdom	223.0	267.5
Europe (excluding UK)		13.0

B. BROOKS (NORWICH) LTD

Secretary and registered office

Mrs S.A. Brooks, Little Melton, Norwich, Norfolk NR9 3NP; (0603) 810137; Fax (0603) 812272

Directors

B.E. Brooks (chairman and managing director), Mrs S.A. Brooks, B.K. Brooks, A.D.C. Lewis, S.A. Barnett (New Zealand), R.A. Burnside (New Zealand)

The company founded by Barry Brooks in 1971, sold a 50% stake to the Primary Producers Co-operative Society of New Zealand (PPCS) in 1988. Shares in Brooks are actually held by the Canterbury Frozen Meat Co., a subsidiary of PPCS. PPCS, which also owns the London-based New Zealand Lamb Co. Ltd, had been a major supplier to Brooks for many years.

The company operates from an EC-approved plant near Norwich. It has recently received EC approval for its cold store and bacon processing unit at St Ives which it bought from Dalgety in 1991. More than 35 000 New Zealand lambs are cut each week to exact customer specifications. Joints can be vacuum- or shrinkwrapped. Beef and pork carcases are also stored, cut and processed.

Beefburgers, lamb-burgers, sausages and a range of frozen value-added meat products for caterers are also produced. Form-fill seal machines produce 'free-flow' frozen minced and diced beef, pork and lamb. The company supplies most of the major supermarket and freezer centres in the UK with own-label goods, but plans to develop Brooks' branded goods more widely in future.

Table 1 Financial review (30/1/93)

	1988	1989	1990	1991	1992	1993
Turnover (£m)	41.7	45.9	52.57	60.07	65.1	73.4
Pre-tax profit (£m)	1.1	1.4	0.27	2.08	1.8	2.2
Employees	601	612	624	651	695	733

Table 2 Turnover (£m) by geographical market

	1992	1993
United Kingdom	62.9	71.0
Rest of EC	2.2	2.3
Rest of world	–	0.1
Total	65.1	73.4

BSN (UK) LTD

Secretary and registered office

R.T. Burdon, 45 Northampton Road, Market Harborough, Leicestershire LE16 9BQ; (0858) 410144

Directors

C.R. Lanbie, J.C. Horen, R.T. Burdon, S.D. Freedman, L. Sauguet, N.J.L. Maulin, L.A.L. Zinsou-Derlin

The BSN Groupe SA is the largest diversified food group in France, Italy, Spain and Belgium and has operations throughout the world. It is actively involved in developing the emerging markets in South-East Asia. Its origins are based on a glass company Boussois-Souchon-Neuvesel (BSN) formed by a merger of two French glass operations in 1966. It is still Europe's second largest producer of glass containers, but also has major interests in dairy products, grocery products, pasta and ready-to-serve dishes, biscuits, beer, and mineral water. Its well-known brands include Danone, Galbani, Panzani, Tuc, LU, Kanterbrau, Evian, Volvic, Jacobs, Lea & Perrins, Daddies, Amoy Soy Sauce and HP Sauce.

In the UK it has five production plants and employs over 4000 people producing HP, Daddies and Lea & Perrins Sauces and Jacobs Bakery Products.

In July 1993, BSN acquired a 40% stake in BE International Foods, a London-based group which imports Indian and Chinese foods. BEI has been the UK's sole distributor of Amoy Foods soy sauces for the last 25 years. Amoy was acquired by BSN 2 years ago. BEI has a turnover of about £47m.

The Jacob's Bakery

(1992, T/O £197.8m, and pre-tax profit £8.7m, 2811 employees) Under its former owners Nabisco, the Jacob's Bakery operations had been segmented by brands: Jacob's, Huntley & Palmers, Peak Frean's and Club. This strategy seemed inappropriate for the 1990s and under BSN the Jacob's Bakery brand is being much more aggressively exploited and aims to become a creditable challenger to McVitie's (UB Group) dominant position. In 1992 UK biscuit sales fell by 4% in volume, with Jacob's market share decreasing slightly; however, sales increased by 2.3% to £226m.

HP Foods Ltd

(1992, T/O £82.9m and pre-tax profit £10.2m, employees 748) A new variety of ketchup with Worcester Sauce was launched in 1992 under the Lea & Perrin's brand. Generally the market for brown sauces and ketchup remained flat with intense price competition. HP sales increased by 2.5% to £83m. HP Foods has two main factories at Market Harborough and North Walsham.

Table 1 Four-year review (31/12/92)

	1989	1990	1991	1992
Turnover (£m)	218.35	2.80	0.20	0.34
Pre-tax profit (£m)	9.92	(5.38)	(0.98)	8.36
Employees	5002	42	–	–

CAMPBELL'S U.K. LTD

Secretary and registered office

D.A. Lacy, Heathrow Boulevard, 284 Bath Road, West Drayton, Middlesex UB7 0DT; 081-2564 8686; Fax 081-2564 8620

Directors

G. Lord, J.L. McKeowin, J.W. T. Mustoe

Campbell's UK is the UK subsidiary of the American soups and grocery company, Campbell Soup Co. Campbells Frozen Foods changed its name to Campbell's U.K. Ltd on 26 March 1990. Its main subsidiaries are Campbell Grocery Products Ltd (canned foods produced at King's Lynn), Campbell Chilled Foods Ltd (factories at York and Peterlee) and Campbell Frozen Foods Ltd (factories at Braintree, Salford, Peterlee and Glasgow). (Despite investment in new plant and equipment for vegetarian ready meals, the Braintree plant closed during 1993.)

Restructuring and rationalization in 1988 returned the UK group to profit during 1991. Slaters (Port Talbot) Ltd, producing chilled ethnic meals, was sold to Madigan Foods. The catering and bakery distribution business was sold to Mill House Foods. Campbell's Freshbake, cut from nine plants to four in 1990, now claims to be market leader in value frozen bakery products. Campbells condensed soups improved sales volumes by 14% in a market which declined by 2.6% in 1992/3.

In March 1993, acquisition of the Fray Bentos brand from Brooke Bond Foods for an undisclosed sum was announced. The Fray Bentos range, worth £35m, will be manufactured at Campbell's King's Lynn factory and will complement its existing range of savoury canned products.

Campbell Chilled Foods Ltd

T/O £10.48m; pre-tax profits £0.3m; employees 203

Campbell Grocery Products Ltd

T/O £49.85m; pre-tax profits £2.5m; employees 485

Campbell Frozen Foods Ltd

T/O £89.96m; pre-tax loss (£4.41m); employees 1525

Campbell Foods plc

This processes frozen and dried vegetables in Ireland. Its main subsidiaries are Premier Vegetables Ltd, Campbell Irish Foods Ltd, Mallow Foods Ltd and Campbell Foods Ireland Ltd.

Table 1 Five-year review – Campbell's UK Ltd (2/8/92)

	1988	1989	1990	1991	1992
Turnover (£m)	47.1	82.5	201.3	186.3	150.3
Pre-tax profit (loss) (£m)	(3.2)	(11.1)	(12.3)	(9.2)	(1.55)
Employees	654	1104	3318	2821	2241

Table 2 Two-year review – Campbell's Foods plc

	1990	1991
Turnover (£m)	79.0	13.3
Pre-tax profit (loss) £m	(10.5)	(1.2)
Employees	2342	376

CARGILL PLC

Secretary and registered office

J.C. Reynolds, Knowle Hill Park, Fairmile Lane, Cobham, Surrey KT11 2PD; (0932) 861000; Fax (0932) 861200

Directors

R. Murray (chairman), D.A. Nelson-Smith, Rt. Hon. Earl of Carrick, A. Blankestijn (Netherlands), J. S. Erikson (USA)

Cargill plc is the UK operating arm of Cargill Inc. USA, a diversified, global group employing 70 000 people in 60 countries. It has interests in commodity trading (grain, oilseed, cocoa, coffee, meat, sugar, cotton, petroleum), international transportation and shipping, the supply of agricultural materials, financial and technical services and food processing.

Cargill plc has 5000 employees in 22 plants and offices throughout the UK involved in agriculture, food processing, trading, risk management and financial services.

The maize milling plant at Tilbury produces starch and glucose used in pharmaceuticals and a wide range of processed foods. In Liverpool and Hull oilseeds (rapeseed and soya) are crushed and the oils refined for use in cooking oils, sauces, salad dressings, margarines and spreads. Oilseed and milling by-products form the basis of animal feeds.

Sun Valley Poultry

This is a leading producer of value-added poultry products as a consumer brand in its own right and as an own-label supplier. At its Hereford sites there are 3500 employees who handle 750 000 chickens/week. The Abergavenny plant processes 85 000 turkeys/week. More than 30 million chickens are processed under the Sun Valley label every year. It is an integrated farm-to-factory operation, with plants in Hereford, Abergavenny and Blaenavon which process chicken and turkey into products such as Chicken Kiev for sale to retail, wholesale and catering outlets in the UK and Europe.

Table 1 Five-year review (31/5/92)

	1988	1989	1990	1991	1992
Turnover (£m)	1588.6	1606.8	1166.8	1256.6	1301.1
Pre-tax profit (£m)	2.787	3.429	6.663	(0.245)	15.452
Employees	4110	4538	4479	5296	6020

Table 2 Financial record by division and geographical area (origin)

	Turnover (£m)		Pre-tax profits (losses) (£m)	
	1991	1992	1991	1992
Division				
Commodity trading	789.0	807.0	(0.935)	6.905
Foodstuff processing and distribution	457.0	479.0	9.048	15.986
Financial services	11.0	15.0	0.522	0.430
Net interest	–	–	(8.880)	(7.869)
Total	1257.0	1301.0	(0.245)	15.452
Geographical area				
UK	1246.0	1287.0	6.937	22.486
Rest of world	11.0	14.0	1.698	0.835
Net interest	–	–	(8.880)	(7.869)
Total	1257.0	1301.0	(0.245)	15.452

CARR'S MILLING INDUSTRIES PLC

Secretary and registered office
R.C. Wood, Old Croft, Stanwix, Carlisle CA3 9BA; (0228) 282291

Directors
I.C. Carr (non-executive chairman), D.B. Armstrong (group managing director), R.C. Wood (finance), J.E. Tudor, C.N. Holmes*, A. R. Heygate*
*non-executive

The company dates from 1831 when Jonathan Carr set up business as a baker and dealer in meal and flour. The Solway Mill at Silloth, opened in 1886, remains a key part of Carr's activities today with a major modernization programme carried out in 1992. The company now operates in two main areas: agribusiness and flour milling and baking. It also has engineering (Keytor) and vehicle body building (Hinds) interests and has recently acquired the remaining 25% of Dickensian Coffee Houses. North Country Eggs has continued to perform poorly due to low egg prices and a decline in consumer demand, leading Carr's to withdraw from egg production and packing in 1993.

Flour milling and bakery continue to be affected by intense price competition, overcapacity in the industry and changes to the Monetary Compensatory Amounts designed to limit the effect of EC currency fluctuations on flour prices. Nevertheless, Carr's has a strong position as a high-quality regional supplier.

In the agribusiness, sales of cattle, pig and sheep feed increased and there was a reduction in the sale and supply of low-margin poultry feeds. Fertilizer blending operations and merchant branches, which include machinery and animal health products increased both sales and profitability.

Trading subsidiaries

B.R.B Trust Ltd	Financial services
Carr's Flour Mills Ltd	Flour milling
Dickensian Coffee Houses Ltd	Coffee houses and restaurants
R. Hind Ltd	Commercial vehicle body builders/repairers
John Stronach (Shipbrokers) Ltd	Shipping agency
Keytor Ltd	Engineering
Carrs Agriculture Ltd	Agricultural feed millers and merchants
Robertson (Bakers) Ltd	Baking
Caltech Biotechnology Ltd	Manufacturers of biotechnological products
North Country Eggs Ltd	Egg producers and packers (closing in 1993)

Table 1 Five-year review (28/8/93)

	1989	1990	1991	1992*	1993
Turnover (£m)	82.1	77.7	69.0	68.2	73.1
Pre-tax profit (loss) (£m)	1.1	0.5	(0.2)	(0.3)	1.12
Employees	1133	1151	965	891	805

* Re-stated according to FRS No. 3

CAVAGHAN & GRAY LTD

Secretary and registered office

J.P. Murphy, Brunel House, Brunel Way, Carlisle, Cumbria CA1 3NQ; (0228) 518200; Fax (0228) 518215

Directors

A.H. Sanderson (chairman)*, H.R. Sims (chief executive), I.D. Husband, P.T.H. Osborne, J.P. Murphy, J.A. Nightingale*, H.A. Vice*
*non-executive

Cavaghan & Gray is a successful, privately owned food manufacturer with a turnover of about £80m. It was established over 80 years ago as a pork and bacon producer and maintained strong links with the British Meat Manufacturers' Association for many years. It has its own farms producing pigs and vegetables which are used in the company's products. Bacon production ceased in 1989 because it had been trading unprofitably for some time.

In 1990, in response to consumer demand for convenience and variety, the company invested £22m in new facilities and now produces over 100 different products, including chilled recipe dishes for Marks and Spencer as well as quiches, pies, sausages, crisp-bake products and a new range of long-life microwaveable products. It is also developing more exotic, Indian products. It operates from five factories and a distribution centre, all within the Carlisle area. Subsidiary, Alex Ross, is a fish merchant employing about 100 people. In 1991 it acquired Calthwaite Dairy Products Ltd for £1. Calthwaite had experimented unsuccessfully with an exotic range of desserts in 1990 using premises belonging to Cavaghan & Gray. It currently produces yoghurts and related products. In 1992, the Emile Tissot ready meals business was bought from Hunter Saphir for £3.75m. Cavaghan & Gray also owns Lakeland Meats Ltd.

Table 1 Five-year review (28/3/92)

	1988	1989	1990	1991	1992
Turnover (£m)	55.395	63.632	71.526	72.903	74.908
Pre-tax profit (£m)	1.913	2.478	1.606	2.189	2.272
Employees	1559	1689	1784	1820	1734

CPC (UNITED KINGDOM) LTD

Secretary and registered office

Mrs S. Flook, Claygate House, Esher, Surrey KT10 9PN; (0372) 462181; Fax (0372) 468775

Directors

P. W. Phillips* (chairman), I.M. Ramsay (managing director), R.A. Crawford, A.C. Krauss*, A.B. Simon*, A.J. Terry
*non-executive

CPC (United Kingdom) Ltd is a group undertaking of CPC Europe (Group) Ltd, both being part of the US parent company CPC International Inc. (CPC Europe has a turnover of about £1.5bn, with 14 000 employees in 20 European countries.) Its main activities are the manufacture and sale of branded food products. The UK operation has factories at Paisley (Knorr soups, sauces, stock cubes, stuffings and casserole mixes), Redditch (Adpac Division), Burton-on-Trent (Marmite and Bovril) and Lifton (Ambrosia desserts and puddings).

Best Foods Division supplies the retail trade with CPC's well-known branded products (Knorr, Aromat, Napolina, Frank Cooper's, Mazola, etc) while the Caterplan Division offers a similar range of the group's products to the catering trade. The Food Ingredients Division supplies specialist seasonings and savoury flavours to food manufacturers under the Garrick Foods and Bovril labels.

Following the failure of a cost-cutting exercise at the Paisley factory, it is likely that production of Knorr products will be transferred to factories in France and Italy with the loss of about 450 jobs at Paisley.

Subsidiaries

Operating
Napolina Ltd
Frank Cooper Ltd
Napolina Italia SRL (incorporated in Italy)

Non-operating
Adpac Ltd
Best Foods Ltd
Brown & Polson
Caterplan Ltd
Corn Products Co. Ltd
Garrick Foods Ltd
Knorr Anglo-Swiss Ltd

Table 1 Five-year review (30/9/92)

	1988	1989	1990	1991	1992
Turnover (£m)	121.3	131.3	182.3	242.5	250.5
Pre-tax profit (£m)	6.45	6.34	(5.30)	5.63	9.55
Employees	1168	1206	1599	1972	1958

DALEHEAD FOODS HOLDINGS LTD

Secretary and registered office

R.M. Adamson, Cambridge, Linton, Cambridge CB1 6JD; (0223) 891000

Directors

A.S. Ambler, P.W. Merrick, A.W. Phillips, J.S. Hughes, C.F.R. Kemp

Dalehead's main activity is the production, processing and packaging of fresh meat, particularly pork, for supermarkets and major manufacturers. Formerly Farmers Pork Distributors, Dalehead has built its expertise in the pork business over the last 22 years, emphasizing quality and efficiency as well as animal welfare. Dalehead has EC-approved abattoirs in Cambridge and Mansfield and factories in Bury St Edmunds and Cambridge. In 1991 it also acquired the Somerwest bacon plant and abattoir in Bristol from GrandMet and the Dixon Pork & Bacon abattoir in Ashton under Lyme from Booker.

Bacon is cured by the traditional Wiltshire method at the Bristol Westerleigh plant. Since the takeover Dalehead has raised hygiene and efficiency standards at the plant, increasing pig throughput from 1500 to 6500/week. Most of the Westerleigh bacon is sliced and packed outside the Dalehead plant although some is packed and sold under the Somerwest brand locally.

Table 1 Four-year review (16/5/92)

	1989	1990	1991	1992
Turnover (£m)	33.33	43.28	50.07	96.92
Pre-tax profit (£m)	1.12	1.52	2.55	1.05
Employees	321	379	438	817

DALEPAK FOODS PLC

Secretary and registered office

P.N. Holley, Dale House, Leeming Bar, Northallerton, North Yorkshire DL7 9DQ; (0677) 424111; Fax (0677) 424443

Directors

M.D. Abrahams (chairman)*, C.R. Ivory (chief executive), P.N. Holley (finance), J. Kerwin, E. Honeywell, R.S.E. Tucknott, T.H.P. Brown*
*non-executive

The company was established in 1976 as a manufacturer of frozen and chilled foods, notably grill steaks. In the late 1980s it acquired Fawcetts Frozen Foods Ltd, Arnold Worsley & Sons (Barnsley) Ltd (1988), Newdale Bakery Co. Ltd, Frozen Foods Ltd and Northallerton Wholesale Bacon. Following further losses, Arnold Worsley was subsequently closed.

Dalepak Frozen Foods

In the year to April 1993, turnover rose by 4% to £34.8m. Sales of grillsteaks increased slightly in a static market. Sales of vegetable grills/burgers grew substantially, with Dalepak now claiming a 30% share. The product range now includes value-added poultry products. A capital investment programme has entailed installation of computer controlled freezing and automated handling plant and a continuous, six-line integrated production system. Delays in commissioning the new plant did adversely affect profits for the year.

Other Divisions

Fawcetts Frozen Foods increased profits on sales up 8%, maintaining its base of catering customers. Dalepak Ready Meals and Northallerton Wholesale foods were profitable despite depressed market conditions.

Table 1 Five-year review (30/4/93)

	1989	1990	1991	1992	1993
Turnover (£m)	24.54	33.71	40.66	40.61	42.23
Pre-tax profit (£m)	1.41	1.92	3.02	3.93	2.21
Employees	416	606	685	632	608

ESS-FOOD (UK) LTD

Secretary and registered office

A. Cowlishaw, 12 The Broadway, Amersham, Bucks HP7 0HP; (0494) 434343

Directors

J. Josefsen, J. Edelmann, C. Mathieson, A. Cowlishaw, C. Christensen

Prior to October 1990, Ess-Food in the UK comprised three companies: the Danish Bacon Co. (wholesaling and distribution with 23 depots), Ess-Food UK (meat trading) and Danepak (bacon processing, slicing and packing for the Danepak brand as well as for retailers' own-labels).

In October 1990, Tulip International A/S (TI) was formed from the processing operations of Tulip, Danepak, Normeat and Jaka. As part of the deal the Thetford Danepak plant became the head office of Tulip in the UK, and Ess-Food acquired shares in the newly formed group. The two Jaka meat slicing units in Liverpool and the Normeat plant at Wellingborough were retained as part of TI.

Already an international operation, the parent company Ess-Food embarked on a joint venture with the Dutch slaughterhouse group Encebe. Also, the merger of Tulip, Wenbo, and Ost Jyske to form Danish Crown brought Danish Crown into the Ess-Food group. For Ess-Food UK it meant that bacon and pigmeat sales are increasingly coming from non-Danish sources *viz.* it sells about 100 000 tonnes of Dutch, Belgian, English, Irish and French meat. In January 1991, it merged operations with Danish Viking Food, closing Viking's Aylesbury depot.

By the end of 1992, Ess-Food had no food manufacturing operations and was involved entirely in sales and distribtion activities. The Danish Bacon Company plc is now the principal shareholder in Ess-Food UK Ltd.

The ultimate parent company of the Danish Bacon Co is Ess-Food (Denmark). Danish Bacon's other immediate subsidiary is Danish Viking Food UK Ltd. Its primary function is the distribution and retailing of Danish Bacon.

Table 1 Ess-Food UK Ltd – Four-year review (3/10/92)

	1989	1990	1991	1992
Turnover (£m)	85.73	73.51	32.76	43.73
Pre-tax profit (£m)	0.28	(0.28)	0.16	(0.01)
Employees	–	14	15	14

Table 2 Danish Bacon Company plc – Four-year review (3/10/92)

	1989	1990	1991	1992
Turnover (£m)	166.93	168.85	195.95	212.75
Pre-tax profit (£m)	0.55	1.32	0.93	(0.24)
Employees	1439	1348	1278	1224

EXPRESS FOODS LTD

Secretary and registered office

A.N. Leigh, Keaton House, Widewater Place, Moorhall Road, South Harefield, Middlesex; (0895) 825999; Fax (0895) 826194

Directors

S.M.D. Oliver (chairman*), J.R. Davenport (managing director), R.E. Trott, J.M. Murphy, A.N. Leigh, (finance), T.D.J. Syder, D.G. Milne
* non-executive

In November 1992, Grand Metropolitan disposed of Express Foods for £96m to a management buyout team headed by Roger Davenport, the current managing director. The core of the business is cheese manufacture and chilled food distribution and the new company currently has five creameries and two distribution depots. Express had been owned by GrandMet since 1969 and at takeover had a turnover of £260m, a trading profit of £13m and 1700 employees. 41% of Express's income was from retail, 42% from food service and 17% from sales to other manufacturers and exporters.

Its trading divisions are: Express Foods Export; Express Foodservice; Express Foods Ingredients; Tuxford & Tebbutt.

In its first 11 months of trading as an independent company (to October 1993), pre-tax profit was £7.48m on sales of £289.84m. Milk prices increased during the year as a result of green pound devaluations. Sales of butter and whey products remained depressed.

With the sale of the Express Dairy and Eden Vale brand names to Northern Foods, the Foodservice Division now supplies the catering, cash and carry and wholesale sectors with products under the Bladen, Cuisine, Bodyline, Tuxford & Tebbutt and Lockerbie brand names.

Claiming to be the largest independent cheesemaker in the UK, Express launched 'The Cheese Company' brand to take over from Eden Vale.

FACCENDA HOLDINGS LTD

Secretary and registered office

Mrs S. Faccenda, Willow Road, Brackley, Northamptonshire NN13 5HB; (0280) 703641

Directors

R.M. Faccenda (chairman), Mrs S. Faccenda, D.M. Williams, J.E. Ames, B. Noble, P.C. Chatterton

Faccenda is a holding company for a number of businesses engaged primarily in poultry breeding, production and processing. In April 1992 it became a subsidiary of Hillesden Investments Ltd.

The group processes over a million fresh and frozen birds a week from three plants. It also operates as a poultry wholesaler supplying many leading manufacturers, runs a number of broiler farms and has a broiler breeding unit at Rochford in Essex.

Hinton Poultry was acquired from Hillsdown in 1990. Based near Bath it has a weekly throughput of 30 000 birds and supplies the major retailers.

Table 1 Five-year review (30/4/92)

	1988	1989	1990	1991	1992
Turnover (£m)	62.0	72.0	96.3	119.2	113.8
Pre-tax profit (£m)	6.65	2.25	0.69	2.41	1.45
Employees	679	755	920	1098	1077

FAVOUR PARKER LTD

Secretary and registered office

B.D. Trollope, The Hall, Stoke Ferry, King's Lynn, Norfolk PE33 9SE

Directors

M.J.B. Parker (chairman), P.C.J. Coles, J.G. Parker, G.G. Allen, K. Boardall, W.B. Holmes, G.T. Fox, S.F. Parker, D. Carter, N.J. Roche

Favour Parker is the holding company responsible for businesses engaged in farming, chicken rearing, feed production, poultry production and processing, and garage services. In the year to February 1992, the animal feeds division traded profitably, but farming again suffered from adverse market conditions. Sovereign Chicken continued to trade unprofitably. The Burwash factory ceased production, and the supply of frozen whole chickens was discontinued.

Sovereign Chicken produces extensively reared, free-range poultry under the Home Farm brand and own-label. It also claims to be one of the first companies to reform white breast meat to produce Chicken Kiev, Chicken Cordon Bleu, Chicken Calabria, etc. Its 5-year plan is to extend its production of value-added products at its Attleborough plant, particularly into private labels. Products will include other meats and vegetable-based dishes. To reflect this change of emphasis, in October 1993 the group announced that Sovereign Chicken would in future be called the Sovereign Food Group.

Table 1 Five-year review (1/2/92)

	1988	1989	1990	1991	1992
Turnover (£m)	108.2	105.1	108.7	109.6	116.1
Pre-tax profit (£m)	2.3	1.7	0.5	0.9	(4.1)
Employees	1754	2043	2065	2121	2073

JAMES FINLAY PLC

Secretary and registered office

R.W. McCracken, Finlay House, 10/14 West Nile Street, Glasgow G1 2PP; 041-204 1321; Fax 041-248 4751

Directors

R.J.K. Muir (chairman), R.G. Capper (deputy chairman), A.C. Afxentiou*, D.W. Forsyth, J.M. Ingleby*, E.J.R. Scott*, R.L. Stone-Wigg, G.D.W. Swire*, M.J.B. Todhunter*
*non-executive

The group trades in five main areas: plantations; trading, manufacturing and merchanting; merchant banking and international confirming; oil and gas; confectionery and beverage manufacturing.

Tea

The group has estates in Kenya, Bangladesh, Australia and Sri Lanka. During 1992, there was a strengthening of the tea market, although there was a decline in prices. The Kenya estates made a profit of £9.8m in 1992, £4.6m more than in 1991. The second phase in the expansion of the instant tea plant in Kenya was completed and worked at full capacity during the year to meet the increasing demand for cold water soluble product used mainly in mixes for soft drinks in the US and for hot water soluble product used for the instant teas in Britain. A small instant tea plant in Bangladesh is also performing well.

The improved quality of Bangladeshi tea allowed sale of substantial quantities in the UK, following a withdrawal from these markets by the Russians. Losses were recorded for tea estates in Australia, and an investment in run down estates in Sri Lanka will require long term commitment by managers to bring them up to standard.

Confectionery and beverage manufacturing

These operations are carried out through George Payne & Co. Ltd. Turnover in 1992 was £66.1m, with operating profits of £4.2m. Comparable 1991 figures were £63.2 and £4.5m.

In 1991, George Payne's tea packing operation was finally moved from the company's original premises in Bermondsey to its new tea factory in Swindon. The group pack tea exclusively for Sainsbury's. Construction of a coffee roasting and packing factory on the same site as the tea packing factory at South Emsall in Yorkshire is also underway. The Hull plant has been successful in developing good quality decaffeinated black teas. The 1993 interim statement (14/9/93) reports increased demand for these with an accompanying improvement in results.

Paynes also make Poppets, Just Brazils and Lift Lemon Tea. With a slight decline in the confectionery market, Payne's did well to maintain trading profits at 1991 levels.

Interim results

In the 6 months to 30/6/93, pre-tax profits increased from £3.6m to £3.8m on turnover up from £70.7 to £76.3.

Five-year review (31/12/92)

	1988	1989	1990	1991	1992
Turnover (£m)	184.6	183.5	149.1	143.7	165.2
Pre-tax profit (£m)	7.0	12.0	9.5	8.9	10.1
Employees	33 967	33 419	33 260	32 354	33 290

FYFFES PLC (FORMERLY FII FYFFES PLC)

Secretary and registered office

J.F. Gernon, 1 Beresford Street, Dublin 7; 010 353 1 730200

Directors

N.V. McCann (chairman), C.P. McCann (deputy chairman), J.E. Callaghan (chief executive), D.V. McCann (managing director), P.G. McNamee, R.E. Benner, A.J. Ellis, D.J. Bergin*, J.F. Flavin*
* non-executive

Fyffes plc is involved in the importing, distribution and marketing of fresh fruit and produce in the UK, Ireland and continental Europe. Although the group reports its results in Irish pounds, most of its business and earnings are in pounds sterling.

Results for 1992 reflected the difficult trading conditions with fruit and vegetable markets seriously over-supplied. The company envisaged that stability would return to the banana market in 1993 as a result of new EC structures introduced. A major capital investment in the replacement of all the UK banana ripening depots has now been completed.

1990 and 1991 saw the acquisitions of Glass Glover Fresh Produce for IR£6.6m, the Padwa Group for IR£8.1m and James Lindsay for IR£11.9m. In 1992 several planned acquisitions did not go ahead, but negotiations with Saba Trading AB of Sweden resulted in a trading agreement through which Fyffes would supply a substantial proportion of Saba's banana requirements. In July 1993 Fyffes acquired 50% of the share capital of BRDR Lembcke A/S for IR£51.7m. Lembcke is the largest Danish importer of fresh fruit and also has major interests in pot plant trading. Lembcke is a wholly owned subsidiary of Axel Johnson AB.

In the UK, Fyffes' subsidiary the Fyffes Group Ltd employs over 1000 people.

Table 1 Fyffes plc – Four-year review (31/10/93)

	1990	1991	1992	1993
Turnover (IR£m)	557.14	607.95	524.31	611.39
Pre-tax profit (IR£m)	26.25	27.11	28.54	31.78
Employees	3260	2446	2411	n/a

Table 2 Fyffes Group Ltd – Two-year review (31/10/91)

	1990	1991
Turnover (£m)*	361.1	218.6
Pre-tax profit (£m)	9.08	4.63
Employees	2571	1056

* pounds sterling

GEEST PLC

Secretary and registered office

Jane K. Scriven, White House Chambers, Spalding, Lincolnshire PE11 2AL; (0775) 761111; Fax (0775) 710445

Directors

J. van Geest (Life President), L.W. van Geest (chairman), D.A. Sugden (chief executive), B.G. Wallace (finance), J.W. Spirling (personnel), M. Dowdall*, A.J. Habgood*
* non-executive

With the sale of its 50% stake in Macfish in September 1992 and two American subsidiaries, Manor Hill Foods and Sidari's Italian Foods in February 1993, Geest has refocused its operations on fresh produce, chilled prepared foods and European markets. It has operating subsidiaries in the UK, Netherlands, Belgium, Windward Islands, Costa Rica and the Bahamas and Guernsey and associated undertakings in the Netherlands, Spain and Italy.

Fresh produce

Geest is Europe's third largest importer and distributor of bananas. It has banana ripening centres in all EC countries and imports 220 000 tonnes from Costa Rica, 270 000 from the Windward Isles and 50 000 from the Canaries. Implemetation of the new EC banana regime in 1993 will restrict imports of 'dollar' bananas and offer some protection to Windward Isles suppliers. Mangoes and pineapples are also sourced from Costa Rica, and there has been development of vegetable pre-pack facilities in Kenya. Geest now has 250 retail greengrocer shops trading under the 'Field Fresh' banner.

Wholesale Services Division

This offers the foodservice industry a wide range of fruit, vegetables and prepared foods from nine distribution depots throughout the UK, under the Cater Fresh brand. The division has a fleet of 140 temperature controlled vehicles.

Food Preparation Sector

This has three divisions. Prepared Foods based in Spalding produces dressed salads, dips and recipe dishes. Convenience Foods includes chilled pizza production at Harrow (Katie's Kitchen), fresh pasta production at the new £6m facility at Barton on Humber (The Pasta Factory) and the new £5m speciality bread facility currently under construction on the same site. Prepared Produce has grown from the single small facility at Bourne acquired in 1989 to four sites (Bourne in Lincolnshire, Deal in Kent, and Shieldness Produce and Caledonian Foods in Scotland). The Bourne site is currently being re-fitted to make it the most advanced salad bagging unit in the UK.

Table 1 Five-year review (2/1/93)

	1988	1989	1990	1991	1992
Turnover (£m)	463.8	499.0	594.6	626.3	625.0
Pre-tax profit (£m)	18.3	20.0	24.8	26.2	20.4
Employees	5142	5304	6246	6740	6998

Table 2 Segmental analysis

	Turnover (£m)		Trading Profit (£m)	
	1991	1992	1991	1992
Fresh produce	508.3	523.2	19.3	15.0
Food preparation	109.7	98.5	5.4	6.2
Other activities	8.3	3.3	1.0	1.0
Total	626.3	625.0	25.7	22.2

GOLDEN VALE PLC

Secretary and registered office
G. Curley, Charleville, County Cork, Republic of Ireland

Directors
D.F. Wallis (chairman), M. Doody (vice chairman), P.J. McGuane (vice chairman), J. O'Mahoney (chief executive), W.J. Irvine (finance), K. Daly, F. Hogan, J. Kennedy, T.J. Lalor, G. Madigan, P. McKenna, P. O'Brien, D. O'Callaghan, J. O'Connor, P. Pyne, L. Slattery, D. Ward

Golden Vale (GV), listed on the Irish Stock Exchange, is Ireland's largest producer of processed cheese, and the fourth-largest cheese manufacturer in Europe. Other products in GV's range include full and low-fat dairy and non-dairy spreads for own-labels and its own Golden Olive low fat spread and Easigold butter with oils. A recent addition has been an enzyme-modified cheese with 25–30 times the strength of normal cheese flavours which is heat- and process-stable. Despite the general decline in butter sales, GV retains substantial markets in Scotland and Northern Ireland.

Since 1989 Golden Vale (GV) has grown rapidly through acquisitions which have included Dairy Produce Packers, the dairy division of Food Industries, several dairies in South Wales, Leckpatrick in Northern Ireland, Vejle Margarinefabrik in Denmark, the Dutch processed cheese manufacturer, Vonk, Haslington Cheese and Ingredients and Cheese Express (UK agent for Vonk and grated cheese supplier). (See also Company Changes, Section 1.3 for more details.) It is predicted that recent acquisitions will increase GV's turnover to over IR£500m in 1993. Half-year results announced in September 1993, showed an increase in pre-tax profits of 18% to IR£9.14m.

During 1992 a new Specialist Food Ingredients plant was established at Charleville which will increase GV's range of dairy based powders, creams and cheeses sold to industrial and catering users.

GV's agribusiness includes the manufacture and sale of animal milk replacers, cattle, pig and poultry feeds, fertilizers, hardware and animal health care products. While sales of animal feeds increased, turnover as a whole in this segment declined by 1.5% in 1992. The loss-making Mullagh feed mill was closed.

In July 1993, Golden Vale UK was set up to handle the UK businesses. These include Golden Vale Dairies Ltd based in Bridgend with a turnover of about £17m employing 160 people and Golden Vale DPP Ltd in Llantrisant and Manchester which supplies processed cheese products to fast food outlets and sandwich makers.

Table 1 Five-year review (31/12/92)

	1988	1989	1990	1991	1992
Turnover (IR£m)	155.3	192.7	209.7	295.5	298.0
Pre-tax profit (IR£m)	8.3	10.1	12.5	14.7	16.7
Employees	n/a	n/a	n/a	939	957

GRAMPIAN COUNTRY FOOD GROUP LTD

Secretary and registered office

Ian Smith & Co., 20 Queens Road, Aberdeen AB1 6YT; (0224) 662040; Fax (0224) 685853

Directors

A.J. Duncan (chairman), A.M.D. Stephen, W.W. Paris, R.G. Barnes, H.J. Tocher, D.P. Witham

1993 saw the development of this fast growing group from a collection of independent operating units to a fully integrated group operating from 'plough to plate' in pigs and chickens. It owns farms, hatcheries, feed mills and both chicken and pork processing units. The group's most recent acquisition is Cymru Country Chickens, Llangefni in Wales, bought from Hillsdown Holdings in October 1993. Cymru includes 22 breeding and growing farms, a hatchery, a feed mill and a frozen oven-ready operation.

Grampian Country Pork

This is Scotland's largest EC-approved pork processor, handling 6000 pigs a week at its modern plant in Buckie. In 1992, it entered a partnership with the Scottish Pig Industry Initiative and Wm. Low supermarkets to promote outdoor-bred, farm-assured Scottish pork. It currently rears 3000 pigs outdoors, with another 1000-sow herd being established at Nairn. In addition to supplying leading supermarkets in the UK, about six container loads a week of quality pork cuts are exported to manufacturers in Holland, France, Denmark and Germany.

Grampian Country Chickens

This handles 1 million birds a week via its five chicken processing operations at Banff and Bucksburn in Scotland, Thorne (Thorne Poultry) and Garstang (North Country Poultry) in Yorkshire and Llangefni (Cymru Chickens) in Wales. All these plants are EC approved and both branded and own-label, fresh and frozen whole and portioned birds are produced. The flagship brand is the Super Roaster 5–9 lb chicken reared for 11 weeks on a special diet. From 1993, all branded chickens and chicken products will be sold under the Grampian logo, with the Thorne and North Country logos being dropped.

The Further Processed Division at Bucksburn will be expanded in 1994 with the opening of a new £3m value-added factory at Ellon, just north of Aberdeen. This is largely in response to a dramatic increase in demand for value-added poultry products from the foodservice sector. At present the range includes formed products: burgers, nuggets and fingers, filled and formed: kiev, cordon bleu, Mexican and Italian and fillets: light and spicy, sweet and sour, etc., and bone-in: drumsticks and nibbles.

Group members

Grampian Country Chickens Ltd, Grampian Country Pork Ltd, Thorne Poultry Ltd, Yorkshire Country Feeds Ltd, Grampian Country Chickens (Bucksburn) Ltd Further Processed Division, Grampian Country Chickens (Bucksburn) Ltd, Grampian Country Feeds Ltd, North Country Poultry Ltd, Vale Royal Hatcheries Ltd, Cymru Chickens Ltd.

Table 1 Four-year review (31/5/92)

	1989	1990	1991	1992
Turnover (£m)	52.8	105.6	158.0	139.9
Pre-tax profit (£m)	(0.22)	8.1	1.2	(0.42)
Employees	1064	1489	2055	1734

GREGGS PLC

Secretary and registered office

N. Calvert, Fernwood House, Clayton Road, Jesmond, Newcastle-upon-Tyne NE2 1TL; 091-281 7721

Directors

I.D. Gregg (chairman), M.J. Darrington (managing director), M. Simpson (finance), C.S. Gregg*, N. Calvert*, S. Curran*, Miss S.I.L. Elkin*
* non-executive

The group manufactures bread, flour, confectionery and savoury products sold through its own retail shops. It operates one of the largest bakery chains in the UK with a total of 497 shops operating in the following areas: Rutherglen, Gosforth, Leeds, Manchester, Birmingham, Enfield and South Wales. During the year 16 new shops opened and three closed. In London there are now 47 shops.

1992 saw turnover increase by 5.7% to over £100m for the first time, with profits increased by 17.6% to £7.1m. Improved performance was attributed to stable ingredient prices and improved efficiency throughout the group. The sandwich business is developing strongly, together with the cakes and savoury products, while intense competition and a price war in the bread sector has adversely affected sales.

The company is currently building the New Enfield Bakery in Mill Marsh Lane near its existing bakery and Ponders End Head Office. The new bakery will cost £9m and cover 30 000 sq.ft, with an extra 10 000 sq.ft of offices. 450 people are employed at Enfield and this is expected to increase when the new bakery is fully operational.

Table 1 Five-year review (26/12/92)

	1988	1989	1990	1991	1992
Turnover (£m)	70.8	76.7	87.0	95.5	100.9
Pre-tax profit (£m)	5.7	6.7	7.0	6.0	7.1
Employees	5328	5639	6220	6045	6075

DAVID A HALL LTD

Secretary and registered office

L. Murray, East Main Street, Broxburn, West Lothian EH52 5AL; (0506) 853300

Directors

J.A. Hall (chairman), L. Murray, D.A.H. Hall, I. Brown, C.L. Smith, S.J. Collins, J.H. Currie, B.E. Sealey

In 1951 the brothers David and James Hall began a small meat product manufacturing operation to supply their father's three butchery shops on the outskirts of Glasgow. Subsequently the shops were sold and the manufacturing operation expanded. The company was incorporated in 1955 and moved to its present site in Broxburn in 1963. It is still an independent family-run business. The premises have been extended several times and now occupy 14 400 sq.m. An additional 9000 sq.m site at Woodburn was burnt down within a few days of opening, but 95% of production was restored within 2 weeks by installing new equipment at the Broxburn site. An extension at Broxburn to replace the fire-damaged plant should be fully operational by mid-1994.

Halls produce mainly pork products including bacon, ham, gammon, sausages, sliced cooked meats, black pudding and patés as well as a range of meat pies, etc. It is Scotland's leading bacon producer with 80 tonnes/week. About 50% of sales are under the Halls brand, distributed through van sales to small retailers. The rest of the business goes to the major supermarkets as own-label products. In January 1991, Halls acquired the brand name and manufacturing rights for McKellar Watt's range of meat products from Campbell's UK.

Halls has animal lairage, an EC-approved slaughterhouse and butchery and curing facilities. It operates six distribution centres in Scotland and one in Tyne Tees, England and a fleet of 100 vans.

Table 1 Five-year review (29/8/92)

	1988	1989	1990	1991	1992
Turnover (£m)	33.1	36.0	42.4	46.9	48.23
Pre-tax profit (£m)	1.61	0.41	2.03	3.76	2.01
Employees	1012	1049	1176	1263	1297

HARRISONS & CROSFIELD PLC

Secretary and registered office

P. Brown, One Great Tower Street, London EC3R 5AH

Directors

J.N. Maltby (chairman), G.W. Paul (chief executive), W.J. Turcan (finance), P.J. Savage, J. Miller, P.G.W. Simmonds, G.E.M. Brown, T.D. Preseton, J.G. Barnard, R.A. Easton, Sir Richard Lloyd*, M.J. Hartnall*, H.S. Mellor*
* non-executive

Harrisons & Crosfield is a diversified international group with interests in chemical and industrial operations, timber and building supplies, plantations, and food and agriculture.

Food and agriculture

This comprises businesses in four market areas. BOCM Pauls Ltd, formed following the acquisition of Unilever's feed business, claims to be the largest manufacturer of top quality compound animal feed. Integration of the two operations led to the closure of 13 mills during 1992 and 1993. It is also the largest pig producer in the UK through Breckland Farms, its joint venture with J. Sainsbury plc.

Pauls Malt is a leading European maltster. The recession led to reduced consumption of beer and whisky worldwide and the closure of maltings at Grimsby and Kirkcaldy. Pauls Food Group manufactures packet soups, sauce mixes, instant drinks and breakfast cereals in the UK and France. Sales and profits increased in 1992, with H & C Cereales in France increasing breakfast cereal sales by over 50%. Telford Foods (the main UK supplier of own-label packet soups) also benefitted from a new breakfast cereal plant. In order to concentrate on its core activities the group sold Soutra Foods (rusk and coatings) to Allied-Lyons and Glentham International Ltd to McCormick (UK) plc.

Edward Baker manufactures complete dry dog and cat foods under the Bakers Complete brand which is now sold by many UK and continental supermarkets.

Table 1 Five-year review (31/12/92)

	1988	1989	1990	1991	1992
Turnover (£m)	1804	1919	1794	1833	1942
Pre-tax profit (£m)	122.9	130.3	94.1	71.2	85.0
Employees	27254	29004	29181	33047	32177

Table 2 Financial record by division and geographical area

	Turnover (£m)		Operating profits (losses) (£m)	
	1991	1992	1991	1992
Division				
Chemical and Industrial	507	516	29.5	27.7
Timber and Building supplies	734	724	22.8	20.3
Food and Agriculture	436	612	28.7	38.6
Plantations	55	72	10.5	20.1
Discontinued operations	101	18	2.3	0.3
Property and other disposals	–	–	(1.4)	3.3
Total	1833	1942	92.4	110.3

H.J. HEINZ COMPANY LTD

Secretary and registered office

M. Cook, Hayes Park, Hayes, Middlesex UB4 8AL; 081-573 7757; Fax 081-848 2325

Directors

J.F. Hinch (chairman), A.L. Barrett, L.H. Balfe, J.A. Connell, P.I. Corddry, R. Elliott, R.J. England, W.A. Fera, R. Millar, A.J.F. O'Reilly, B.R. Purgavie, A.J.H. Skrimshire, R. Stewart

Following the year-end in April 1993, the ultimate parent company, H.J. Heinz Company, Pittsburgh, USA announced its intention to implement a £15m restructuring plan for the UK operation during 1993/94. At present, however, the company continues to be involved in the manufacture, processing, growing and distribution of food with factories and farms at Harlesden in North London, Kitt Green and Chorley in Lancashire, Poling in Sussex, Horley, Woking and Camberley in Surrey and Deeside in Clwyd. Future developments are likely to focus on new recipe dishes. The Weight Watchers range is now the largest brand in the low calorie and healthy food category with 80 products in 15 markets. Nine Lives and Award petfood brands have recently been introduced, as well as canned tuna. For Heinz babyfoods, ringpull ends and plastic lamination on cans are the latest developments.

During the year Sous Chef was acquired (although it is currently not trading). It specializes in the use of ohmic heating and aseptic packing techniques to produce shelf-stable catering foods. Custom Foods Ltd completed a new factory at Dundalk in Ireland, to supply the UK group with pizza products for UK and European markets. The frozen and processed division of Blue Prince Mushrooms was sold to a management team (now called Orchidwood Mushrooms), with Heinz retaining the fresh mushroom division, but closing two farms at Angmering and Bradford-on-Avon. Halls European Foods, a manufacturer and distributor of pizza products joined the group in 1991. Montrose Foods Ltd sells packaged food products mainly in Germany. In 1992 Heinz acquired a 22.8% stake in Industrias De Alimentacao Ltda in Portugal and H.J. Heinz Company (Ireland) Ltd took over responsibility from Erin Foods for trading in Heinz products in Eire and Northern Ireland.

Table 1 Five-year review (24/4/93)

	1989	1990	1991	1992	1993
Turnover (£m)	423.2	455.1	484.8	489.4	485.8
Pre-tax profit (£m)	44.6	45.7	43.4	31.9	45.6
Employees	4564	4386	4117	3681	3486

HYGRADE FOODS LTD

Secretary and registered office

P.A. Bigwood, 43 Upper Wickham Lane, Welling, Bexleyheath, Kent DA16 3AJ; 081-304 7829

Directors

Hilda M. Bigwood, J. Bigwood, R.B. Kerr (USA), R.S.W. Eades, G.W. Finley, R.J. Backler

The company, founded by Bernard Bigwood, developed from a family butchers business. The manufacturing side soon outgrew the retail side and the shops were sold. The company moved to the premises in Latona Road, London in 1962. The parent company is Latona Foods Ltd.

Hygrade Chilled Foods Ltd operate from the original Latona Road site. In 1983 the former Chard Meat Co. factory was acquired and now houses Hygrade Foods and Oscar Mayer Quality Foods, which makes frozen ready meals for the major retailers. The old Wall's meat factory at Chippenham was also bought by Hygrade in the early 1980s and provides a slicing and packing operation.

The most recent acquisitions include Ferndale Foods Ltd, bought from Union International Group for £0.9m in 1992, and Borthwicks Consumer Products purchased in February 1993 for £275 000 in cash. Prior to takeover, the Borthwicks business had made a pre-tax loss of £152 000. It manufactures fresh, ready-to-cook meals for supermarkets and caterers. With the addition of the these two companies the Hygrade Foods Group as a whole now employs about 1000 people and is predicted to turn over around £120m in 1993.

Table 1 Five-year review (28/3/92)

	1988	1989	1990	1991	1992
Turnover (£m)	34.1	44.7	47.9	52.2	58.7
Pre-tax profit (£m)	0.71	0.88	0.10	2.19	0.71
Employees	777	949	895	900	951

WILLIAM JACKSON & SONS PLC

Secretary and registered office

P. Farley, 40 Derringham Street, Hull, HU3 1EW; (0482) 224131; Fax (0482) 24526

Directors

P.B. Oughtred (chairman and joint managing director), R.H. Mountfield (vice-chairman), C.M. Oughtred (joint managing director), P.W. Farnsworth, P. Farley, J.A.M. Oughtred, M.N. Oughtred

A family business established in 1851, the Hull-based group of companies has a diverse range of interests that includes retailing, food manufacture and catering. Recent results for the group have been disappointing and a number of businesses have been sold or closed.

Retailing

A number of the larger supermarkets trading under the Grandways name were sold: nine to the Argyll Group for £23m and a total of 19 to Kwik Save for about £14.5m (seven in 1991 and a further 12 in 1992). This leaves the group with 33 existing stores operating as Jacksons Family Food Stores.

Manufacturing

Production lines for morning goods at Jackson's Bakery were improved and a new bakery is being built at Hull. At Tryton Foods, frozen and chilled businesses were integrated and the meat processing operations at the Cottingham factory rationalized.

Catering

The catering division has been slimmed down and commitments at the Market Rasen Racecourse relinquished.

Table 1 Five-year review (24/4/93)

	1989	1990	1991	1992	1993
Turnover (£m)	180.92	196.00	207.76	205.79	151.29
Pre-tax profit (£m)	4.17	4.22	2.98	(1.00)	0.89
Employees: full-time	2411	2403	2487	2171	1801
Employees: part-time	2455	2437	2422	2116	1841

JLI GROUP PLC

Secretary and registered office

P. Newman, JLI House, Guildford Street, Chertsey, Surrey KT16 9ND; (0932) 569600; Fax (0932) 569585

Directors

J.M. Alexander (chairman)*, Y. Gottesman (chief executive), I.W. Bayer (finance), B. Land*, P.M. Ainsworth*, R. Temple*
* non-executive

From small beginnings in the Covent Garden fruit market, Jack L. Israel Ltd has developed into a group with a turnover of £104.7m in 1993. It operates in three divisions: snacks, ingredients and distribution.

Snacks

(Turnover £32.8m; 365 employees at units in Bootle and Leeds.)

Tee Gee Snacks Ltd produces salty snacks, nuts, savoury mixes and popcorn and was formed from the merger of Playtime Foods (acquired 1991) and Golden Twin and TG Smith (Fine Foods) (both acquired in 1992). It supplies both retail, foodservice and sport and leisure outlets.

Ingredients

(Turnover £31.2m; 658 employees at units in Pocklington, Birmingham, Boston and Skelmersdale.)

The division includes Sundora Foods (dried fruits), Cadec Ltd (nuts for confectionery, bakery and cereal customers), Jack L. Israel (Norfolk) Ltd (brined vegetables for use in pickles, relishes etc.) and Langwoods (fresh vegetables for retailers and caterers).

Distribution

(Turnover £40.7m; 115 employees.)

Jack L. Israel Ltd, Chertsey, Surrey and Brookerpaks Ltd, Newington, Kent supply supermarkets, caterers and manufacturers with canned goods, concentrates, pulps, honey, nuts, garlic, ginger, corn and beetroot. Brookerpaks Ltd was acquired in September 1992.

Recent developments

In October 1993, JLI bought Turban Foods from Field & Co for £50 000, including intercompany debts of £1.7m. Turban processes and distributes nuts, dates and other dried fruits at Martlesham Heath in Suffolk and has a turnover of £4.9m and operating profits of £200 000. It will be integrated with Sundora Foods. JLI is also paying up to £3m for Frozen Herbs of Suffolk which made £0.5m profit on sales of £1.84 in 1992.

Table 1 Five-year review (31/3/93)

	1989	1990	1991	1992	1993
Turnover (£m)	63.4	82.5	90.1	84.3	104.7
Pre-tax profit (£m)	1.9	2.0	2.3	3.2	5.2
Employees	390	567	542	769	1148

KELLOGG COMPANY OF GREAT BRITAIN LTD

Secretary and registered office

H.I. Keepe, The Kellogg Building, Talbot Road, Manchester M16 0PU; 061-869 2000; Fax 061-869 2100

Directors

T.A. Knowlton (chairman and managing director), G. Lord (deputy managing director), J.A. Chappell, H.I. Keepe, B.R. King, R.S. Morgan, J.A. Newhouse, R.C. Taylor, G.W. West, T.P. Mobsby, W.F.T. Humes

Ownership

The company, which was registered in England in 1924, is a subsidiary of the Kellogg Co., USA.

Principal activities

Kellogg has held brand leadership in the cold ready-to-eat breakfast cereal market for many years. Its overall market share of the £895m cereals market is estimated to be 45%. Kellogg's Cornflakes alone hold 13%. The company's product range includes Cornflakes, Frosties, Special K, Ricicles and Bran Buds.

Of the major producers of breakfast cereals in the UK – Quaker Oats, Weetabix and the former RHM breakfast cereals business (now operated by Cereal Partners) – Kellogg is the largest, and most profitable. Unlike its competitors, Kelloggs continues to withstand pressures to manufacture under own-label.

Subsidiaries

Cereal Packaging Ltd (sold to Low & Bonan plc for £33m in March 1993)
Kellogg Co. of Ireland Ltd (Eire)
Kelloggs Espana SA (Spain)
Seaforth Corn Mills Ltd

Table 1 Five-year review (31/12/92)

	1988	1989	1990	1991	1992
Turnover (£m)	377.6	431.1	479.3	523.1	578.5
Pre-tax profit (£m)	58.5	58.6	72.9	85.5	90.7
Employees	3030	3006	2926	2964	2870

Table 2 Turnover (£m)

	1989	1990	1991
United Kingdom	348.5	389.5	418.0
Rest of Europe	80.8	87.8	103.1
Other	1.8	2.1	2.1

KENNERTY FARM DAIRIES LTD

Secretary and registered office

A. Cochran, 6 Bon Accord Square, Aberdeen AB9 1XU; (0224) 890444; Fax (0224) 890402

Directors

P.J. Harrison

Kennerty Farm Dairies is Scotland's largest independent dairy company. In 1992 it acquired Norfarm Dairies from the Aberdeen-based Northern Co-operative Society. It now has five production and packaging centres and 25 distribution depots. Its new range of 'Cool' milks is intended to help consumers identify various types of milk more easily according to their fat content: Cool 100 is full-cream milk, Cool 50 is semi-skimmed milk and Cool 2.5 is skimmed milk. In addition to milk and cream, Kennerty also manufactures yoghurt at its Edinburgh plant. Fresh and long-life ambient yoghurts suitable for schools, hospitals and independent distributors are available. Its immediate subsidiary is the Forthill Dairy Ltd.

Table 1 Four-year review (2/1/93)

	1989	1990	1991	1992
Turnover (£m)	39.06	41.23	40.86	43.16
Pre-tax profit (loss) (£m)	4.78	1.19	0.39	(0.62)
Employees	715	725	721	813

THE KERRYGOLD COMPANY LTD

Secretary and registered office

J.M. Paterson, Sunnyhills Road, Leek, Staffordshire ST13 5SP; (0538) 399111; Fax (0538) 399918

Directors

Dr N. Cawley (chairman), N. Coakley (managing director)

An Bord Bainne – The Irish Dairy Board

The Kerrygold Company (formerly Adams Foods) is a wholly owned subsidiary of An Bord Bainne. Board members include manufacturing co-operatives and dairy companies in Ireland. The Board's product portfolio is exported through an integrated network of subsidiaries, agents, distributors and the Kerrygold Trading House Ltd. It currently markets its members' products in 80 countries, and has turnover of IR£1.23bn and pre-tax profits of IR£16.2m. It operates in three main areas: Kerrygold branded products, food ingredients, and commodity dairy markets. Its UK dairy ingredients subsidiary is Adams Food Ingredients Ltd.

In May 1993 it was reported that An Bord Bainne had taken a majority share in a new company, Dairy Ingredients (UK), formed by a management buyout of Ernest George, a supplier of butter and dairy products.

Kerrygold in the UK

The original family business of Adams Foods started in 1922 and remained independent until being acquired in 1972 by the Irish Dairy Board. In 1985 the company sold its bakery interests to Northern Foods for £11m. This was followed by the sale of its fruit juice interests, Just Juice, to RHM, allowing the Leek-based company to concentrate on its dairy interests. The decision to concentrate on dairy products resulted in Adams adopting the title Kerrygold Co. in 1989. The Kerrygold brand was first launched in the UK in 1962.

1991 saw the first year of operation from a new purpose-built factory in Leek. Significant savings have resulted from the consolidation of manufacturing activities which has also led to a substantial reduction in the workforce, now standing at about 350. In the year ended 31/12/92, Kerrygold at Leek increased both turnover and profits. The cheese packing operation, which can accommodate 40 000 tonnes annually, ran at 90% capacity and a new cheese grating facility was installed. Against a background of increased retail prices and declining consumption, Kerrygold butter maintained 1991 volume levels, particularly in own-label business.

During 1992 Kerrygold disposed of surplus land and buildings in Leek and Cuffley, but the depressed property market prevented the sale of other surplus assets.

Table 1 Five-year review (28/12/91)

	1987	1988	1989	1990	1991
Turnover (£m)	128.2	165.9	184.7	151.8	127.8
Pre-tax profit (£m)	1.25	0.67	0.46	(0.3)	(0.13)
Employees	822	900	651	500	405

KERRY GROUP PLC

Secretary and registered office

B. Durran, Prince's Street, Tralee, Co. Kerry, Eire; 010 353 66 22433; Fax 010 353 66 22353

UK Head Office

Thorpe Lea Manor, Thorpe Lea Road, Eghham, Surrey TW20 8HY; (0784) 430777

Executive directors

M. Hanrahan (chairman), D. Brosnan (managing director), H. Friel (deputy managing director), D. Cregan (deputy managing director), M. Griffin, J. Houlihan

Non-executive directors

D. Buckley, M. Fitzgerald, R. Fitzgerald, M. Harty, E. Hayes, P. Healy, S. Kelliher, M. Leslie, E. McSweeney, J.J. O'Brien, P. O'Connell, P. O'Connor, A. O'Donnell, D. O'Donoghue

In the last 2 years Kerry has continued to grow organically and by acquisition. It has subsidiaries incorporated in Eire, Northern Ireland, the UK, Germany and the USA. The business is in four divisions: Kerry Ingredients; Kerry Foods; Meadow Meats; and Kerry Agribusiness. Group turnover more than doubled between 1988 and 1992 (IR£397m to IR£827m). Pre-tax profits increased by 19% from IR£24.1 in 1991 to IR£28.7 in 1992.

Kerry Foods

In the UK, Kerry Foods Ltd now employs over 3000 people in seven manufacturing units. It has 22 depots and 200 refrigerated vans. It is a major supplier of own-label, low-fat spreads, and is dominant in the milk substitutes market with its Kerry Plus Pints bottled powders. Robirch and Miller offer a range of high-quality sausages, pies and savoury products from factories in Burton on Trent and Poole in Dorset. In white meat products, Kerry operates through the A.E. Button poultry business in Norfolk and a cooked meats business in Blackpool. In July 1992, Buxted Duckling joined the group. The group is a major supplier to retailers, contract caterers, hotels and pub chains.

Kerry Ingredients

Reduction in specialist departments within major companies has allowed Kerry to develop its specialist ingredients business alongside its customers in a key R & D role. Products include Kerrynor Proteins, Beatreme fat powders, Kerrykreme creaming agents, Bealite emulsifiers, Cheesetreme cheese powders, Whiptreme whipping agents, Nature-tones natural flavourings, BeaTrim fat replacers, Dell seasonings for snack foods and Ultrasavours spray dried ingredients. Recent acquisitions include Dairyland Products (USA) and Eastleigh Flavours in the UK.

Meadow Meats

Kerry's beef and lamb business is a leading supplier of value-added beef and lamb cuts to UK and mainland Europe food manufacturers and retailers. It has successfully developed markets in Germany, Netherlands and Italy with progress also underway in France. Acquired from Guinness in 1991, Meadow Meats has processing plants in Ireland at Monaghan, Cork, Laois and Waterford and in the UK at Chard in Somerset (South West Meats) and Thirsk in Yorkshire.

Kerry Agribusiness

This division covers milk assembly and farm inputs trading, provender feed milling, a cattle breeding service and pig breeding and production.

Principal subsidiaries

Dairy products
- North Kerry Milk Products Ltd
- Kerrykreem Food Products Ltd
- Kerrykreem Ltd
- North Kerry Milk Exports Ltd
- Kerry Milk Exports Ltd
- Kerry Foods Ltd (GB)

Food ingredients
- Kerry Proteins Inc. (USA)
- Beatreme Foods Inc. (USA)
- Kerry Food Ingredients GmbH (Germany)
- Kerry Ingredients (UK) Ltd (GB)

Liquid milk
- Kerry Dairies Ltd
- Kerry Co-operative Dairies Ltd

Agricultural trading
- Kerry Creameries Ltd
- Kerry Farm Supplies Ltd

Meat products
- Henry Denny & Sons (Ireland) Ltd
- Henry Denny & Sons (NI) Ltd (N. Ireland)
- Kerry Holdings (Ireland) Ltd
- Duffy Meats Ltd
- Meadow Meats Ltd
- K.M.P. (Exports) Ltd
- K.M.P. Co-operative Society Ltd
- South West Meat Ltd (GB)
- Robirch Ltd (GB)
- W.L. Miller & Sons Ltd (GB)
- Grove Turkeys Ltd
- Grove Farm Ltd
- Ballyfree Farms Ltd
- Glenealy Farm (Turkeys) Ltd
- A.E. Button & Sons Ltd (GB)

Table 1 Five-year review (31/12/92)

	1988	1989	1990	1991	1992
Turnover (IR£m)	397.0	560.0	584.0	755.0	827.0
Pre-tax profit (IR£m)	12.3	17.1	19.3	24.1	28.7
Employees	2615	3372	3747	6431	6279

Table 2 Turnover (IR£m) by division and geographical area

Division	1991	1992	*Geographical area*	1991	1992
Kerry Ingredients	271.1	288.5	Ireland (inc. N. Ireland)	271.1	278.2
Kerry Foods	276.8	296.3	GB and Rest of Europe	287.5	330.9
Meadow Meats	177.0	203.7	USA	186.1	203.4
Kerry Agribusiness	30.0	38.2	Rest of world	10.2	14.2
Total	754.9	826.7		754.9	826.7

KITCHENS OF SARA LEE – UK LTD

Secretary and registered office

T.A. Schouten, Lancaster Road, Carnaby Industrial Estate, Bridlington, Humberside YO16 1AA; (0262) 678212

Directors

D. Baines, B.W. Stevens, M. Woodall, J. Woodbridge, T.A. Schouten, A.M. Dunstan

The Kitchens of Sara Lee are owned by the Sara Lee Corporation in the USA. They are a major manufacturer of branded and own-label luxury frozen cakes and desserts. The company claims market leadership in the £218m UK retail frozen confectionery market with a 17.5% share. Core lines include chocolate gateaux, lemon meringue pies and Danish apple bars as well as hot-eating products like strudels, puddings and crumbles. A new range of 'lighter' Tortes are also selling well.

*The contracts (see table) for all the employees at the Carnaby bakery are held by another Sara Lee company, Sara Lee Bakery (Employees) Ltd. This has been the situation since July 1991, and is reflected in the table as a zero in the number of employees for 1992. In 1992 there were 888 employees at the bakery.

Table 1 Four-year review (26/6/92)

	1989	1990	1991	1992
Turnover (£m)	32.60	44.60	52.88	58.49
Pre-tax profit (£m)	0.04	(1.77)	(0.45)	(4.98)
Employees	716	699	973	0*

KRAFT GENERAL FOODS LTD

Secretary and registered office

C.K. Cash, St George's House, Bayshill Road, Cheltenham, Gloucester GL50 3AE; (0242) 236101; Fax (0242) 512084

Directors

A.M.F. Anscombe (K), R.J.S. Bell (K & G), P.C. Beloe (K), S.T. Colman (K), S.D. Gray (G), R.J.D. Halliday (G), G. Jardine (K & G), J.M. Keenan (USA, K & G), P.A. Smith (K & G), N. Thompson (K) (K = Kraft Foods Ltd, G = General Foods Ltd)

Kraft General Foods (KGF) was formed in December 1989 following the $13bn acquisition of Kraft Foods Inc. by Philip Morris Companies Inc. which also owns General Foods. In 1990, Jacobs Suchard, based in Switzerland, was added to the group. During 1991, a major restructuring of its food operations worldwide led to the disposal of unprofitable and non-core businesses. Globally the food business now comprises KGF North America, KGF Commercial Products and KGF International, which includes Jacob Suchard.

In March 1993, the Terry's confectionery group was bought from United Biscuits for £220m and was merged with the existing Suchard business to form the Terry's Suchard Division.

Coffee and dessert products

A factory in Banbury (the former Head Office of General Foods) produces instant coffees, dessert mixes and custard powders under many well-known brands including Maxpax, Cafe Hag, Kenco, Maxwell House, Bird's Angel Delight and Dream Topping.

Dairy products, spreads and meat products

Factories in Bristol and Liverpool make well-known branded products including Dairylea, Philadelphia, Golden Churn and Vitalite. Premium meat pies are marketed under the Cotswold Pie Co. and Brains Frozen Foods.

Confectionery

The Terry's Group provided synergy with the Jacobs Suchard brands already imported into the UK including Toblerone, Milka, Cote d'Or and Twilight. Terry's has factories in York and Bridgend (Callard & Bowser) employing about 1700, a turnover of about £177m and an operating profit of £13.4m (1992 figures).

Table 1 Five-year review (14/12/92)

	1988	1989	1990	1991	1992
Kraft Foods Ltd					
Turnover (£m)	143.1	138.9*	180.6†	204.0‡	205.88
Pre-tax profit (£m)	10.1	13.7	14.4	20.8	25.42
Employees	1024	906	900	777	758
General Foods Ltd					
Turnover (£m)	269.4	282.8	253.6	266.7	n/a
Pre-tax profit (£m)	18.8	3.0	25.0	39.3	n/a
Employees	2015	2008	1632	1546	n/a

* 47 weeks; † 53 weeks; ‡ 54 weeks

LLOYD MAUNDER LTD

Secretary and registered office

W.C. Tucker, Willand, Cullompton, Devon EX15 2PJ; (0884) 820534; Fax 0884 821404

Directors

R.L. Maunder, J.L. Maunder, D.H.L. Maunder, G.R.L. Maunder, R.L. Maunder, I.R.E. Maunder, A.J.L. Maunder

Lloyd Maunder is a family business operating poultry and meat processing plants in Devon. The company also has a hatchery, a number of poultry rearing units, a feed mill and laboratories. Poultry provides the major part of the company's activities. 200 people work in the poultry division which handles 14 million birds/year. A highly automated plant was built from scratch in 1992 after fire destroyed the 6-year-old poultry processing plant.

Lloyd Maunder also processes up to 100 beef carcases a week, 1500 pigs and 12 500 lambs. All meats are sourced from within the UK and are mostly of local origin. Fresh pork and lamb come in sizes ranging from primals to tray overwrapped retail packs. The company currently exports lamb to Belgium, France, Spain and Germany.

A wide range of traditional and new value-added recipe dishes are also manufactured at Cullompton. Many of these are developed for the major multiples, while others are sold through independent butchers mainly in the London, the South West and Wales and Lloyd Maunder's eight own butchers shops.

Table 1 Four-year review (26/09/92)

	1989	1990	1991	1992
Turnover (£m)	62.23	63.29	60.91	59.82
Pre-tax profit (loss) (£m)	(0.42)	2.33	(0.25)	1.92
Employees	836	831	926	861

MARSHALL FOOD GROUP LTD
(formerly D.B. Marshall (Newbridge) Ltd)

Secretary and registered office

R.J. Preston, Newbridge, Midlothian EH28 8SW; 031-333 3341; Fax 031-335 3270

Directors

E.C. Tarr (chairman), W.M. Marshall (chief executive), R. Stewart, R.J. Preston, J.J. Hogarty, K.S. Kirkpatrick, K.M.M. Marshall, S.P.J. Smith

In 1991, the group changed its name and restructured operations into three divisions: Marshall Agriculture (from feedmills through to the delivery of live chickens to the processing units); D.B. Marshall (responsible mainly for supplying Marks & Spencer); Marshall Foods (dealing with major retail and food service customers).

The group is a fully integrated operation having its own hatcheries, feed milling and pasteurizing plants, farms and breeding stock, as well as poultry factories at Newbridge, Cambuslang, Coatbridge and Coupar Angus. Weekly production is 1.3 million birds. In 1992, £10m was invested in development kitchens and value-added processing lines at the Coatbridge plant. Operations were further expanded in February 1993, by the acquisition of Premier Poultry from Hillsdown Holdings. The deal increased Marshall's turnover by 25% and added 12 breeding farms, 15 broiler units, a hatchery, 500 employees and 93 000 sq.ft of processing plant at Sandycroft in Wales (fresh, whole and portioned birds).

Its subsidiaries include Robert Miller (Denny) Ltd (which produces poultry houses), Grain Brokers (Leith) Ltd and a 50% holding in Douglas Miller (Transport) Ltd (road haulage).

Table 1 Five-year review (30/9/92)

	1988	1989	1990	1991	1992
Turnover (£m)	143.8	132.6	141.3	141.0	146.7
Pre-tax profit (£m)	(1.42)	(5.19)	5.36	0.83	2.55
Employees	4041	4143	3947	3955	3764

BERNARD MATTHEWS PLC

Secretary and registered office

J.G. Brown, Gt. Witchingham Hall, Norwich NR9 5QD; (0603) 872611; Fax (0603) 871118

Directors

B.T. Matthews (chairman), D.J. Joll (managing director), N.F. Bartram (operations and finance), J.G. Brown, G.L.J. Hayes, A.M. McFarquhar, C.J.H. Simpson*
* non-executive

The business was established in the early 1950s by the current chairman, Bernard Matthews. Production initially concentrated on whole birds at the company's headquarters, Witchingham Hall, bought in 1955. It has an integrated operation with a breeder unit, hatcheries, feed mill and about 20 farms in Suffolk and Norfolk. It has four factories on two sites, all EC-approved. Two factories close to Witchingham Hall produce oven-ready birds and further processed turkey products; the others at Halesworth in Suffolk produce value-added turkey products and red meat products. The company has also recently trialled fish products under the Norfolk Bay brand and Dinosaur-shaped turkey burgers were successfully launched.

During 1991, low-priced poultry imports from the EC and USA, coupled with falling demand for whole turkeys, adversely affected profits which fell by 75% in 1992 to £3.3m, compared with £13.2 for 1991. Sales of red meat roasts also declined. In addition heavy costs were incurred in launching a range of turkey and fish products into French markets through its French subsidiary, Bernard Matthews SA.

In 1992, the company continued its European development by acquiring 90% of the share capital of Savari Baromfiipari Rt., the leading poultry company in Hungary which processes and markets turkeys, chickens, geese and ducks in Hungary, Switzerland and Austria.

Interim results

Interim results for September 1993 show a recovery in pre-tax profits to £4.6m (1992, £2.13). Increased turnover (20%) came mainly from value-added products and a new range of cooked meats whose sales are likely amount to £20m/year. Sterling's devaluation also helped to stem the flow of cheap imports from France and the USA. Savari contributed £464 000 to profits and the 1993 Christmas market for whole birds improved significantly.

Recent development

In December 1993, Bernard Matthews bought Turner's Turkeys from Unigate for £18m in cash.

Table 1 Five-year review (3/1/93)

	1988	1989	1990	1991	1992
Turnover (£m)	135.2	135.8	150.1	148.4	144.2
Pre-tax profit (£m)	10.2	9.1	15.5	13.2	3.3
Employees	2586	2678	2529	2544	2478

Table 2 Turnover (£m) by geographical destination

	1991	1992
United Kingdom	133.9	130.3
Europe	14.5	13.9
Total	148.4	144.2

McCAIN FOODS (GB) LTD

Secretary and registered office

H. Kinder, Havers Hill, Eastfield, Scarborough, North Yorkshire YO11 3BS; (0723) 584141

Directors

H.H. McCain (chairman, Canada), G.W.F. McCain (Canada), C.M.A. McCarthy, H. Kinder, A.D. McCain, R.D. Frawley, I.S. Cameron, P.J. Burman, G.B. Calder

McCain is a subsidiary of McCain Foods Ltd (Canada). It has major interests in frozen potato chips and frozen pizzas, both conventional and microwaveable. Despite tough competition in 1993 it claims to have retained its position as the UK's largest supplier of frozen chips with a 38% brand share of sterling sales. It also holds almost 50% of the oven-ready chip market, a market which it pioneered in earlier years. Microchip, a product developed for the microwave, is said to have an 86% share of a market that has grown to £10m in the last 2 years.

Dansco-Tolona

In 1990, the company announced the formation of Dansco-Tolona whose combined turnover is in excess of £30m. Dansco, acquired in 1988, makes mozzarella cheese, used in pizza products. Since its acquisition, £5.5m has been invested equipping and modernizing Dansco's factory in Dyfed which is a major supplier to most of the UK's high street pizza chains in the UK, Europe, the Middle East and the Far East. Tolona Pizza Products, based in Skelmersdale, supplies pizza crusts and other pizza products.

McCain pizza perfection claims to be the top selling brand with 19% of the large single sector with a 19% value share.

Trading subsidiaries

PAS (Grantham) Ltd	Manufacturers of frozen French fries and potato products
McCain Potatoes (GB) Ltd	Traders of seed potatoes and contract potato growers
SSHP Holdings Ltd	Holding company
Dansco Dairy Products Ltd	Manufacturers of mozzarella cheese
Tolona Pizza Products Ltd	Manufacturers and distributors of pizzas and pizza catering equipment
Britfish Ltd	Manufacturers of frozen fish products

Table 1 Five-year review (30/6/92)

	1988	1989	1990	1991	1992
Turnover (£m)	145.1	153.9	187.8	197.5	211.4
Pre-tax profit (£m)	15.8	7.2	11.9	22.3	22.6
Employees	1880	1968	1991	2036	1969

Table 2 Turnover (£m) by geographical destination

	1988	1989	1990	1991	1992
United Kingdom	137.4	144.2	179.1	186.9	197.3
Rest of Europe	6.0	7.0	3.7	10.6	11.4
Rest of world	1.7	2.7	5.0	0.008	2.7
Total	145.1	153.9	187.8	197.5	211.4

McCORMICK (UK) PLC

Secretary and registered office

C.J. Mann, Owen House, 8 Bankhead, Crossway North, Edinburgh EH11 4BR; 031-556 8555; Fax 031-442 4924

Directors

Dr J.J. Albrecht (USA), N.S. Garrow (managing director), R. Keefe, C.D.F. Savage, A.J. Sykes, W. Sinclair

The company is the UK subsidiary of McCormick & Co. Inc., one of the world's largest spice, seasoning and flavouring companies. Following rationalization by the parent company in 1988, Paterson Jenks plc changed its name to McCormick (UK) plc. McCormick (UK) has three principal divisions: Jenks Group, McCormick Foods and McCormick Stange.

Jenks Group

This provides a marketing and national sales, distribution and finance service to UK and overseas manufacturers. It currently handles many well-known brands including Highland Spring mineral water, Ocean Spray cranberry sauce, Realemon, Kikkoman, Hermesetas, Milupa and Bick's relish.

McCormick Foods

Based at Thame, this is responsible for the manufacture and marketing of all retail lines, principally the Schwartz range of which there are more than 80 herbs, spices and seasonings and dry casserole mixes.

McCormick Stange

This is responsible for the supply of bulk ingredients and flavours to companies involved in fast food, ready meals, desserts, confectionery, sauces, soups and dressings. A £6m investment programme at Stange's Ellesmere Port plant has quadrupled capacity and established the site as McCormick's centre for research and development in Europe. The company claims to be the first UK herb and spice manufacturer to achieve BS 5750 accreditation.

In February 1992, McCormick Stange bought Glentham International from Harrisons & Crosfield for £15.5m which included Glentham's operation in South Africa. The new business, McCormick Glentham, has factories in Northampton and Ellesmere Port.

Table 1 Five-year review (31/10/92)

	1988	1989	1990	1991	1992
Turnover (£m)	73.7	72.9	79.6	91.5	105.0
Pre-tax profit (£m)	4.8	3.2	5.2	5.8	4.5
Employees	510	531	563	576	664

Table 2 Financial record by geographical area

	Turnover (£m)		Profits (losses) (£m)	
	1991	1992	1991	1992
United Kingdom	86.3	96.2	5.5	4.9
Other Europe	5.2	6.5	0.3	0.4
Rest of world	–	2.3	–	0.3
Other income			0.9	0.8
Interest			(0.9)	(1.9)
Total	91.5	105.0	5.8	4.5

MD FOODS PLC

Secretary and registered office

D.A. Firth, Craven House, Kirkstall Road, Leeds lS3 1JE; (0532) 440141; Fax (0532) 448643

Directors

F.G.B. Blake (chairman), A. Allinson, C.F. Hall, G.O.W. Williams, D.A. Firth, K.B. Anderson, N. Freeman

In December 1990, Associated Fresh Foods was taken over by MD Foods a.m.b.a. of Denmark, becoming MD Foods plc in March 1992, a subsidiary of MD Foods International A/S. Despite difficult trading conditions in the UK dairy sector significant progress has been made towards MD's stated aim of becoming one of the UK's leading dairy companies. Following the additional purchase of the CRS dairy interests in June 1992 for £65m, MD claims to be fourth in the liquid milk market in the UK with a 10% share and a turnover of £400m.

Processing and distribution dairies operate from Newcastle, Leeds, Bristol, Oakthorpe and London. The Accrington bottling plant was closed in May 1993 with the loss of 190 jobs and production transferred to the plant at Bamber Bridge recently acquired from Dairy Crest.

MD also markets Lurpak butter in the UK; sales of this product have been in decline for several years. Market share in 1992 was maintained by widespread TV advertising in the run-up to the Barcelona Olympics. In the cheese sector MD Foods (UK) in Redhill, Surrey has reversed several years of stagnation, with volume growth in danablu, mozzarella and cream cheeses. The group is also aiming to strengthen its position as a supplier to the major retailers in the UK, and preparing for the dissolution of the Milk Marketing Board in 1994.

The fruit juice and fruit drinks business of Cravendale Foods was strengthened by refitting the former Scottish & Newcastle Brewery at Blackburn, Lancashire. Citrus Grove, Southern Gold and Cravendale brands will be produced on this site.

MD Foods cheese division has worked closely with Tesco on a pilot scheme using electronic data interchange based on bar codes which allows MD to assess the retailer's requirements on a daily basis directly from the point of sale.

Subsidiaries include Lakeland Creameries Ltd, Lakeland Ice Cream Co. Ltd and Lakeland Dairy Products (Kendall) Ltd.

Table 1 Two-year financial record (26/9/92)

	1991 *	1992
Turnover (£m)	259.2	217.9
Pre-tax profit (loss) (£m)	10.6	(0.6)
Employees	1839	2201

* 17 months

MILK MARKETING BOARD FOR NORTHERN IRELAND

Secretary and registered office

S.A. Agnew, 456 Antrim Road, Belfast BT15 5GD; (0232) 770123

Board members

W.R. Morrow (chairman), W.E.S. Fullerton (vice-chairman), W.M. Carson, Mrs A.L. Dickson, W.R.M. Farrell, O.R. Gault, J.R.H. Hamilton, J.B. McAllister, W.E. Rankin, S.A. Robinson, S.A. Shanks, J.A. Walker, J. Warden

The MMB for Northern Ireland was set up in 1955 and, with four other Boards (England & Wales, Scottish, Aberdeen & District and North of Scotland) is responsible for marketing milk in the UK. With the end to statutory milk marketing schemes in 1994, it is proposed that the Board in Northern Ireland will change to a producer co-operative which all milk producers will be able to join. The co-operative will market members' milk supplies, returning a pooled price to producers. It will also transport milk, assure its quality and provide services to producers. Dromona Quality Foods, the wholly owned subsidiary of the Board will become a limited company owned by the co-operative.

In 1992/3 a total of 1171 million litres of milk were purchased from 5900 producers, 83% of this milk was used for manufacture of dairy products and 17% sold as liquid milk. With increased demand from consumers for low-fat milks, 36% of liquid milk sales were low fat. Milk for manufacturing is now being used to a greater extent to produce whole milk powder and cheese, with much less being used for butter. Total income from milk sold by the Board was £247.6m.

Dromona Quality Foods

This was formed in 1989 to undertake manufacturing, processing and transport operations of the Board. It has manufacturing sites at Bangor, Tassigh, Dromona, and Dunmanbridge. It had a turnover of £78.3m in 1992/3 and profits of £3.4m. It produces and markets a range of dairy products under Spelga and Dromona brands as well as own-label products.

Associated undertaking

Nidco Foods Ltd (50% owned) – sale of dairy products.

Subsidiary undertaking

Halib Foods International Ltd (100% owned) – sale of dairy products mainly into export markets.

During the year, the Board sold its 50% holding in Ballyshrane Foods Ltd.

Table 1 Five-year review (31/3/93)

	1989	1990	1991	1992	1993
Income from sales of milk by Board (£m)	227.1	245.5	213.3	225.3	247.6
Turnover for Dromona Quality Foods (£m)	59.3	73.5	63.5	68.2	78.3
Dromona Quality Foods – profit (£m)	1.58	1.02	1.50	2.65	3.36
Employees – Board	962	948	872	813	814
Employees – Dromona	665	652	583	535	522

MOY PARK LTD

Secretary and registered office

I. Brewster, Royal Oak Building, High Street, Crewe, Cheshire CB2 8BL

Directors

W.M. Carson (chairman), R.T. Campbell (managing director), I. Brewster, J. O'Toole, E. Reid, P.D. Wolfe, K.J. Baird, J.N. French, S. Rooney

The principal activity of the group is the rearing, processing and marketing of poultry. It is a major supplier of own-label poultry to major supermarkets in the UK. It also sells goods under the Moy Park brand. Its associated company in France, BMP SA, set up as a joint venture with Bourgoin SA, produces speciality cooked and flavoured poultry products for the European market. BMP commenced production in September 1991.

The group also has interests in a food technology consultancy (Verner Wheelock Associates), agricultural and food consultancy (Agra Food Consultants) and computer software systems for monitoring food product quality and safety (MCP Consultancy and Park Applications Computer Engineering Ltd). Moy Park International Ltd is responsible for export marketing activities as well as providing a business performance improvements consultancy for food manufacturers.

Turnover for 1992 was £113.1m, an increase of 9.5% on the 1991 figures. However, despite continued development of value-added products and increased efficiencies, there was a fall in pre-tax profits to £2.4m (£3.5m in 1991).

Recent development

The Coolhill processing plant in Dungannon, Co. Tyrone won the 'Factory of the Year' award from the journal *Management Today*.

Table 1 Five-year review (31/3/92)

	1988	1989	1990	1991	1992
Turnover (£m)	61.6	68.16	82.77	103.3	113.1
Pre-tax profit (loss) (£m)	0.22	(1.13)	2.23	3.48	2.42
Employees	1919	1822	1888	2200	2317

NEWMARKET FOODS LTD
(formerly Haverhill Meat Products)

Secretary and registered office

D. Everitt, Little Wratting, Haverhill, Suffolk CB9 7TD; (0440) 704444; Fax (0440) 62120

Directors

P. Clark (managing), P.D. Collins, R. Cooper, D.E. Henson, D.M. Hickman, C. Roberts, R.A. Clark

NewMarket Foods is a wholly owned subsidiary of J. Sainsbury plc, with activities on a single site ranging from pig slaughtering, bacon curing and packaging, meat product manufacture and packaging, the sale of fresh pork and the processing of by-products.

With unsatisfactory results in the last 2 years the company has been restructured and its customer base widened in the UK and Europe. In 1990, its subsidiary J.S. (Market Drayton) Ltd, formerly Palethorpes Ltd, disposed of its business and assets to Northern Foods for £14m. In April 1993 it also changed its name, £5m has been spent on a new abattoir (able to process 20 000 pigs/week) and a further £2m on upgrading facilities to meet EC and USDA standards.

It has also extended its range of products and now produces pizza toppings, sandwich fillings, marinaded lines and products made to customer specifications.

Table 1 Five-year review (14/3/92)

	1988	1989	1990	1991*	1992†
Turnover (£m)	97.35	98.40	108.26	112.56	119.9
Pre-tax profit (loss) (£m)	7.026	5.20	0.010	13.04	(1.44)
Employees	2261	2308	2348	1855	1865

* 52 weeks ending 17/2/91; † 56 weeks ending 14/3/92

THE NICKERSON GROUP ROTHWELL LTD

Secretary and registered office

P.R.C. Braithwaite, Villa Office, Rothwell, Lincoln LN7 6BR; (0472) 371271; Fax (0472) 371545

Directors

W.C. Thompson (chairman), A.H. Westropp, M.S. Edmundson, M. Kerrigan, J.S. Denton, T.N. Wood (finance)

The Nickerson group of companies, built up by the late Sir Joseph Nickerson, is involved in farming and agrogenetics connected with the breeding, rearing, processing and sale of ducks and pigs. Subsidiaries include Cherry Valley Farms, Cotswold Pig Development Co., and North Lincolnshire Farms Ltd. The group has a 20% holding in Maple Leaf Farms, the largest duck producer in the USA, acquired in exchange for exclusive rights to Cherry Valley breeding stock. It also holds a 30% interest in Palotasi Kacsa KFT (Hungary) and 50% of Robert Nickerson Farms UK partnership. The Cotswold Pig Development Co's pig breeding business has performed well in home and export markets. A notable success has been the development of a boar free from the halothene gene which causes pale soft exudative pork and sudden death in pigs.

Weakened demand for whole ducks in the UK and USA resulted in the closure of Concord Farms, Inc. in the USA in December 1991 and concentration of its UK activities on value-added duck products. To this end, Cherry Valley Farms have just completed the refurbishment of the development kitchens and factory at Cleethorpes as part of a £6m investment in the group's farms, growing accommodation, research division, hatchery and processing facilities. An innovative, premium duck sandwich filling has been developed as well as a new chicken recipe dish – Chicken Fajitas – which complements the successful Crispy Peking Duck range.

Despite the complications associated with the 'Grune Punkt' packaging legislation, Cherry Valley have increased the export of whole ducks and several value-added products to Germany and are trialling the supply of whole ducks to the Belgian Delhaize supermarket chain.

Table 1 Five-year review (25/4/93)

	1989	1990	1991*	1992	1993
Turnover (£m)	51.99	59.97	50.64	50.94	54.50
Pre-tax profit (loss) (£m)	(1.82)	1.53	2.85	2.51	3.08
Employees	1328	1406	1150	1073	1095

* 1991 figures re-stated after the closure of Concord Farms, USA on 7/12/91

Table 2 Turnover by geographical market (£m)

	1991	1992	1993
United Kingdom	37.95	36.92	36.23
Rest of EC	9.47	10.50	12.93
Non-EC countries	3.22	3.52	5.34
Total	50.64	50.94	54.50

G.W. PADLEY (HOLDINGS) LTD

Secretary and registered office

L.J. Wilkinson, Anwick, Sleaford, Lincs NG34 9SL; (0526) 832661; Fax (0526) 833180

Directors

D.G. Padley (chairman), Mrs P. Padley

Padley's is one of the UK's leading privately owned food processing companies with a turnover of £150m, producing an extensive range of high-quality branded and own-label chicken products for supermarkets, caterers, fast food operations and airlines. Anwick, the largest factory in the group, employs 800 people. The Prepared Foods Division and Padley's Foodservice are based at Grantham; cooked meats are at Wisbech.

Padley's plans to develop its range of value-added products and uses its market research team to constantly review eating habits in pubs and restaurants in the UK and abroad. Recent additions to the range include Chick'n Quick and a microwaveable chicken burger in a bun. It also has a range of exotic Mexican, Louisiana smoked, Tikka and Chinese diced roast chicken for filling rolls, nachos and pittas.

Table 1 Five-year review (1/8/92)

	1988	1989	1990	1991	1992
Turnover (£m)	107.6	195.2	123.4	136.6	150.0
Pre-tax profit (£m)	3.24	1.05	4.83	4.62	3.658
Employees	2700	2771	2999	3237	3334

PERKINS FOODS PLC

Secretary and registered office

P.R. Henson, Trinity Court, Trinity Street, Peterborough PE1 1DA; (0733) 555706; Fax (0733) 558499

Directors

A.M. Davies (chairman*), H.G. Phillips (chief executive), D.W. Ball*, F.R. Black*, I.M. Blackburn (finance), T.J. Finn*, J.L. Van Den Heuvel, M.P. Lawrence, A.D. Malpas*, I.L. Shearer*
* non-executive

Perkins Foods is a North European group engaged in the manufacture of frozen and chilled foods and the distribution of fresh produce. It supplies supermarkets and caterers in the UK, Netherlands, Germany, France and Scandinavia. Operations in the Netherlands contribute about 65% of group profits compared with 20% from the UK.

Frozen Foods

This is the group's largest division, accounting for 42% of turnover (£148.4m) and 48% of operating profit (£14.2m). It manufactures value-added convenience foods in the UK, Netherlands, Germany and Luxembourg. A European network of sales companies supports these manufacturing operations.

Chilled Foods

These account for 18% or turnover (£64.5m) and 23% of operating profit (£4.8m). Chilled meat, chicken and turkey products for retail and wholesale customers are manufactured in the UK and Netherlands. The five production units for red meat products have EC licences. G.G. Baxter's factory was closed during 1992 and production transferred to other lower cost facilities within the group.

Fresh Produce

Fruit and vegetables contribute 34% to turnover (£121.1m) and 25% to operating profit (£6.7m) and are imported from all over the world and supplied to supermarkets and wholesalers in the Netherlands.

Mushrooms

This is the smallest division in the group, accounting for 6% of turnover (£19.7m) and 4% of operating profit (£0.99m). Chilled, frozen and canned mushrooms are produced. In 1992 the market suffered from price competition and over-capacity, a stagnant market in Germany and cheap French imports.

Recent developments

In November 1993, Perkins bought Cromer Crab in a share deal, and in December it disposed of its Dutch mushroom processing interests to a new joint venture in which Perkins will take a minority shareholding.

Trading subsidiaries

Frozen foods

Anchor Seafoods Ltd (UK)	manufacturing
Rowleys Frozen Foods (UK)	"
Cogel SA (Lux)	"
Enko BV (NL)	"
Fellside Foods Ltd (UK)	"
KB de Souffle Specialist BV (NL)	"

Peppino's Pizza (Germany)	"
Van der Made Potato Products BV (NL)	"
Comigel SA (France)	distribution
G+K Teifkuhlprodukte GmbH & Co. KG	"
Nuka BV (Netherlands)	"
Sunbird Foods Ltd (UK)	"
Chilled foods	
Bakker Lekkerberk Holland BV (NL)	manufacturing
G.G. Baxter Ltd (UK)	"
Studleigh-Royd Ltd (UK)	"
Fresh produce	
Groenten-en Fruitbedrijf C. den Hollander Barendrecht BV (NL)	distribution
Hage International BV (NL)	marketing
Vebruco Import BV (NL)	"
Mushrooms	
Champex BV (NL)	distribution
Champifri Holland BV (NL)	manufacturing
Holland Champignons BV (NL)	manufacturing

Table 1 Five-year review (31/12/92)

	1988	1989	1990	1991	1992
Turnover (£m)	64	134	196	260	354
Pre-tax profit (£m)	2.8	9.8	18.1	24.3	22.0
Employees	769	597	–	1382	1980

PIGS MARKETING BOARD (NORTHERN IRELAND)

Secretary and registered office

W.H. Hirst, Bridgewater House, Bridge Street, Lisburn, County Antrim BT28 1XY; (0846) 677070; Fax (0846) 660008

Board members

W.H. Robson (chairman), J.F. Beggs, H. Gabbie, R.A. Attwell, T.J.A. Barclay, S.A. Anderson, O. Priestly, J.W.D. Telford, P. Killen

Directors of Bridgewater Country Foods Ltd

W.H. Robson, J.F. Beggs, J.W.D. Telford

The Pigs Marketing Board of Northern Ireland is an organization that controls about half of the pigs slaughtered in Northern Ireland through Unipork and the Ulster Farmer Bacon Co. Ltd. Producers faced 18 price changes in 1991, high feed costs, and blue ear disease. Processors have faced the costs of licence upgrading in line with EC requirements, erratic pig supplies, margins cut by the major retailers in the UK and increased costs associated with veterinary inspection and by-product disposal.

For the year ended 2/1/93, the Board purchased a total of 401 934 pigs from enrolled producers, a decrease of 24% on the previous year. To contain costs, the killing line at Enniskillen was closed, and Unipork operations were concentrated at the abattoir and butchery at Cookstown. Curing and slicing continued at Enniskillen.

In July 1991, a new subsidiary was created under the title of Bridgewater Country Foods Ltd (BCFL). Early in 1992, the Board transferred its entire shareholdings in both Unipork (Holdings) Ltd and Ulster Farmer Bacon Co. (Newry) Ltd to BCFL in return for an identical number of shares in the new subsidiary. BCFL also invested £250 000 in a joint venture with Ferne Foods Ltd to manufacture bacon and egg-based products.

Table 1 Pigs Marketing Board (N.I.) – Three-year review (2/1/93)

	1990	1991	1992
Sales of pigs (£m)	39.92	34.35	28.75
Supplementary income (£m)	0.85	0.31	0.22
Purchase of pigs (£m)	(40.05)	(34.39)	(28.73)
Transport (£m)	(0.035)	–	–
Total	0.69	0.27	0.24
Employees	13	12	10

Table 2 Bridgewater Country Foods Ltd – Three-year review (12/1/93)

	1990*	1991	1992
Turnover (£m)	59.38	54.95	50.68
Pre-tax profit (loss) (£m)	1.38	(0.52)	(1.04)
Employees	955	985	868

* pre-merger trading companies Unipork (Holdings) Ltd and Ulster Farmers Bacon Co. (Newry) Ltd

PORTFOLIO FOODS LTD

Registered office

Pasadena Close, Hayes, Middlesex UB3 3NQ; 081-573 1010; Fax 081-848 4282

Directors

B. Stokes (chairman)*, P. Courtenay-Luck (chief executive), S. Joseph (finance), I. Parham*, A. Gibbons*
* non-executive

Portfolio Foods has now completed 2 full years of trading since its £59m management buyout of the snacks and confectionery interests of Hazlewood Foods in 1990. Results to year-end in March 1993 show small increases in turnover from £77.9 to £79.7 and in operating profit from £7.7 to £8.1m.

Confectionery and popcorn

Cravens of York, one of the companies acquired in the initial buyout has now been integrated with Barker & Dobson, Keiller, Bensons and Milady, bought from the receiver of Alma Holdings for £3m. Sweetmate produces both branded and own-label items for the bagged sector. The German division includes Dragess aus Weseke and Heidrich Confiserie. Portfolio claims to have 35% of the high-quality polished dragee market in Germany.

At the end of 1992, Butterkist was acquired from Hunter Saphir for £2.9m and combined with Bard Bros operations, bringing its turnover to about £20m. New equipment from Butterkist's Dagenham plant has been transferred to Bard's London site and the Dagenham site closed. A new microwaveable Butterkist popcorn has been introduced. Bard's has also maintained its 27% share of the table jelly market, mainly in own-label.

Snack foods

In the UK, Red Mill is the market leader in the economy sector of the children's market. Its factory at Wednesbury in the Midlands has been extended and a smaller site at Upton on Severn closed. Well-known brands include Snack Mate, WWF Superstars, Millies, Tangy Toms and several others. Red Mill NV in the Netherlands markets Millies and Snack D'Or in Benelux, Germany and France. The Golden Twin own-label nut processing business was sold to the JLI Group in April 1992 for £4m.

Table 1 Financial results (31/3/93)

	1992	1993
Turnover (£m)	78.0	79.7
Pre-tax profit (loss) (£m)	2.1	(1.3)
Employees*	1500	1300

* approximate figures for full-time equivalent, calculated from turnover/employee

WILLIAM PRICE & SONS LTD

Secretary and registered office

A.G. Price, 38 Raleigh Street, Walsall, W. Midlands WS2 8RB; (0922) 25566

Directors

W.B. Price, A.G. Price, J.D.W. Price, M.R. Price, S.A. Hele, R.M.D. Price

William Price & Sons Ltd are bakers and confectioners. They are the holding company for a group comprising Primebake, New Primebake, Harvestime and Boboli Pizzas.

Primebake supplies specialist bakery products like garlic bread to the multiples. In October 1993 it acquired the Dairy Crest Lymeswold factory in Nantwich, Cheshire. This is a modern plant, designed for high-risk foods, and should allow Primebake to develop a range of novel bread products for both retail and catering markets. The first of these 'Crusti' has already been launched. It is a hand-held snack with a bread base and meat, vegetable or cheese toppings.

Boboli cheese crust pizzas are based on an authentic Italian feasting bread and are made at Price's site at Walsall, West Midlands. Boboli (UK) Ltd is also a subsidiary of William Price & Sons.

New Primebake, operating from premises in Telford, Shropshire produces an extensive range of part-baked Continental style frozen doughs, while Harvestime makes bread, rolls, baps and other morning goods at the Walsall Bakery.

Boboli (UK) Ltd

Turnover £2.63m; pre-tax profit £0.24m; employees 61 (2/1/93)

Harvestime Ltd

Turnover £23.29m; pre-tax loss (£0.21m); employees 374 (2/1/93)

New Primebake Ltd

Turnover £5.12m; pre-tax profit £0.13m; employees 143 (2/1/93)

Table 1 William Price & Sons Ltd – Four-year review (2/1/93)

	1989	1990	1991	1993
Turnover (£m)	20.25	22.42	27.74	31.47
Pre-tax profit (loss) (£m)	0.83	0.32	(0.18)	0.32
Employees	358	472	551	605

Note: No figures are available for 1992

PRIDE FOODS LTD

Secretaries

Hanover Management Services Ltd

Registered office

16 Hans Road, London SW3 1RT; 071-581 1477

Directors

M.S. Gorvy (chairman), Z.H. Cohen, R.L. Wunsh, A.J. Leibowitz

Pride Foods Ltd (J. Gerber & Co. Ltd until 1985) is a holding company for a group of companies better known as Gerber Foods. It is a subsidiary of a larger investment, trading and management services group – Hanover Acceptances Ltd. The ultimate holding company is Quadriga Holdings SA (Luxembourg).

Gerber Food Holdings Ltd

- Gerber Foods International Ltd – a sourcing, sales and marketing operation
- Gerber Frozen Foods Ltd – markets value-added ready meals, vegetables, meats, speciality ice cream and gateaux
- Delican Ltd – supplier of juices and drinks to small independent off-licences, wholesalers and retailers as well as airlines, hospitals and hotel groups; depot in South London
- Givco BV – a joint venture with Ilio srl, Italy and Vienna Commerz; based in Rotterdam, distributes food products throughout Europe and the Middle East
- Gerber Foods Manufacturing Ltd – based in Bridgwater in Somerset, provides development and production of 70% of the food group's total output
- Gerber Foods Manufacturing (Hellas) Ltd – juice production in Greece
- Frigopak – canning, frozen fruit and vegetables in Turkey
- Gerber Frozen Foods Manufacturing Ltd – frozen ready meals manufactured in a purpose-built factory in Kettering, Northants

Commercial overseas offices

Italy (Eurofrutta), Spain (Frucotrade), Israel (Medijuice) and Greece (Gerber Foods (Hellas)).

Gerber claims to be the largest producer of long-life juices and drinks in the UK with an annual output of 300 million units in packs from 200 ml to 5 litres. Substantial investment in processing and packaging technology at the Bridgwater plant has ensured product quality. Soft drinks and fruit juices are marketed under the Sun Pride brand. Canned goods are produced for own-labels and the Gerber Pride brand. Gerber Frozen Foods, which acquired the factory and workforce of Prime Foods, Kettering, from the receiver in 1991, is targetting local authorities and catering organizations as well as retailers in a bid to increase its share of the £2bn frozen food market. Frequently operating through joint ventures with other companies, Gerber has recently developed vegetarian ready meals with the small health food producer Berrydales (see Section 2.3).

In February 1992, Gerber Foods International set up Gerber Confectionery Division to penetrate the own-label chocolate confectionery market. Manufacturers represented by Gerber include Rademaker, Netherlands, Carl Brandt, Germany, Belgapraline, Belgium Produktion, Switzerland and Van Meer and Holgates in the UK.

Table 1 Pride Foods Ltd – Five-year review (28/12/91)

	1987	1988	1989	1990	1991
Turnover (£m)	106.9	118.9	140.7	165.5	164.8
Pre-tax profit (£m)	2.0	6.2	5.8	7.9	6.4
Employees	755	1075	688	601	1193

QUAKER OATS LTD

Secretary and registered office

D.C. Edwards, PO Box 24, Bridge Road, Southall, Middlesex UB2 4AG; 081-574 2388; Fax 081-574 6615

Directors

G. Sewell, J. Atkins (USA), G.G. Jewell, R.G. Lagden, B.J. Nolan, P. Farrand, P. K. Lawton

The ultimate holding company is The Quaker Oats Co., Chicago, Illinois, USA, with net sales of $5.6bn in 1992 and net income $247.6m. European operations turned over $1.4bn in net sales and generated an operating income of $41.2m. These are divided between pet foods (60% of sales) and human foods (40% of sales). Primary lines of business are pet foods, cereals, Gatorade health drink, and edible oils. The largest businesses are in Italy and the UK, followed by France, Benelux, Germany and Scandinavia, with distribution arrangements in Austria, Switzerland and Spain.

In June 1991, Fisher-Price, Quaker's toy and juvenile products subsidiary was spun off to shareholders, creating an independent company Fisher Price, Inc. In the UK, 1990 figures were re-stated to take account of the demerger of Fisher Price (including Kiddicraft Ltd).

UK operations

The company has manufacturing sites at Southall near London and Cupar in Scotland.

Petfoods

Quaker, currently the UK's third-largest petfood manufacturer, produces Chunky, Minced Morsels, Felix, Delikat and Meowmix. It has recently relaunched its Chunky dog food with a revision of product, packaging and promotion with the aim of doubling the brand share by 1997. New equipment has recently been installed on lines for Felix cat food and Chunky dog food.

Breakfast cereals and snack bars

Quaker Oats is also a major producer of breafast cereals, with a strong presence in the hot cereals market with Quaker and Scotts porage oats. Ready-to-eat brands include Sugar Puffs, Puffed Wheat and Harvest Crunch. Harvest Chewy Bars are also produced in a variety of flavours.

Sutherlands Foods

Originally part of Edwin Sutherlands' potted beef business established in 1947, Sutherlands was acquired by Quaker Oats in 1964. However, in January 1993 Quaker announced the sale of Sutherlands to its main rival C. Shippam for an undisclosed sum. Sutherlands holds about 25% of the spreadable markets, 20% branded and 5% own-label, and has a turnover of £12m.

All turnover and profits originate in the UK, being the sole location of company assets.

Table 1 Five-year review (30/6/92)

	1988*	1989*	1990†	1991	1992
Turnover (£m)	156.3	158.9	185.8	170.3	166.4
Pre-tax profit (£m)	13.5	12.1	7.7	2.2	6.3
Employees	3200	1545	1746	1344	981

* consolidated accounts; † restated following Fisher Price demerger

RAINES DAIRY FOODS LTD

Secretary and registered office

L.Y. Jacobs, Raine House, Crown Road, Enfield, Middlesex EN1 1TX; 081-804 8151; Fax 081-805 4211

Directors

R. Raine, L.Y. Jacobs, N.A. Raine, M.I. Jacobs, P. Moys, J.G. Guested

Raine's Dairy Foods was established in the East End of London in 1920 by Lewis Raine, father of the present chairman. Today, Raines remains a major player in fresh dairy foods employing over 900 people in dairies in Basildon, Enfield, Wantage, Yeovil and in the Head Office at Enfield. The company has two working parts: Raines Dairy Foods is the marketing and sales company; Basildon Dairy Foods is the production company.

The company processes over 150m litres of milk/year, turning out the equivalent of 450 million units of the following dairy products; yoghurts, fromage frais, desserts, sour cream, soft cheese, cottage cheese, salad dressings and dips, ice cream. Major customers include all the major retailers as well as Boots, United Biscuits, Sara Lee, Sun Valley and Ross Youngs.

Raines has a fleet of 15 refrigerated articulated vehicles and operates 15 van sales rounds to delicatessens, bakers and caterers from depots in London and Birmingham.

Table 1 Raines Dairy Foods – Four-year review (26/9/92)

	1989	1990	1991	1992
Turnover (£m)	63.99	66.18	77.18	86.46
Pre-tax profit (£m)	0.25	2.57	3.95	2.79
Employees*	108	102	131	138

* Figures for Enfield site. As noted above, group as a whole employs over 900 people

Table 2 Basildon Dairy Foods Ltd – Four-year review (26/9/92)

	1989	1990	1991	1992
Turnover (£m)	15.81	17.95	21.27	23.15
Pre-tax profit (£m)	4.95	4.10	3.91	3.81
Employees	643	663	697	739

JOHN RANNOCH LTD

Secretary and registered office

A.F. Wilcox, Haughley Park, Stowmarket, Suffolk IP14 3JZ; (0359) 40472

Directors

R.J. Williams, R.A. Kent, A.J. Williams, P.R. Kerridge, A.F. Wilcox

Despite the fact that the turnover of the poultry business was up 8.9% on 1991 figures, total turnover for the group was 4.1% down – reflecting its withdrawal from the egg business in January 1991. Floor space released by the closure of the egg business has been redeployed in developing the poultry business.

Table 1 Five-year review (3/5/92)

	1988	1989	1990	1991	1992*
Turnover (£m)	24.82	24.95	24.23	24.48	23.47
Pre-tax profit (loss) (£m)	0.62	(0.08)	(0.68)	0.69	0.68
Employees	654	657	577	568	525

* 53 weeks

CHRISTIAN SALVESEN PLC

Secretary and registered office

G.W. Horne, 50 Fettes Avenue, Edinburgh EH4 1EQ; 031-552 7101; Fax 031-552 5809

Directors

Sir Alick Rankin (chairman), Dr C. Masters (chief executive), G.O. Tourlamain, W.R. Irving, B.H. Fidler (finance), A.C. Salvesen*, R.S. Salvesen*, R.A.B. Miller*
* non-executive

Christian Salvesen is a Scottish company with interests in 16 countries. It operates in three main areas: distribution; food services; specialist hire.

Distribution

This encompasses operations throughout the UK, Europe and North America and contributes 44% of group trading profits. The UK operation generated £20.5m in 1993, up 3.5% on 1991/2. In the grocery sector increased volumes were handled for J. Sainsbury, Safeway and Kwik Save. Services for Asda and Tesco are planned. Retail support services include tray washing for major supermarkets. New contracts have been gained with food manufacturers for chilled and frozen foods, notably Tudor Ice Cream and Pillsbury and with fruit importers like Primafruit at Evesham.

Specialist hire

This includes Light & Sound Design, Salvesen Oilfield Technology and Aggreko, a company able to supply temporary sources of power for events (e.g. Barcelona Olympics) or in remote locations. It similarly rents temperature control equipment.

Manufacturing

This includes Salvesen Brick and Vikoma (oil pollution control equipment).

Salvesen Food Services

This offers complete cold chain management from the raw materials through to the packed products delivered to wholesalers or retailers. It is responsible for processing 30% of the UK's output of frozen vegetables, packing almost 40% of own-brand labels. A new £8m packing factory at Easton, Lincolnshire, has been commissioned, the facilities at Grimsby upgraded and older plants at Eyemouth and Lowestoft have been closed. Processing facilities at Spalding and Lowestoft have been rationalized. In addition to green vegetables, the spectrum of products handled continued to widen include petfood constituents, and roots crop with 2000 tonnes of frozen rice processed at Lowestoft, 2500 tonnes of fish products at Grimsby and Edinburgh and increased soft fruit volumes at Dundee. It also provides a distribution service for Tendafrost, Birds Eye Wall's and Findus.

Table 1 Five-year review (31/3/92)

	1988	1989	1990	1991	1992
Turnover (£m)	312.0	380.7	422.5	498.3	528.0
Pre-tax profit (£m)	41.3	60.9	52.4	46.3	75.9
Employees (UK)	5429	6349	6803	7850	7497
Employees (overseas)	3381	3756	3438	3643	3601

SAMWORTH BROTHERS LTD

Secretary and registered office

M.R. Allen, 200 Medeline Road, Beaumont Leys, Leicester LE4 1EX

Directors

D.C. Samworth, J.C. Samworth, R.G. Samworth, L. Samworth

Perhaps the best-known company in the Samworth Group is Ginsters Cornish Pasties Ltd. Ginsters (set up in 1968 by Geoffrey Ginster), with a turnover of £43.4m and pre-tax profit of £2.19m in 1992, now employs about 900 people in two factories at Callington in Cornwall. Ginsters produces over 1 million pasties a week which are split fairly evenly between branded and own-label.

Walker & Son (Leicester) Ltd employs 390 people in the production of pies and sandwiches from units in Leicester. Sandwiches are supplied to major retailers from a £7m purpose-built factory at Beaumont Leys. Turnover for 1992 was £19.5m and pre-tax profits £1.2m.

Moss Side Products Ltd provides the Group with wholesale depots and distribution services and Samworth Bros (Leicester) Ltd provides management services to Walkers.

In March 1992, Samworth acquired Dickinson & Morris, a manufacturer of Melton Mowbray pork pies.

Table 1 Four-year review (31/12/92)

	1989	1990	1991	1992
Turnover (£m)	38.59	44.42	53.59	62.20
Pre-tax profit (£m)	4.04	4.37	5.82	5.67
Employees	1055	1209	1301	1372

SCOTTISH MILK MARKETING BOARD

Secretary and registered office

Marie I. Aitken, Underwood Road, Paisley PA3 1TJ; 041-887 1234

Board of Scottish MMB

A.L. Howie (chairman), J.A. Brown (vice-chairman), F.R. Evans, A.M. Gilchrist, A.Gray, E. Mundell, A.D.S. Rolland, R. Simpson, W. Weir, I. Wilson

Scottish Pride Group

I. Laidlaw (executive chairman), A. McCartney (managing director)

In March 1993 there were 2178 milk producers registered with the Scottish Milk Marketing Board (SMMB). Total milk sales off farms for the year amounted to 996.48 million litres (1012 million litres in 1990/91 and 1008.9 million litres in 1991/2). The largest percentage goes into liquid milk (545 million litres), followed by cheese (283.9 million litres) with smaller amounts going to condensed milk and chocolate crumb, cream, butter and other products.

With the imminent demise of statutory milk marketing boards in 1994, SMMB has announced plans for the creation of Scottish Milk Ltd, a new milk co-operative which will come into existence when the statutory responsibilities of the Board end.

In 1987, the Board separated its commercial division – Scottish Pride Quality Dairy Foods and William W.Stevenson Ltd (Scottish Farm) – from its milk marketing function. In 1992, these two functions were merged under one managing director into Scottish Pride Ltd. Proposals for Scottish Pride to take over the fresh milk business of CWS were rejected by the MMC in December 1992. Scottish Pride is now one of Scotland's largest food manufacturers employing over 1000 people in five liquid milk units, two transport depots, and five creameries. It is a major supplier of UHT milks and sales of its Fresh'n'Low pasteurized skimmed milk have topped £7m.

Principal subsidiaries

Kintyre Creameries Ltd
Campbeltown Creameries Ltd
Scottish Pride Daries Ltd

Principal associates

Galloway Cheese Company Ltd
Green Valley Products Ltd
Perratt & Sons Ltd
Cuthbertson Foods Ltd

Table 1 Three-year review (31/3/93)

	1991	1992	1993
Milk fund			
Total value of milk sold (£m)	216.5	221.6	225.7
Average value of milk sold (ppl)	21.07	21.77	22.95
Total payments to producers (£m)	204.71	207.35	210.24
Average payments to producers (ppl)	20.22	20.54	21.46
Cattle breeding			
Sales (£m)	2.15	2.62	2.81
Trading surplus (£m)	0.30	0.40	0.33
Scottish Pride 'Group'			
Turnover (£m)	164.3	184.9	166.9
Pre-tax profit (£m)	1.9	3.1	3.4
Employees – SMMB	282	291	401
Employees – Scottish Pride Group	1076	1085	974

C. SHIPPAM LTD

Secretary and registered office

M.E. Parker, East Walls, Chichester, Sussex PO19 1PQ; (0243) 788373; Fax (0243) 773278

Directors

C.N.G. Shippam, S.W. Young, R.L. Mittelbusher, E.A. Haberli

Shippams was a family business until in 1975 it was taken over by Wm Underwood, of Boston Massachusetts. In 1982, Underwoods were acquired by IC Industries and Shippams became an important part of the international operations of its subsidiary Pet Inc. In 1988, IC Industries changed their name to the Whitman Corporation, and following a further change in strategy, Pet Inc. was demerged from Whitman in 1991. Pet is now an independent, publicly quoted company which claims to be the 14th-largest packaged food company in the US.

Shippam still has its Head Office in Chichester, Sussex in the premises acquired by the Shippam family in 1919. Its main activities include the manufacture of meat and fish pastes and spreads in glass jars, the canning of chicken products, and the sales and marketing throughout Europe of Pet's Old El Paso brand of Mexican food products.

Shippams has factories in Chichester, West Sussex (mainly pastes and spreads) and Crediton in Devon (poultry slaughtering, processing and chicken canning). For many years its also had fish processing and canning operations at Newlyn in Cornwall but these were disposed of in 1989. It also has a distribution depot in Chichester and a freezing plant and cold store in Penzance in Cornwall.

Shippam first introduced its canned Chicken Supreme in 1960, making it under the St Michael brand label for Marks & Spencer in 1961. As the largest UK canner of chicken, Shippam processes over 70 000 birds per week for all the major retailers and the Shippam brand.

Since 1984 Shippams has been responsible for importing and marketing Pet's Old El Paso products to retail, catering and foodservice sectors. This it now does primarily through Old El Paso Foods B.V. at Amsterdam in the Netherlands.

Already a market leader with a 47% share of spreadables, in 1993 Shippams took over the business of Sutherlands Foods from Quaker Oats Ltd. Sutherlands had been with Quaker since 1964 and had a turnover of £12m and 25% of the spreadables market.

Table 1 Four-year review (30/5/92)

	1989	1990	1991[a]	1992[b]
Turnover (£m)	35.09	38.31	19.66	39.07
Pre-tax profit (£m)	5.62	6.22	3.10	6.15
Employees	568	594	625	621

[a] From 24/11/90 to 25/3/91 (4 months); [b] From 25/3/91 to 30/5/92 (14 months)

SIMS FOOD GROUP PLC

Secretary and registered office

D.R. Brady, Sims House, Sims Food Park, Sherbourne Drive, Tilbrook, Milton Keynes MK7 8BS; (0908) 270061; Fax (0908) 270063

Directors

J.M. Stone (chairman*), B.P. Ford (chief executive), D.R. Brady (finance), E.J.R. Canvin, D.B. Gunner, J.W. Lowndes, C.D. Lenox-Conyngham*, N.J. Iren*
* non-executive

The Sims Group has seen major changes in the last few years, transforming it into one of the UK's largest red meat organizations. In 1986/87 the group, then known as Sims Catering Butchers, employed just over 100 people and had a turnover of £13m, supplying meat to catering outlets, mainly in small tonnages. Since then the group has grown rapidly through a large number of relatively small acquisitions giving it a turnover of £312.6m in 1993. Nevertheless the group has been badly hit by the recession and changes in EC legislation affecting the meat industry and has undertaken a disposal, closure and cost-cutting programme. This has included disposal of J. Redmond (food machinery) and Country Feasts.

Retail Division

This comprises Canvin International, Ziff Meats, Lincs Turkeys, General Provisions, and Buitelaar Boston and generated sales of £197m in 1993. Buitelaar Boston acquired in July 1992 for £1.25m specializes in the export of cow beef to Europe. Canvin International bought the Pure Meat Co. when it went into voluntary liquidation following the sudden death of its owner. Pure Meat was supplying conservation grade meat to Asda.

Slaughtering capacity has been reduced by 40% by closing the Bow Abattoir, transferring some Boston abattoir business to Carnaby, closing three Aristocracy meat processing plants at Milton Keynes and consolidating all Safeway business at Cardington. Other retail business is now centralized at the Food Park in Milton Keynes.

Catering Division

This includes Russell Hume, W.D. Brabin, Woodhouse Hume and W & W Meat Supplies. Sales increased by 2% during 1992/3 despite the closure of depots at Motherwell, Southampton and Cambridge. Six depots remain. Catering turnover was £55m.

Manufacturing Division

In June 1992 the sausage and burger manufacturing business of Puritan Maid was acquired for £1.1m, in April 1993 followed by the Oakland Fast Food burger business of GrandMet for £4.2m. Both have been successfully integrated into J.L. Quality Foods which has a factory near Bristol. The division achieved sales growth of 31% to £17m.

Trading Division

This disposed of its shares in T S & W and in January 1992 set up First City Trading to handle UK exports and international meat trading. Turnover was £44m.

Table 1 Five-year review (31/3/93)

	1989	1990	1991	1992	1993
Turnover (£m)	143.3	167.9	228.0	251.4	312.6
Pre-tax profit (loss) (£m)	6.1	8.1	8.2	6.8	(0.6)
Employees	1504	1602	1830	2090	1956

SMITHKLINE BEECHAM PLC

Secretary and registered office

I.M.F. Balfour, New Horizons Court, Brentford, Middlesex TW8 9EP; 081-975 2000

Directors

H. Wendt (chairman), R.P. Bauman (chief executive), H.R. Collum (finance), H.C. Groome, Dr J-P. Garnier, J. Leschly, Dr G.H. Poste, A.R.F. Buxton, Sir Robert Clark (non-executive vice chairman), W.R. Grant (non-executive vice-chairman), Sir Christopher Hogg*, D. McHenry*, R.A. Pfeiffer, Jr.*, Sir Peter Walters*, J.White*
* non-executive

SmithKline Beecham plc was created in July 1989 from the merger of the SmithKline Beckman Corporation and the Beecham Group plc, forming one of the world's major pharmaceutical, healthcare and toiletry companies. It has operations in 30 countries, with substantial interests in the UK, USA and throughout continental Europe. Business sectors in which it operates are pharmaceuticals, animal health, consumers' brands and clinical laboratories. Of these, pharmaceuticals and consumer brands have the largest turnovers (£2889m and £1384m, respectively). The long-term strategy of the group is to move away from toiletries and personal care products and focus more strongly on consumer medicines and healthcare products. To this end, in 1992 SKB disposed of men's personal care products in North America and Manetti Roberts toiletries in Italy. More recently (June 1993) it announced the sale of a batch of leading toiletry brands (Brylcreem, Badedas, and Body Mist) to the Sara Lee Corp. and hair care lines (Silvikrin, Vosene, Bristow's and Amami) to Wella.

SB Consumer Brands accounted for 27% of sales (£1384m) and 21% of trading profit (£244m) in 1992, and include well-known brands such as Macleans, Aquafresh, Tagamet, Tums, Oxy, Ecotrin and Contac together with health drinks Ribena, Lucozade and Horlicks. Figures for this sector for the 9 months to October 1993, showed a rise in sales to £964m (£790: 1992) but a fall in trading profit to £109m (£112m: 1992).

Ribena and Lucozade are produced at the Royal Forest Factory in Coleford, Gloucestershire and Horlicks is produced at the Slough site. The Coleford factory is one of the UK's largest drinks manufacturing plants employing over 600 people in fruit processing and production of drinks in bottles, cans, aseptic cartons and pouches.

In December 1992, SmithKline Beecham sold the rights to manufacture, distribute and sell Schloer and PLJ fruit drinks to Merrydown plc for £8.5m.

Table 1 Five-year review (31/12/92)

	1988	1989	1990	1991	1992
Turnover (£m)	4302	4897	4764	4685	5219
Pre-tax profit (£m)	564	724	860	1002	1115
Employees	61 900	62 800	–	54 000	53 700

Table 2 Turnover (£m) by business sector and geographical location

Business sector	1991	1992	*Geographical location*	1991	1992
Pharmaceuticals	2470	2889	United Kingdom	504	508
Animal health	318	337	USA	2078	2361
Consumer brands	1358	1384	Continental Europe	1273	1398
Clinical laboratories	539	609	Rest of world	830	952
Total	4685	5219	Total	4685	5219

SWIZZELS MATLOW LTD

Secretary and registered office

N.S. Menzies, Carlton House, Albion Road, New Mills, Stockport, Cheshire, SK12 3HA; (0663) 744144; Fax (0663) 742800

Directors

J.D. Dee, M. Matlow, T.J.L. Matlow (joint managing director), M.N. Dee (joint managing director), A.N. Dee, B.N. Dee, A.C. Matlow

Swizzels, based in Stockport, has been involved in the sugar confectionery trade since 1928 when it was founded by Alfred and Maureen Matlow and David Dee. Well known for its Love Hearts, it also makes chews, sherbet dips, marshmallows, fun gums as well as chocolate coated brazil nuts and travel tins of fruit and mint drops.

Trading subsidiaries

Printo Wrappings Ltd
Clevedon Confectionery (Blackpool) Ltd (50% held by parent company, 45% by subsidiaries)
Castleford Ltd (Eire)
Swizzels Matlow (USA) Ltd (USA)

Table 1 Five-year review (31/12/92)

	1988	1989	1990	1991*	1991†	1992
Turnover (£m)	24.9	24.7	26.1	30.6	24.9	37.7
Pre-tax profit (£m)	0.86	0.32	0.99	0.92	0.55	2.14
Employees	758	743	701	737	765	785

* Year ended 31/3/91; † 9 months ended 31/12/91

THORNTONS PLC

Secretary and registered office

M.S. Thornton, Thornton Park, Somercotes, Derbyshire DE55 4XJ; (0773) 540550; Fax (0773) 540757

Directors

C.J. Thornton (chairman and chief executive), M.S. Thornton (deputy chairman and chief executive), A.D. Goodwin (finance), D.J. Mitchell (MD Thornton's UK), J. Coyle (MD Thornton's Europe), J.G. Culverhouse (personnel), D.C. Samworth*, J.W. Seabright*, M Tuttle*
* non-executive

Thorntons is a manufacturer and retailer of high-quality chocolate confectionery and ice cream with divisions focused on retail, manufacturing, and commerical activities.

Retail

Thorntons now has a total of 511 outlets, 451 in the UK, Eire and Channel Islands and 60 in France. Outlets include retail, franchise, and mini-franchises and chill cabinets. During 1993, 39 new shops and franchises were opened in the UK, several within major supermarkets (e.g. Asda). A successful Christmas in 1992 gave rise to UK retail and franchise sales of £66.9m.

Restructuring of the French business and a write-down of goodwill resulted in a pre-tax loss for the year of £4.8m. As a result about 25 shops are to be sold, and the remaining operations focused around profitable and well-located sites near Paris and Ile de France. In France shops trade under the name of Martial or Sunset.

Manufacturing

The activities of the five manufacturing plants (Derby, Belper, Flixborough, Antwerp and Clermont-Ferrand) have now been closely integrated and new plant and equipment installed for lollipop and chocolate moulding, chocolate coating and improving boiled sweet capacity and product range. In Belgium, Gartner's factory in Antwerp continues to supply specialist products to Thorntons and Martial shops and in Germany Gartner's products are being trialled through the Hussel chain. The ice cream factory at Clermont-Ferrand in France makes ice cream for both retail and commercial customers. Two new luxury ice creams modelled on successful confectionery lines – Apricot parfait and Cherry Truffle – were introduced.

Commercial

The Division makes premium chocolates for Marks & Spencer own-labels and a range of diabetic chocolates and sugar-free lines for Boots. Sales turnover was £15m in 1993.

Table 1 Five-year review (26/6/93)

	1989*	1990†	1991	1992	1993
Turnover (£m)	63.92	76.17	79.91	84.33	92.48
Pre-tax profit (loss) (£m)	10.09	11.31	11.88	9.20	(4.80)
Employees	2450	–	2667	2619	2775

* 56 weeks; † 53 weeks

Table 2 Turnover (£m) by geographical destination

	1992	1993
United Kingdom and Belgium	75.6	81.9
France	8.7	10.6
Total	84.3	92.5

TULIP INTERNATIONAL (UK) BACON DIVISION LTD

Secretary and registered office

M. Forbes, Caxton Way, Thetford, Norfolk IP24 3SB

Directors

J.W. Anderson, F. Lindelor, G. Kjeldsen, P. Rasztar

Tulip International A/S (TI) was formed in October 1990 by merging the processing operations of Tulip, Danepak, Jaka and Normeat. The new group has 12 plants, seven in Denmark, four in the UK and one in Germany. UK plants are at Thetford (Danepak) and Wellingborough (Normeat) and two in Liverpool (Jaka). About 40% of turnover comes from the UK operation, also 40% of the workforce.

Results for 1991 showed losses totalling £18.8m (DkK 211m) on a turnover of £500m (DkK 5600m). Subsequent half-year results reported a further deficit of £11m and turnover down to £222m. Poor results were blamed on recession and over-capacity in the UK, sharp rises in raw material and pigmeat prices in the US, and unfavourable trends in exchange rates. A restructuring programme designed to cut costs led to the loss of 75 jobs at Thetford.

Despite a decline of 10% in the bacon market, Danepak and Tulip brands gained market share. During 1992 'Lean & Low' – premium quality bacon aimed at the health conscious consumer – increased sales by 50%. Value-added products such as 'Classic Dry Cure' and 'Free Range Bacon' from TI's joint venture with Friland Foods in Denmark, were introduced in 1993.

Tulip International (UK) Cooked Meat Division Ltd (formerly Jaka) is based in Merseyside. It supplies cooked and canned meats from Denmark, South America, Italy, France and Germany. It has 370 employees and had sales of £88.5m in 1992, on which it made a pre-tax loss of £3.71m.

In November 1992, Tulip International put in a bid for Plumrose chilled and canned meat which was put up for sale by the owners East Asiatic Company. Tulip's bid was subsequently rejected as inadequate.

1993 results

In the year to the end of September 1993, Tulip International made pre-tax profits of £15.1m on sales of £480m. The UK Bacon division at Thetford made a pre-tax loss of £1.7m on sales of £97m and the cooked meats operation increased sales by 10% and turned 1992's loss into a profit of £400 000.

Table 1 Four-year review (26/9/92)

	1989	1990	1991	1992
Turnover (£m)	88.06	86.92	97.79	102.44
Pre-tax profit (loss) (£m)	(3.44)	0.86	1.61	(5.84)
Employees	1041	940	903	874

THOMAS TUNNOCK LTD

Secretary and registered office

H. McKellar, 36 Old Mill Road, Uddingston, Glasgow G71 7HH; (0698) 81355

Directors

A.B. Tunnock, M. Tunnock, E.A. Tunnock

Tunnock is a family business, established as a retail business in 1890 by the grandfather of the company's present chairman, Boyd Tunnock. In terms of turnover, 90% of total sales are from biscuits, principally Chocolate Caramel Wafers, Snowballs, and Chocolate Teacakes. In addition, the company produces a limited amount of meat products – Scotch pies, sausage rolls and steak pies – and various cakes, including wedding cakes. Recent additions to the product range include Ditto, a chocolate-covered wafer and caramel biscuit and Milk Chocolate Ginger Perkins, chocolate ginger biscuits.

Its van sales force began after the second world war and now serves about 600 customers in Scotland from small shops to large wholesalers. Most of the major multiples stock Tunnock's biscuits which are also exported to over 30 countries. New markets have recently emerged in Portugal, Lebanon and Hong Kong.

Tunnock's Glasgow factory has recently completed the last phase of a 5-year expansion plan.

Table 1 Four-year review (28/2/93)

	1990	1991	1992	1993
Turnover (£m)	18.43	21.17	20.80	23.16
Pre-tax profit (£m)	1.55	1.15	0.90	1.56
Employees	722	729	709	692

WALKERS SMITHS SNACK FOODS LTD

Secretary and registered office

P.S.E. Valentine, 1600 Arlington Business Park, Theale, Reading, Berkshire RG7 4SA; (0734) 306666; Fax (0734) 303152

Directors

Smiths Crisps: J.S. Pingel (chairman), P.B. Robinson, A.M. Gardiner, N. Harrison, T. Raven, D. Fennell, A. Hunt, A. Garland, J. Lawson

Walkers Crisps: J.S. Pingel (chairman), T. Charlesworth, J.R. Cullip, N.B. Hazeldine, P.S.E. Valentine, J. Benton, D.F. Morrison, J.P. Collins, T.C. Wentworth

In July 1989, the French food group BSN sold the Smiths and Walkers snack food companies to Pepsico for $1.35m only a few weeks after buying it from buyout specialist Kohlberg, Kravis Roberts. Pepsico, Inc. is a global corporation operating in three main areas: beverages, snack foods and restaurants. The beverage segment markets Pepsi, Diet Pepsi, Mountain Dew and 7Up (outside the US). Concentrates are sold to franchises with bottling operations worldwide. The snack food business operates through Frito-Lay with other major businesses in Mexico, Canada and the UK. Restaurant chains include Pizza Hut, Taco Bell, and Kentucky Fried Chicken.

In 1992, Pepsico spent $63.9m on consolidating and streamlining the Walkers and Smiths businesses in the UK, which included integration of their commercial functions at a new Head Office in Reading. Part of Pepsico Foods International, Walkers Smiths Snack Foods is held by Frito-Lay Holdings Ltd in the UK. It has factories at Thurmaston, Leicester, Beaumont Leys, Fleetwood, Bracebridge, Liverpool, Peterlee and Swansea.

Between them Walkers and Smiths have many of the top selling branded crisps and snacks in the UK including Big D, Planters, Crinkles, Quavers, Squares, Cheese Moments, Tubaloops, Looney Tunes, Bacon & Scampi Fries and Walkers and Smiths crisps. They are constantly competing with the other major players, Dalgety (Golden Wonder) and United Biscuits (KP) for market share. At the end of 1992, Walkers claimed its largest market share ever with 37.8%.

Table 1 Financial information

	1990	1991*
Smiths Crisps Ltd		
Turnover (£m)	161.3	243.7
Pre-tax profit (£m)	13.9	4.7
Employees	3434	3149
Walkers Crisps Ltd		
Turnover (£m)	146.3	247.1
Pre-tax profit (£m)	37.0	71.0
Employees	1713	1722

* 18-month period from June 1990 to December 1991

WARBURTONS LTD

Secretary and registered office

R.N. McMullan, Back o' th' Bank, Blackburn Road, Bolton, Lancashire BL1 8HJ; (0204) 31004; Fax (0204) 23361

Directors

W.R. Warburton (chairman), R.J.L. Abbott (deputy chairman), W.B. Warburton, J. Warburton, B.E. Sealey*, J.P. Speak*
* non-executive

In March 1991, the assets and business of the Bolton Cake Bakery, part of Warburtons (Soreen) Ltd were sold to Northern Foods plc, providing a profit of £2.9m on the sale. Since then the malt bakery of Warburtons (Soreen) has continued to trade from the Trafford Park bakery as an operating division of Warburtons Ltd. The US Associate, Warburtons, Inc. (52% holding) was disposed of during 1991.

Bread Division

1992 was a year of overall growth, despite intense price competition in bread markets. Increases in sales in the North East and Yorkshire were offset by declines in the company's traditional stronghold of Lancashire. The Newburn Bakery moved into profit in its first full year of trading.

Retail Division

1990's good results, enhanced by the acquisition of Sayers Confectioners, were not repeated in 1991, and in 1992 none of the retail companies maintained 1991 sales volumes.

Meat Division

Peter Hunt's Bakery Foods opened a new £7m, 88 000 sq.ft EC-approved factory in Kearsley in September 1993. It is producing over 60 chilled and frozen lines (meat pies, pasties and savoury products) for own-label as well as for Warburtons' 250 retail outlets. The division maintained its profitablility despite a decline in van sales. The Farnsworth bakery has been closed.

Subsidiaries

Warburtons Fylde Ltd
Warburtons Pennine Ltd
Warburtons (Colne Valley) Ltd
Warburtons (Yorkshire) Ltd
Warburtons (Tyne & Wear) Ltd
Warburtons Estates Ltd
Sayers Confectioners Ltd
Burtons (Blackpool) Ltd
Hampson (Bolton) Ltd
Parker Bradburn Ltd
F. Spinks Ltd

Table 1 Five-year review (30/9/92)

	1988	1989	1990	1991	1992
Turnover (£m)*	87.3	100.3	115.1	144.5	149.1
Pre-tax profit (£m)	4.94	8.0	8.3	13.1	11.6
Employees	3433	3660	4162	5517	5430

* All the group's turnover is now within the UK

WARNER–LAMBERT (UK) LTD

Secretary and registered office

K.A.M. Sykes, Lambert Court, Chestnut Avenue, Eastleigh, Hampshire SO5 3ZQ; (0703) 620500; Fax (0703) 629812

Directors

A.W. Yates, A.K.I. Walker

The company is a subsidiary of Warner–Lambert Co., USA, one of the world's largest confectionery groups. In November 1991, the parent company announced major rationalization of UK operations as part of a wider global restructuring and the formation of a new European headquarters in Brussels.

In June 1992, Hall Brothers (Whitefield) Ltd, part of Warner–Lambert Confectionery was placed in members' voluntary liquidation. Hall's had been part of the group for about 27 years.

Well-known brands manufactured include Hall's Mentholyptus, Hall's Soothers and Clorets. The group also distributes Dentyne and Stimorol chewing gums and Van Melle's Fruit-Tella, Dummys and Mentos.

Warner–Lambert Health Care markets a range of mouthwashes, suntan lotions and self medication products.

Subsidiaries

- Lambert Chemical Co. Ltd – marketing, Middle East and Africa
- Parke Davis & Co. Ltd – pharmaceuticals, healthcare products, toiletries and confectionery

Table 1 Five-year review (30/11/91)

	1987	1988	1989	1990	1991
Turnover (£m)	64.5	68.7	77.0	141.1	151.2
Pre-tax profit (loss) (£m)	0.5	2.7	1.3	1.7	(2.3)
Employees	907	671	625	2014	1967

WATERFORD FOODS PLC

Secretary and registered office

M.J. Dempsey, The Friary, Main Street, Dungarvan, County Waterford, Eire; 010 353 58 41033

Directors

J. Barry (chairman), J. Dowley (vice-chairman), S. O'Connor (managing director), T. Corcoran, P.Fitzgerald, M. Flood, C. Hill, M. Kiely, B. Maye, W. Moore, D. Shanahan, J. Walsh, T. Walsh, B. Whelan

Waterford Foods is a leading Irish dairy and consumer foods group with major businesses in Eire, Northern Ireland and the UK and Galloway-West in Wisconsin, USA.

Dairy Division

The Dungarvan processing plant produces butter, milk powder, casein and specialist food ingredients. Sales of lactic butter to Germany have increased, winning the 'Bronze Cow' award for the fourth year running. Kilmeaden, one of the largest cheese production units in the UK or Ireland, makes cheese and whey products for domestic and export markets. Premier Dairies, acquired in 1992, has plants in Dublin and Dundalk. It supplies a range of milk and dairy products under the Premier and Dealgan brand names. It also supplies Virginia Milk Products Ltd with milk for further processing.

Galloway West

This manufactures bulk sweetened condensed milk for the chocolate and confectionery industries, and hard and semi-hard continental cheese (particularly Swiss cheese) and supplies cheesemaking ingredients including phage-resistant starter cultures. A new science and technology centre provides resources which allow product functionality to be tailored to customer requirements.

Waterford Food Ingredients

In 1992 the group acquired Virginia Milk Products Ltd and Express Foods Northern Ireland Ltd whose core businesses involve the production of cream base for Irish cream liqueurs, specially formulated milk powders, cheddar cheese, mozzarella and whey protein concentrates. Increased demand for mozzarella has led to the upgrading of the plant at Magheralin for this purpose.

Consumer Division

The division manufactures, markets and distributes branded goods including Yoplait, Petits Filous, Snowcream and Castlelyons brands. A new fromage frais plant was opened at Inch, County Wexford in December 1992. There is a sophisticated hard and soft cheese packing operation at Corsham in Wiltshire supplying most of the major multiples in the UK.

Waterford Foods (UK)

A single dairy business named Waterford Dairies, has been created out of Heald's, Rigby's and United Co-Operative Dairies (Daisy Dairy), supplying over 25% of the liquid milk market in North West England. Plans are in hand to fully franchise the doorstep delivery business. The juice business has also been rationalized with the closure of Heald's Maidstone plant.

Trading Division

This sells Gain animal feeds and supplies a range of farm hardware, fertilizers, crop protection materials and veterinary supplies to the agriculture sector.

Table 1 Five-year review

	1988	1989	1990	1991	1992
Turnover (IR£m)	238.4	278.9	439.1	423.8	551.5
Pre-tax profit (IR£m)	7.9	11.3	11.3	15.9	20.4
Employees	1170	n/a	n/a	2371	3224

WEBBS COUNTRY FOODS LTD

Secretary and registered office

B.W. Watkins, Moorside Road, Winnall, Winchester, Hampshire SO23 7RX; (0962) 68046; Fax (0962) 68421

Directors

C.G. Webb, L. St J. Webb, N.E. Caffyn, Mrs I.J. Webb, D.A. Price, B.R. Taylor, H. Gallagher

Company history

The original business was established just after the second world war by Percy Webb, father of the company's present chairman and managing director. Following a series of acquisitions – J.G. Read, Jaywill, Stonegate Farmers and Wiltsdown Ducks – the company is now one of the UK's largest independent poultry processors, handling 750 000 birds/week and employing about 1500 people at its six factories. It also has poultry dealing, exporting and engineering interests, 31 farms and its own feed mill at Westbury in Wiltshire.

Following the acquisition of Wiltsdown Ducks from Unigate in 1988, permission for a new £1.5m plant on the same site was refused by the landlords. It was subsequently closed with the loss of 45 jobs.

Principal activities

The company operates a total of six production facilities in the south of England: at Okeford Fitzpaine, Lymington, Winchester, Chichester, Corby and Sutton Benger near Chippenham. Processed fresh birds, portions and added-value dishes are prepared at Okeford Fitzpaine, the company's first custom-built broiler operation opened in 1977. The Lymington site employs 350 people and processes 10 million broilers each year into fresh and chilled products. Production at Winchester is devoted to the processing of hen meat for sale to food manufacturers, while the £4m purpose-built further processing factory at Chichester produces added-value products such as sausages, burgers and chicken rolls. The company also operates a fifth plant in Corby, which was bought jointly with Faccenda Chicken in 1985. The Northampton-based plant now provides excess capacity for primary processed chicken ranges and further processed products.

In September 1992, Webbs bought an 11-acre site at Sutton Benger near Chippenham from Associated Poultry, part of Hillsdown Holdings.

Webbs currently supplies fresh chicken to the leading retailers – Marks & Spencer, Sainsbury, Waitrose, Asda and Safeway – and also supplies bulk chicken meat to manufacturers such as Heinz, Ross Foods and Campbells.

Tabıe 1 Five-year review (30/9/92)

	1988	1989	1990	1991	1992
Tunover (£m)	52.4	56.6	66.0	70.8	74.8
Pre-tax profit (£m)	3.1	3.4	3.3	3.6	4.1
Employees	1141	1120	1256	1286	1477

Table 2 Divisional analysis – 1992

	Turnover (£m)	Pre-tax profit (£m)
Food processing	69.6	3.6
Poultry dealing	5.2	0.4
Engineering	0.02	0.1*
Total	74.8	4.1

* Includes £0.078m profit from associate companies

WEETABIX LTD

Secretary and registered office

I.P. Clarke, Weetabix Mills, Burton Latimer, Kettering, Northamptonshire NN15 5JR; (0536) 722181; Fax (0536) 726148

Directors

R.W. George (chairman and managing director), H.R. Reynolds, J.H. Carver, T. Briggs, I.F. Comley, T.A. Hart

Weetabix is primarily engaged in the manufacture of breakfast cereals in the UK. Other members of the group also handle sales promotion schemes, and carton and package manufacture. It also has subidiaries in USA, Canada and France. In 1991, its two Canadian subsidiaries merged to form one company, Weetabix of Canada Ltd.

1992 saw an overall profit increase of 22% despite difficult trading conditions. The apparent decline in overseas business was entirely related to the high value of sterling at the end of the year. In the UK Weetabix has faced intense competition from Kelloggs and Cereal Partners (Nestlé and General Mills joint venture), whose total advertising spend is far in excess of that of Weetabix. Nevertheless, Weetabix has used both national and regional television to promote Weetaflakes, Weetos, Oat and Wheat Bran, and Ready Brek. Weetabix whole wheat bisks, the group's staple brand, has attempted to halt its declining market share using the children's TV series Stingray in its recent promotions. Acquisition of Ready Brek from Allied-Lyons in 1991 increased Weetabix's share of the overall cereals market. It is now produced at one of the Corby factories. Alpen is market leader in branded muesli products, but its derivative Cruesli sold poorly and was discontinued in 1991.

In 1991, Weetabix won a Queen's Award for Exports largely due to its European sales which have been growing at the rate of 40% a year, particularly in France.

Vibixa, the group's packaging subsidiary increased profits and expanded sales to companies outside the group.

Table 1 Five-year review (31/7/92)

	1988	1989	1990	1991	1992
Tunover (£m)	123.5	138.8	150.3	184.4	184.3
Pre-tax profit (£m)	12.5	12.9	15.7	19.7	24.0
Employees	2295	2299	2326	2500	2466

Table 2 Geographical origin of turnover and pre-tax profit (31/7/92)

	Turnover (£m)		Profits (losses) (£m)	
	1991	1992	1991	1992
United Kingdom	128.3	129.1		
Rest of EC	15.2	17.3		
Other	2.8	3.8		
Sub-total	146.3	150.3	18.6	21.9
North America	38.1	34.0	2.1	2.6
Net interest			(0.9)	(0.4)
Total	184.4	184.3	19.7	24.0

WHITWORTHS HOLDINGS LTD

Secretary and registered office

R. Shortland, Victoria Mills, Wellingborough, Northamptonshire NN8 2DT; (0933) 76351

Directors

R.W. George (chairman), R.V.J. Cadbury (managing director), M.C.M. Evans, D.A. George, A.J. Reynolds, H. Reynolds, M.J.B. Ward

The principal activities of the group during the year were flour milling and agricultural merchanting, production of bread and flour confectionery, sales and servicing of motor vehicles, and other activities including farming, estate management and forestry.

Bakeries

This is the largest part of the group, employing about 1500 people. The Wholesale Bakery made a pre-tax loss of £3.7m in 1992 attributed to a costly and protracted move to a new bakery in Leicester and the well publicized bread price war. Retail bakeries performed better and 16 new in-store units were added during the year, and profits increased. Following the acquisition of the master franchise for Delifrance, a French-style bakery and cafe, the first unit was opened at Peterborough followed by two at Gatwick Airport.

Whitworth Bros.

Sales increased to both external customers and the bakery division. The purchase of Quality Bakery Supplies Ltd for £188 000 in cash broadened the range of products offered to the bakery trade and enhanced flour sales. Production of bread crumb for food coatings was commenced as part of a strategy to develop more value-added products.

Related companies

While Ryecroft and Hill Biscuits performed satisfactorily, Napier Brown Holdings increased pre-tax profits by 9%. Napier Brown is involved in trading and packing sugar, dried fruits, nuts and other produce. In 1992 it generated a pre-tax profit of £11m on a turnover of £388.4m. It employs 1100 people.

Food-based subsidiaries

- Fresha Bakeries Ltd, Leicester
- Don Millers Hot Bread Kitchens Ltd, Luton, Bedfordshire
- Rising Dough Ltd, Hove, Sussex
- Whitworths Bakeries Ltd (sub-group management)
- Whitworth Bros. Ltd, Wellingborough, Northamptonshire (flour milling, wheat processing and other agricultural activities)

Related companies

- Hill Biscuits Ltd, Ashton-under-Lyne, Lancashire (20%)
- Napier Brown Holdings Ltd, London (52.6%)
- Ryecroft Foods Ltd, Ashton-under-Lyne, Lancashire (20%)

Table 1 Five-year review (31/3/92)

	1988	1989	1990	1991	1992
Turnover (£m)*	231.4	150.7	130.4	117.4	109.0
Pre-tax profit (loss) (£m)	2.15	2.79	4.31	6.09	(0.4)
Employees	2963	1990	1908	1821	1937

* All turnover of the group is generated in the UK

THE WRIGLEY COMPANY LTD

Secretary and registered office
A.C. Ogle, Estover, Plymouth, Devon PL6 7RR; (0752) 701107; Fax (0752) 778850

Directors
P. Dixon, W. Wrigley, A.C. Ogle, D.S. Barrie, M.J. Geraghty, P.G. Hamilton

This Plymouth-based manufacturer of chewing gum is the UK subsidiary of Wm. Wrigley Jnr Company (USA), the world's third-largest producer of sugar confectionery with an annual production of 140 000 tonnes.

The UK operation, established in 1921, is brand leader in the chewing gum sector of the confectionery market, holding 87%. Over the past few years the market for chewing gum has been increasing, since dentists recommend that chewing sugar-free gum after meals can help prevent tooth decay. Wrigley's two sugar-free gums, Extra and Orbit, are worth £66m – more than half the UK chewing gum market.

Table 1 Four-year review (31/12/92)

	1989	1990	1991	1992
Turnover (£m)	51.97	58.86	68.00	89.49
Pre-tax profit (£m)	7.12	6.36	8.46	12.93
Employees	503	544	588	622

YORKSHIRE FOOD GROUP PLC

Secretary and registered office

R. Atkinson, Carter mills, 146 Cleckheaton Road, Bradford, West Yorkshire BD12 0HF; (0274) 691285; Fax (0274) 604971

Directors

M.R. Firth (chairman and chief executive), P.S. Haley (finance), A. Giddings, D. Holden, P.A. Merchant, L.M. White, C.J.M. Haines*, Sir Marcus Fox*
* non-executive

Incorporated in 1986 and re-registered on 17 February 1993 as a public limited company. In March 1993, shares were placed on the Official List.

It has grown from a very small operation packing sugar for E. Billington and then Prewitts Muesli for Holland and Barratt, through the acquisition of several small companies (Hunni Foods, a rice supplier, Freshroast Coffee and Coffee Mac, and two bakeries – Crossfields in Blackburn and Forshaws in Preston, and Pie Tom's).

Ingredients

This division includes Scotia Haven (acquired from Berisford in 1991 for £12.8m), William Brock and Hunni Foods. They handle 50 000 tonnes/year of beans, pulses, rice, dried and glace fruits and porridge oats under the Great Scot brand name. A new selection of premium culinary nuts has recently been added to the range. Customers include all the major supermarkets, large wholesalers and food manufacturers.

Bakery

Forshaws (Confectioners) Ltd was closed with the loss of 70 jobs and the business sold. Crossfield Foods, operating from a factory in Blackburn, was successful in winning a major contract to supply a range of pies to Euro Disney.

Beverages

This division has been particularly successful. Coffee Mac Ltd opened its fifth branch at Brentford, and now has a national distribution system for installing beverage dispensers and supplying ingredients. Fresh Roast Coffee Co. Ltd has added freeze-dried coffee, drinking chocolate and cappuccino topping to its range. Both operate from purpose-built premises in Halifax.

USA

The group owns Rio Del Mar Foods Inc., a San Francisco-based raisin, prune and walnut trading business. In 1992 it acquired Treehouse Farms from Berisford. Treehouse, based in California, is an almond processor. 1993 saw the purchase of an 80% holding in Del Monte Dried Fruit, part of the American Del Monte Corporation for £18m. This business should make pre-tax profits of $2.4m on sales of $58m.

Table 1 Two-year financial record (31/12/92)

	1990	1991	1992
Turnover (£m)	15.1	49.3	63.2
Pre-tax profit (£m)	0.3	1.5	2.7
Employees	–	685	722

SECTION 2.3 SMALL COMPANIES

Companies with approximately 500 employees or less engaged in food manufacture.

A

Abacus Frozen Foods Ltd
Abel Eastern Foods Ltd
Aberdeen & District Milk Marketing Board
Aberdeen Meat Marketing
Allan & Dey Ltd
Alpine Soft Drinks (UK) Ltd
Anchor Foods Ltd
Armitage Bros. plc
Attwell Farms Ltd
Aulds (Foods) Ltd

B

John C. Bain & Son Ltd
A.J. & R.G. Barber Ltd
Bart Spices Ltd
Baxters of Speyside
Beacon Foods Ltd
Beck Food Group Ltd
Belgravia Salads Ltd
Ken Bell (International) Ltd
Bendicks (Mayfair) Ltd
Beni Foods Ltd
Bennetts of Doncaster Ltd
Berrydales Ltd
B.G. Foods Ltd
Edward Billington & Son Ltd
Blue Bird Confectionery Ltd
G.D. Bowes & Son Ltd
Boxford (Suffolk) Holdings Ltd
BP Nutrition (UK) Ltd
Brandway Group Ltd
Bridge Farm Dairies Ltd
Bristow's of Devon (Holdings) Ltd
Brown Brothers Manufacturing Ltd
Sarah Brownridge
Buchan Meat Producers Ltd
James Budgett & Son Ltd
Buitelaar (Kilmarnock) Ltd
Butcher & Baker Foods Ltd

C

Cambrian Soft Drinks Ltd
Canadian Pizza plc

Carters Drinks Group Ltd
Case & Sons Ltd
Castle Dairies Ltd
Cauldron Foods Ltd
C & D Continental Patisserie Ltd
Central Midlands Co-operative Society
Cheale Meat Ltd
Cheshire Wholefoods (Cereals) Ltd
Chitty Group plc (The)
T.W. Clothier & Sons Ltd
Conrad & Essenhigh Ltd
Cooke Bros. (Tattenhall) Ltd
Coombe Farm Foods Holdings Ltd
Cow & Gate Nutricia Ltd
Cox Seafoods Ltd
Cranswick plc
D.S. Crawford Ltd
Crooks of Harleston Ltd
Crown Chicken Ltd
Crystal Drinks Ltd

D

Dairyfresh Desserts Ltd
Daloon Foods (UK) Ltd
Jeffrey Davies & Davies Ltd
Dawn Foods (UK) Ltd
Dawnfresh Seafoods Ltd
Dell Foods (Products) Ltd
Del Monte Foods (UK) Ltd
De Roma Ice Cream Ltd
Devon Quality Foods Ltd
D.M.J. plc
Dove Valley Poultry Ltd
F. Duerr & Sons Ltd
Duncan's of Scotland Ltd
Dunhills (Pontefract) plc

E

Elisabeth the Chef Ltd
Elsenham Quality Foods Ltd
Emley Farm Chilled Food Ltd
Eryri Food Products
Everest Foods plc

F

Farley Health Products Ltd
Farmers Boy Ltd
Faroe Seafood Company Ltd

Fletchers Bakeries Ltd
Food for Thought (U.K.) Ltd
Forth Valley Foods Ltd
Fredericks Dairies Ltd
Fresh-Pak Eggs Ltd
Fribo Foods Ltd
Furniss Group Ltd

G

J.W. Galloway Ltd
Gianova Foods Ltd
Giles Foods Ltd
F.A. Gill Ltd
Gillard Holdings plc
Glendale Frozen Foods Ltd
Global Group plc
Gold Crown Foods Ltd
Golden West Foods Ltd
Goodlife Foods Ltd
Granelli McDermott Ltd

H

Halal Meat Co. Ltd
Haldane Foods Ltd
Hall & Woodhouse Ltd
Halo Foods Ltd
H. Hargrave & Co. (Wholesale Pork) Ltd
Heritage Food Group (UK) Ltd
Hey UK Ltd
Highlander Snacks Ltd
Hitchen Foods plc
Peter Hollingsworth Ltd
Horizon Farms
Horlicks Farms & Dairies Ltd
Hyman Foods Ltd

I

Icelandic Freezing Plants Ltd
Ilchester Cheese Co. Ltd
International Cuisine Ltd
Inverurie Scotch Meat Co. Ltd

J

Jacksons Foods Ltd
W. Jordan (Cereals) Ltd
Jugg Foods plc

K

Kavli Ltd
Kays Cooked Meats Ltd
Kibun Co. (UK) Ltd
Kinnerton (Confectionery) Co. Ltd
Kitchen Range Foods Ltd

L

Lagan Meat Co. Ltd
Lancashire Dairies Ltd
Lark Valley Foods Ltd
Laxgate Ltd
Mr Lazenby's Traditional Sausages
Leaf United Kingdom Ltd
John J. Lees plc
Le Pain Croustillant
Lion Mark Holdings Ltd
Livera Foods plc
Lofthouse of Fleetwood Ltd
Long Clawson Dairy Ltd
Lord Rayleigh's Dairies Ltd
Lovell's Confectionery Ltd

M

Macfish Ltd
M & W Mack Ltd
Mackie's Aberdeen Dairy Co. Ltd
Macphie of Glenbervie Ltd
Marlow Foods Ltd
Mathers (Inverurie) Ltd
E.C. Matheson & Son Ltd
Matthews Foods plc
Maxway Food Products Ltd
McIntosh Donald Ltd
McIntosh of Dyce Ltd
McKey Food Services Ltd
McLaren Foods Ltd
McQuillan Meats Ltd
Meadow Vale Foods Ltd
Meltis Confectionery Ltd
Meridian Foods Ltd
Merrydown plc
John Millar & Sons (1844) Ltd
Millway Foods Ltd
Moray Seafoods Ltd
Morning Foods Ltd
Müller Dairy (UK) Ltd

N

Needlers & Co. plc
Henry Newbould Ltd
New Covent Garden Soup Company
New England Food Company
J.N. Nichols (Vimto) plc
Nimmo (Holdings) Ltd
Noon Products plc
North Downs Dairy Co. Ltd
North of Scotland Milk Marketing Board
Northern Co-operative Society Ltd
Northumbrian Fine Foods plc

O

Oak Forest Foods Ltd
Oak Lodge Salmon Ltd
Orchard House Foods Ltd

P

Paradise Cuisine
Parkham Foods Group
Pasta Reale Ltd
Patak's (Spices) Ltd
Paulig UK Ltd
Pershore Poultry Ltd
Perimax Meat Co. Ltd
Phoenix Foods Manufacturers Ltd
Pilgrim Food Group Ltd
Pinegain Ltd
Plumrose Ltd
Plumtree Farms Ltd
Pollard Confectionery Ltd
Princes Ltd
L.E. Pritchitt & Co. Ltd
Pukka Pies Ltd
Puredrive Ltd
Puritan Maid Ltd

Q

J.G. Quicke & Partners

R

Rahbek-Food Ltd
Rayner & Co. Ltd
Rea Valley (Canned Meats) Ltd

J.F. Renshaw Ltd
Ringtons Ltd
Roach Foods Ltd
Robertson & Baxter Ltd
Rubicon Products Ltd

S

S & A Foods Ltd
Saxby Bros. Ltd
Schwan's Europe Ltd
Sco-Fro Foods Ltd
Scotch Premier Meat Ltd
R & W Scott Ltd
Benjamin Shaw & Son Ltd
Shearway Foods Ltd
Sidoli Ice Cream Ltd
Silver Spring Mineral Water Co. Ltd (The)
Simpson Ready Foods Ltd
A.C. Skelton & Sons Ltd
F. Smales & Son (Fish Merchants) Ltd
Walter Smith (Birmingham) Ltd
Snack Factory Ltd (The)
Solway Foods Ltd
Speciality Foods Ltd
R & R Spink & Sons (Arbroath)
Stella Products Ltd
A.G. Stewart & Co. (Catering)
St Merryn Meat Ltd
Stonegate Farmers Ltd
Strathmore Foods Ltd
Surrey Hills Foods Ltd
Sweet 'n Savoury Frozen Foods Ltd
William Sword Ltd

T

Taveners plc
Tendafrost Frozen Foods Ltd
Thames Valley Eggs Ltd
Thayers Ice Cream Ltd
Thornley & Son Ltd
Tillery Valley Foods Ltd
Tinsley Foods Ltd
Top Hat Holdings Ltd
Town & Country Petfoods Ltd
Treats Ice Cream Ltd
Tucker Foods Ltd
Tudor Dairies Ltd

U

ULN (UK) Ltd
Unique Cuisine Ltd

W

Waldens Wiltshire Foods Ltd
Walkers Nonsuch Ltd
Walkers Shortbread Ltd
Wander Ltd
Webbs Poultry & Meat Group (Holdings) Ltd
Wells Soft Drinks Ltd
Westlake Foods Ltd
Westler Foods Ltd
Wilkin & Sons Ltd
R & K Wise Ltd
Witwood Food Products Ltd
John Wyeth & Brother Ltd

XYZ

Yeo Valley Farms (Production) Ltd
Zest Foods Ltd

A

A

Abacus Frozen Foods Ltd

T/O £7m; pre-tax profit £75 160; Emps = 112 (31/10/92)

Abacus is a family-owned business producing fish blocks as well as a range of value-added, coated and sauce-based fish products (oven crisp cod and haddock, fish kievs, cordon bleu cod, etc.) from a plant at Mintlaw, Peterhead in North-East Scotland. The company serves both retail and foodservice markets with products in the Diners Club brand.

Abel Eastern Foods Ltd

T/O £4.4m; pre-tax profit £0.49m; Emps = 115 (30/6/92)

Abel, which claims to be the largest manufacturer of nan bread in the UK, began trading in 1984. In its first year of trading its sales reached £300 000 and it has grown rapidly since then to achieve a turnover which is expected to be around £6m in 1993. It claims an 80% share of the nan and pitta bread market. It produces branded and own-label goods and has recently gained contracts from multiples and distributors in Ireland, France, Belgium and Germany in addition to those in the UK. Gas-flushing techniques are used to give the bread a shelf-life of up to 16 days. It is about to begin construction of a new 40 000 sq.ft factory.

In 1993, Abel won the Scottish round of the 'Business in Britain' Award.

Aberdeen & District Milk Marketing Board

T/O £26.1m; surplus prior to taxation £0.3m; Emps = 103 (31/3/93)

Commercial division: sales of milk £13.7m; pre-tax (loss) £0.1m (31/3/93)

This is the smallest of the five MMBs, with particular responsibility for marketing milk in Aberdeenshire, Kincardineshire and Banffshire. In 1992/93 the Board hauled 101.1 million litres of milk at an average price of 21.00 ppl. The Twin Spires creamery (part of the commercial division) produces quality dairy products for industrial and catering markets. In 1992 and 1993, the commercial division made pre-tax deficits and this necessitated a 34% reduction in the workforce at Twin Spires.

In June 1991, the Board disposed of its cattle breeding division.

Aberdeen Meat Marketing Ltd

T/O £27.2m; trading profit £664 000 (1992)

The meat processing and marketing subsidiary of the ANM Group (formerly Aberbeen and Northern Marts). ANM, while primarily involved with livestock marketing, also has interests in estate agency, electronic auction systems, catering and car sales. It is Scotland's largest farmer co-operative with more than 8000 members and a throughput of £160m. Aberdeen Meat Marketing claims to be one of the most innovative and profitable meat companies in Scotland with strong sales in continental Europe, particularly Italy. It operates at the premium end of the market offering a premium of £3/beast for top quality livestock. ANM has recently acquired the Inverurie and Edinburgh abattoirs from Hillsdown Holdings and these will be merged with and managed by Aberdeen Meat Marketing from a new head office in Inverurie under construction before the takeover from Hillsdown. The new enlarged company is called Scotch Premier Meat Ltd. Aberdeen Meat's Banchory site is to be sold. (See also Westlake Foods Ltd.)

Albrighton plc – see Lovells Confectionery Ltd

Allan & Dey Ltd

T/O £14.56m; pre-tax loss £0.2m; Emps = 165 (31/12/92)

Owned by J.W. Holdings (Aberdeen) Ltd, Allan & Dey was founded in 1980 by William

Allan who subsequently went into partnership with Peter Dey. The company now produces smoked, fresh and frozen fish and marinated salmon, with 70% of its turnover coming from exports to Europe and the USA. In the 1970s it became involved in salmon farming and now produces 1500 tonnes of salmon annually from its six salmon farms. It is best known for its high-quality smoked salmon and it is in this area where the company plans to expand with different salmon cuts and a hot smoked salmon. It has recently moved to new EC-approved premises in Sinclair Road, Aberdeen and incorporated the activities of its latest acquisition, Polarfish, into the new plant. Polar's old plant is to be refurbished and used for value-added lines in fresh wet fish.

Alpine Soft Drinks (U.K.) Ltd

T/O £7.2m; pre-tax loss £1.3m; Emps = 154 (30/4/92)

International Media Communications (IMC) acquired Alpine Soft Drinks from Farringford plc (formerly the Alpine Group plc) in January 1992. Corona and Alpine soft drinks were originally distributed by roundsmen who were employees. Following substantial losses, the delivery rounds were franchised or closed during 1990, and the group returned briefly to profit before reporting further trading losses in June 1991. Alpine now operates from factories in Birmingham and Manchester producing carbonated beverages under the Alpine brand as well as producing Pepsi and Corona under licence from Britvic. IMC plan to introduce new consumer products for Alpine and to improve the profitability of its franchisees.

Anchor Foods Ltd

T/O £181.9; pre-tax profit £1.4m; Emps = 410 (31/5/93)

Anchor Foods, based in Swindon, is a subsidiary of the New Zealand Dairy Board, held through MPH (Europe) Ltd. It is responsible for packaging, distributing and marketing a range of dairy products. The Anchor label is a brand leader in butter, pre-packed mature Cheddar cheese (38% share by volume) and UHT aersosol cream, which has a 74% market share. It also markets Anchor Spreadable Butter and a range of savoury butters, which are now being made at the Trowbridge factory. It has just introduced a new aerosol custard topping into the £143m creams, toppings and custards market.

Anglian Tendabeef Co. Ltd – see Fribo Foods Ltd

Armitage Bros. plc

T/O £21.75m; pre-tax profit £0.85m; Emps = 326 (30/5/92)

Armitage claims to be Europe's largest manufacturer of pet products and accessories, currently holding 22% of this £225m retail market. Products include pet treats and dog chews (Good Boy and Good Girl treats), veterinary products like flea collars and powders, and squeaky toys.

Attwell Farms Ltd

T/O £3.1m; pre-tax profit £0.29m; Emps = 82 (31/12/90)

Attwell Farms is an independent, privately owned company which has established itself as a major eggs and poultry company over the last 25 years. Its recently installed processing plant near Redditch can handle up to 2000 birds/hour. The birds – chicken and turkeys – are reared on the adjacent farm and sold fresh.

The second major arm of the group, Oakland Eggs, packs 4000 cases/week – from its own flocks – making it one of the largest egg producers in the Midlands.

Aulds (Foods) Ltd

T/O not disclosed; Emps = 275

Parent Co.: T/O £11.2m; pre-tax profit £0.8m; Emps = 721 (31/3/92)

Parent company Thomas Auld & Sons Ltd, founded 90 years ago, operates from a

30 000–sq.ft bakery in Greenock. Aulds Foods produces 120 different luxury frozen desserts and gateaux for catering and retail markets. It currently has a frozen food plant at Hamilton and has recently invested £3m in a new 58 000-sq.ft factory at Inchinnan in Renfrewshire where it plans to develop its meringue range. 270 jobs should be created over the next 5 years.

B

John C. Bain & Son Ltd

John C. Bain, established in 1955, operates from a modern factory in the Grampian region, supplying beef, venison and game (pheasant, partridge and duck) to catering butchers and freezer suppliers throughout the UK and Europe. The company also operates a fleet of lorries. It is a member of the National Association of Catering Butchers. Its brand name is Bain of Tarves.

A.J. & R.G. Barber Ltd

T/O £35.1m; pre-tax profit £0.75m; Emps = 233 (31/3/92)

Barber is a long-established farming company which has been based in Somerset for over 100 years. Known primarily as a farmhouse cheesemaker, in 1991 the company completed a new EC-approved abattoir and bacon factory at a site in Wiveliscombe originally intended for an expansion of cheese production. A member of the Charter Quality British Bacon scheme, it uses traditional Wiltshire methods to cure premium carcases. Up to 4000 pigs/week from local farms are handled at the plant. An extensive range of pork and bacon products are available from whole sides and primals to catering and consumer packs. The company runs a van delivery service in the West Country.

Bart Spices Ltd

T/O £3.2m; pre-tax profit £0.24m; Emps = 44 (15 months to 31/3/92)

Taken over by West Trust for £2.5m in 1992. Barts produces a wide range of herbs and spices for major supermarket own-labels and small retail jars under the Bart Spices brand.

Baxters of Speyside Ltd

T/O £29.2m; pre-tax profit £3.18m; Emps = 507 (16/5/92)

Baxters is a family-owned, independent manufacturer of high-quality canned soups, jams, marmalades, and chutneys. It claims market leadership in premium soups and pickled beetroot. It exports to over 50 countries, with soups accounting for 80% of exports and jams and marmalades the remainder. 10–12% of turnover is generated through exports. It relies on in-house expertise to develop new products and is planning to invest £8m over the next 3 years on new plant, equipment and technical support services. The company also has a visitors' centre and restaurant.

Beacon Foods Ltd

A small company based in Brecon, Powys which produces chopped and pureed garlic, onion and ginger products for use in exotic ethnic foods.

Beck Food Group Ltd

T/O £32.16; pre-tax profit £0.013m; Emps = 235 (2/5/92)

Beck Foods was formed as a result of a management buyout of Frans Buitelaar. It operates one of Britain's largest beef slaughtering and processing plants at Boston in Lincolnshire. In November 1993, it bought the Perth abattoir from Hillsdown. The Perth abattoir slaughters 3500 lambs and 700 cattle a week, and together with its ready meal manufacturing operations, employs 230 people. The new company will be called the Perth Fresh Meat Company.

Belgravia Salads Ltd

T/O £2.5m; pre-tax profit £6000; Emps = 52 (31/3/92)

Belgravia Salads of North London which supplies speciality and delicatessen lines under the Orexis label took over Yours Tastefully in 1991. Yours Tastefully, a distributor of chilled and frozen foods to caterers, included starters, entrees, patisseries, desserts and Belgravia's salads among its range.

Ken Bell (International) Ltd

T/O £25.6m; pre-tax profit £0.5m; Emps = 214 (31/3/93)

Ken Bell is a processor, wholesaler and distributor of seafoods. It produces crumbed and battered fresh and seawater prawns and a number of gourmet lines including seafood kebabs.

Bendicks (Mayfair) Ltd

T/O £13.7m; pre-tax profit £0.29m; Emps = 211 (31/12/92)

After a series of takeovers the company is now owned by the West German company Storck, who bought the specialist chocolatier from Nabisco in 1988. Its prestigious own-label customers include Harrods though most of its annual 12 000 tonnes production is under the Bendicks label. Bendicks will continue to operate its five specialized shops which provide a showcase for its experience in the production of hand-made chocolates. The acquisition will provide Storck with a vehicle to market its own brands, Toffiffee, Merci and Werther's Original though production will remain in Germany. Werther's Original 150g bag currently claims to be the biggest selling bag in the UK sugar confectionery market. Its immediate subsidiary is Newmans Chocolates Ltd. The West German parent company has worldwide sales of over £300m.

Beni Foods Ltd

T/O £40.6m; pre-tax profit £5.7m; Emps = 277 (31/12/92)

Suppliers of cooked meats and delicatessen products to customers who include Harrods.

Bennetts of Doncaster Ltd

T/O £3.5m; pre-tax loss £0.1m; Emps = 46 (31/12/91)

The company is a catering butcher which operates from MLC-approved premises. Fresh Seal Foods, a division of the company set up to develop a range of value-added meat products, won Pork Product of the Year in 1993 in a competition jointly sponsored by the MLC, Food from Britain and Meat Manufacturing and Marketing Magazine. The winning product was Pork à la Greque.

Berrydales Ltd

T/O £37 140; pre-tax loss £15 670 (31/3/92)

The company produces a novel range of vegetarian ready meals and soya based ice creams. In 1992, it joined forces with Gerber Foods to launch a range of frozen, vegetarian ready meals.

B.G. Foods Ltd

T/O £21m; pre-tax profit £1.52m; Emps = 102 (1992)

B.G. Foods has been trading for 8 years during which time it has grown very rapidly. For the 5 years up to June 1991, thc company grew at an average rate of 67% per annum. Initially the company marketed high-quality dairy and bakery products sourced from France, Germany and Holland. Subsequently it purchased Devon Desserts in Newton Abbot, where demand soon outstripped production capacity and a new 28 000-sq.ft production unit was built and commissioned in August 1990. The new unit produces hot-eating microwaveable and ovenable desserts and puddings, cream slices, and fresh cream layered pot desserts. Brands include Kool Indulgence Desserts, Old Fashioned Puddings, Creme Caramels, Big Junior set milk desserts and French Experience Petit Pain and Croissants.

Edward Billington & Son Ltd

T/O £102.9; pre-tax profit £1.21m; Emps = 483 (31/4/92)

The Billington group of companies includes a number of businesses involved in the manufacture and merchanting of meat, sugar, agricultural products and packaging materials. Carousel Holdings, in which Volvo Foods of Sweden has a 25% shareholding, has three plants: an EC-approved deboning plant at Kirby (Carousel Meat Packers), a cooked meats factory at Runcorn, and a company in Skelmersdale handling 30–40 tonnes of meat per week sold to ready meals manufacturers and fast food outlets.

The Billington group is also a major merchanter of sugar, handling 23 000 tonnes of unrefined sugar per annum. In 1992, Billington Food Services (a subsidiary of Edward Billington (Sugar) Ltd) opened a new, purpose-built food packing factory in Ipswich capable of handling not only various types of sugar but any dry foods. Ingredients such as rice, pulses, dried fruits, sugars can be sieved, blended, screened, milled and packed in form, fill seal or block bottom packs.

Blue Bird Confectionery Ltd

T/O £3.0m; pre-tax loss £0.61m; Emps = 74 (31/3/93)

This long-established manufacturer of toffees and high boilings was bought by Hillsdown in 1986. Subsequently a management team purchased the company from Hillsdown in 1989, but 2 years later it was taken into receivership. Singapore-based Jack Chia–MPH Group have since bought it and invested heavily at the Hunnington factory. New products, flavours and packaging designs have been introduced and as a result sales have increased by 75%. Hi-Sours in lemon and apple flavours, originally imported from the Far East by the parent company are now made at Hunnington. A recent promotion teamed a tin of sweets with three top-quality golf balls under the logo – Foursomes.

G.D. Bowes & Son Ltd

T/O £43.3m; pre-tax profit £0.75m; Emps = 458 (31/3/92)

Based upon a farming enterprise founded in 1936, the company commenced pig slaughtering in 1957 and now handles about 9000 pigs per week at its Watton Abattoir in Norfolk. About 30% of its throughput comes from its own farms. The company, trading as Bowes of Norfolk, is a member of the Charter Quality British Bacon scheme. Customers include high street retailers, export traders and bacon curers.

Boxford (Suffolk) Holdings Ltd

T/O £1.9m; pre-tax profit £0.1m; Emps = 170 (30/4/92)

Boxford (Suffolk) Farms (owned by the Peake family) manufactures English fruit juices, primarily apple, under the Copella brand name. In 1989 the company was bought by Taunton Cider. This gave Boxford sufficient cash to acquire Galleywood Fruit and Storage business in Chelmsford, Essex. Taunton Cider was then the subject of a management buyout when a decision to concentrate on cider was taken. In May 1992 the original owners bought the business back for an undisclosed sum. Boxford has recently acquired the Cawston Vale brand from Robert Marshall, offering Copella opportunities to expand its activities in new trade sectors. The Cawston Vale range includes five pure fruit juices and a traditional lemonade, all packed in the tall, slim 250ml cans.

Copella juices are sold through wholesalers, multiple retailers (35% of sales), cross-channel ferries, pubs and restaurants and health food shops (25% of sales). A small amount is exported to Germany, Belgium the Far East and the US.

BP Nutrition (UK) Ltd

T/O £128.1m; pre-tax loss £6.07m; Emps = 767 (31/12/92)

BP Nutrition is based at Northwich in Cheshire and is a subsidiary of British Petroleum Co. plc. It is involved in producing speciality feeds for agriculture, fish farming and pets.

The pet food operation was substantially enlarged in 1988 by the acquisition of Luda Petfoods.

In 1992, the Sara Lee Corporation of the USA bought BP Nutrition's Consumer Food Group. In November 1993, Paragon Petcare was sold to Dalgety.

Brandway Group Ltd

T/O £16.91m; pre-tax profit £2.6m; Emps = 256 (31/12/92)

Brandway produces a wide variety of home baking and cake decorating aids, breadcrumbs and dried vegetables under its Supercook label. In 1992 Lyons-Tetley sold its Lyons Home Classics range of cheesecake and fool mixes, and chocolate baking aids to the Brandway Group Ltd. The mixes are produced together with Supercook's existing range at Sherburn-in-Elmet, Leeds and the chocolate at Brandway's Beech's subsidiary at Preston.

In February 1993, Brandway acquired the gelatine business of Davis Industries UK Ltd for £334 000.

Bridge Farm Dairies Ltd

T/O £20.4m; pre-tax profit £1.76m; Emps = 239 (3/4/93)

A leading manufacturer of long-life, ambient yoghurts sold under the Dairybell brand in low-fat and thick and creamy varieties. It is owned by Lavenham Investments Ltd.

Bristows of Devon (Holdings) Ltd

T/O £3.7m; pre-tax profit £36 000; Emps = 84 (31/12/92)

The subsidiary Bristows of Devon (Manufacturing) Ltd is involved in the manufacture of various confectionery items including toffees, fudges, and Devonshire butterscotch. The parent company also has interests in wholesaling industrial packaging materials and in printing. In 1993, it bought Pollards Confections of Newton Abbot from the receivers. Bristows will continue to manufacture the Pollards and Tuckers ranges of sweets.

Brown Brothers Manufacturing Ltd

T/O £4.5m approx.

With an eye on expansion into the multiples and supermarket sector, Brown Brothers bought Kerr Little Meats from the receivers. Kerr Little went into receivership almost exactly a year after their new £1m plant at Kirkconnel was opened by Ian Laing, the Scottish Secretary. Brown Brothers cure bacon and also produce a range of Continental style cooked meats. The new Kerr Little plant offers the sort of high-quality manufacturing plant and equipment required by the major supermarket chains.

Sarah Brownridge, Whole Country Food

Sarah Brownridge was set up in 1989 by a private British food group and operates from a purpose built, 55 000-sq.ft factory in Tadcaster, North Yorkshire. It manufactures a range of high-quality cooked meat pies and pasties and ready meals for airlines. About 40% of sales are as own-labels for the major multiples with a further 30% made up of fresh meat supplies and meat pies and caterers.

The company says it is independent and will disclose no further information about itself.

Buchan Meat Producers Ltd

Buchan Meats is a farmer-owned co-operative with 2500 members and established about 18 years ago. It supplies premium quality beef and lamb, up to 95% of it purchased direct from members' farms. It can supply a full range of meat products from carcases to value-added retail packs. Packaging options include vacuum packed, fresh, frozen and modified atmosphere. 60% of sales are to the major supermarkets chains. Buchan belongs to the Guild of Scotch Quality Meat Suppliers and the Farm Assured Scotch Livestock scheme.

James Budgett Sugars Ltd

T/O £224.5m; pre-tax profit £4.61; Emps = 66 (31/3/92)

James Budgett is a subsidiary of E.D. & F. Man Ltd who are financiers and cocoa and sugar brokers and international commodity traders. It is the second-largest sugar trader in the UK. Following an agreement with the Irish Sugar Co., which acquired a one-third stake in the Chelmsford-based sugar merchant in 1989, Budgett will distribute Irish Sugar products in the UK, providing a guaranteed outlet for 20 000 tonnes of sugar per annum.

In 1990, Budgett acquired the sugar merchanting business of John Thomas Sugar of Bristol and Manchester, which will continue to trade under its existing name and management.

Buitelaar (Kilmarnock) Ltd

T/O £19.53m; pre-tax profit £0.06m; Emps = 35 (27/4/91)

The Buitelaar group of companies has interests in farming, livestock and game, pet treats, meat and meat products and smoked salmon. Until 1992, Buitelaar had three abattoirs, one in Kilmarnock and two in Boston, Lincolnshire. In July 1992 Buitelaar (Boston) Ltd was sold to the Sims Food Group for £1.25m. This abattoir specialized in the export of cow beef to France, Italy, Germany and Spain. The market for beef in France was adversely affected by the BSE scare in 1990.

Frans Buitelaar (Meat) in Boston was the subject of a recent management buyout by the Beck Food Group (see above). Buitelaar (Kilmarnock) is owned by Holbeck Holdings Ltd.

Butcher & Baker Foods Ltd

T/O £ not disclosed; pre-tax profit £0.11m; Emps = 145 (27/7/92)

After a period of sustained growth, Butcher & Baker has invested £3m in a new purpose-built plant which will double its current capacity. The company's growth has been based on the sale of its chief products – meat pies, pasties and scotch eggs – to frozen food wholesalers which supply caterers, local authorities and hospitals on a national basis. The company also supplies caterers and fast food operators around its Midland base via its own van sales service, Grangeway Meat Products. The factories are BMMA accredited and EC-approved.

C

Cambrian Soft Drinks Ltd

T/O £28.15m; pre-tax profit £0.14m; Emps = 242 (1990)

Cambrian is part of the Greenalls Group plc. It is a manufacturer and contract bottler of soft drinks. It also packs beer and cider.

Canadian Pizza plc

T/O £15.2m; pre-tax profits £2.21m; Emps = 260 (1992)

The Canadian Pizza Group manufactures high-quality pizza crusts and topped pizzas and sells pizza topping ingredients and ancillary products like cooking equipment and packaging. The business was established in 1982 by a Manchester baker, Harry Kent, who imported pizza bases from an Italian Canadian, Gino Molinaro. By 1985, a British manufacturing operation had been set up, and in 1992 Kent and his fellow managers bought out their Canadian partners for £15.5m.

In the 3 years to 1992, sales rose at an average of 39%/year and profits at 61%/year. With 45% of sales being to export markets (EC, Australia and the Middle East) the company won a Queen's Award for Exports in May 1993. In November 1993, the company was floated on the stock market, with a view to raising cash to pay off debts left from the buyout and fund an expansion programme. Sales are predicted to top £16m for 1993, with profits at over £3m.

The company operates from a £2.5m purpose-built factory in Salford making 40 million pizza crusts a year for retailers, caterers and industrial customers.

Carters Drinks Group Ltd

T/O £58.36m; pre-tax profit £5.1m; Emps = 583 (31/1/92)

The group, based in Derby, includes Carters Gold Medal Soft Drinks, Carters Packaging and Carters Vending. It claims to be the fourth-largest soft drinks manufacturer in the UK making mainly own-label lines. In 1992, Carters was taken over by the European Hero Group, which has a turnover of about £400m. Financial backing from Hero has allowed extension of the Kegworth factory. Hero's range of high-quality, sparkling fruit drinks and juices in six flavours is currently distributed by Carter's Gold Medal Soft Drinks.

Case & Sons Ltd

T/O £12.6; pre-tax profit £0.23m; Emps = 112 (31/3/93)

Case and Sons has been involved with pig meat in one way or another since its formation in 1877 and is a founder member of the Charter Quality Bacon Scheme. Since 1967 the company has been located near Shaftesbury in Dorset which houses its slaughtering, butchery and processing facilities. It has a throughput of about 140 000 pigs per year, using traditional curing methods for whole sides of bacon. Case supplies primals and pre-packs to a range of customers including the major multiples.

In March 1992, Case set up a technical services division based on the business of J.J. Mansour, which it took over. It now offers spare parts for Swissvac brine injection machines, on-site laboratory services offering microbiological and chemical brine analysis and a product development consultancy.

Castle Dairies Ltd

T/O £6.67m; pre-tax profit £0.3m; Emps = 40 (31/10/92)

This is a family-run dairy company manufacturing cheese, milk powders, bulk and packet butter.

Cauldron Foods Ltd

T/O £1.7m; pre-tax profits £0.1m; Emps = 37 (31/12/91)

Now part of the Hero Group, Cauldron has recently expanded into a new manufacturing unit at Portishead near Bristol which includes a £1.5m investment in a line capable of producing 1000 tonnes of tofu/year. It is a leading producer of plain, smoked and marinated tofu and chilled vegetarian products including chickenless and beefless burgers, meatless sausages and luxury vegetable pâtés.

C & D Continental Patisserie Ltd

T/O £1.8m; pre-tax profit £0.094m; Emps = 67 (27/11/92)

A leading independent supplier of premium quality frozen gateaux and pastries primarily to the catering trade. Its recent expansion into the retail sector contributed to an increase in turnover of 37% during the first 6 months of 1992. It supplies cream eclairs, jam doughnuts, strawberry mousse tarts, etc. to retailers for sale from in-store patisseries.

Central Midlands Co-operative Society Ltd

T/O £335m; trading profit £8.6m; Emps = >5000 (30/1/93)

The Central Midlands Co-operative is the second-largest independent Co-operative Society in the UK, with trading centred around Birmingham and Derby. The business includes 150 food outlets including six superstores. In addition there are 46 produce units in the Midlands trading under the Fruity Fruits and Mills names. It operates the largest dairy business in the Midlands turning over >£100m.

Cheale Meat Ltd

T/O £45.8m; pre-tax profits £1.4m; Emps = 64 (31/8/92)

Cheale Meats has an abattoir at Brentwood in Essex approved for the export of meat. The company was involved in exporting pork and beef to Europe even before Britain joined the EC and today exports account for about 90% of the company's turnover. The abattoir has a weekly throughput of 2500–3500 sows and boars, plus 500–600 cows. The meat, in carcase form, is transported in the company's fleet of 12 refrigerated container vehicles.

Cheshire Wholefoods (Cereals) Ltd

T/O £12.23m; pre-tax profit £1.5m; Emps = 106 (31/12/91)

Cheshire Wholefoods claims to be the UK's largest producer of natural mueslis in the premium muesli market. Since its inception in the mid-1970s the company has grown rapidly and joined the Unlisted Securities Market in November 1985, only to be acquired by the Dutch food group, Koninklijke Wessanen just 21 months later. At the time of the takeover the company produced about 550 tonnes of muesli per month, two-thirds of which was for own-label customers. The products are currently sourced through P.A. Ross (Food Brokers) Ltd.

The Chitty Group plc

T/O £30.2m; pre-tax profit £0.025m; Emps = 111: Chitty Abattoir, T/O £2.8m; pre-tax loss £0.095m; Emps = 48: Chitty Wholesale, T/O £29.0m; pre-tax profit £0.12m; Emps = 54 (30/10/92)

The Chitty Group is a leading slaughterer and wholesaler of beef, pork and lamb, operating an EC-approved abattoir and cutting plant in Guildford. It also has a fleet of refrigerated lorries and its own butcher's shop in Dorking. It has now opened its newly acquired and refurbished packing plant at Horndean in Hertfordshire where CAP packs of mince, boned and rolled joints and steaks are prepared for supermarkets.

T.W. Clothier & Sons Ltd

A Somerset-based family-owned farming business which broke away from the MMB in 1980. The family owns 2000 acres of land in Somerset and Dorset supporting 1300 dairy cows, and 3600 sows producing about 100 000 bacon pigs/year. Young pigs are fed whey produced by Clothier's associate company, Wyke Farmhouse Cheese Co. Ltd (T/O £17.58m, pre-tax profit £30 000, with 42 employees). An investment of £1.5m at Wyke's White House Farm site means that the company can now process 8000 gallons of milk/hour to make 3.5 tonnes of cheddar. The modernization and expansion includes cheese-forming towers and a cheese portioning line at the Wincanton cheese conditioning store. Wyke is one of the UK's largest independent farmhouse cheesemakers. It markets direct to retailers from the farm gate.

Conrad & Essenhigh Ltd

No accounts yet published.

Conrad & Essenhigh, based in Somerset, is an independent manufacturer of Bavarois mousses and gateaux. In October 1992 it acquired the traditional pudding business of Docker Foods so adding summer pudding and treacle sponge pudding to its product range.

Cooke Bros. (Tattenhall) Ltd

T/O £14.5m; pre-tax profit £0.43m; Emps = 225 (30/9/92)

The company's better known subsidiary is Calypso Soft Drinks Ltd. It is an independent manufacturer of soft drinks, now in its 104th year of operation. In 1989 it bought Tip Top Soft Drinks from Hazlewood Foods and in December 1992 it bought Scotts Soft Drinks. Both these acquisitions have provided Calypso with more capacity to meet growing demand for its product range. Over the last 30 years the company has focused on children's soft drinks, now producing 300 million single-serve cartons a year using the Flintstones logo. There are plans to introduce a similar Tom and Jerry range. It also produces freeze-pops and claims market leadership in cup drinks.

Coombe Farm Foods Holdings Ltd

T/O £116.5m; pre-tax profit £1.51m; Emps = 395 (31/3/92)

Coombe Farm is a holding company for a group of companies engaged primarily in marketing dairy products. Subsidiaries include Mendip Foods Ltd (55%, cheese marketing), Coombe Castle Fine Foods Ltd. (fine wine trading) and Amermart Ltd (USA, cheese importing). Principal associates are Malvern Farm Foods Ltd (50%, cheese trading) and Oliver Investments Ltd (33%, property).

Mendip Foods markets mature cheddar under the Cathedral City brand and a range of other farmhouse cheeses (Cheddar, Cheshire, Lancashire, Red Leicester, and Double Gloucester) for the Farmhouse Cheesemakers Co-operative under the Centurion brand.

Of the total turnover of £116.5m for 1992, £9.5m was generated through export markets.

Country Style Foods Ltd – see Meadow Vale Food Group

Cow & Gate Nutricia Ltd

T/O £90.7m; pre-tax profits £9.05m; Emps = 405 (2/1/93)

Cow & Gate changed its name to Cow & Gate Nutricia in February 1993. It is the UK's brand leader in baby foods and has a comprehensive product range covering infant milks, pre-prepared meals (including the successful Olvarit range), yoghurts, fruit juices and herbal drinks. During 1992, it re-launched many of its well-known products under a new pan-European brand, Nutrilon. The parent company is Nutricia of Holland in which Unigate holds a minority stake.

Cox Seafoods Ltd

T/O £ not disclosed; pre-tax profits £0.14m; Emps = 115 (31/5/92)

Cox Seafoods, located in Buckie, is strategically situated close to the major Scottish fishing ports of Petershead, Fraserburgh, Aberdeen and Whitehills. The company is a major producer of own-label fish (plain, breaded and smoked), scampi, shellfish and value-added lines for supermarkets, freezer centres and catering distributors. It plans to begin exporting its frozen fish and shellfish lines.

Its parent company is Mancunian Mercantile Investments Ltd.

Cranswick plc

(formerly Cranswick Mill Group plc; name changed 10/12/92)

T/O £109.6; pre-tax profit £2.2m; Emps = 243 (31/3/93)

Cranswick is now a vertically integrated meat business with interests ranging from livestock production and feed manufacture through to value-added meat products in consumer packs. Group companies include: Cranswick Mill Ltd (grain, feeds, livestock sales), Cranswick Country Foods Ltd (supplier of pork products), Charter Pork Cuts Ltd (pig abattoir), F.T. Sutton & Sons (Rossendale) Ltd (manufacturer of cooked meats and pork products) and Cranswick Tuck Marketing Ltd (51%). In April 1993, a 60% interest in George Buckton Ltd (producer of pigeon corn and bird food) was acquired for a maximum of £1m.

Integration of the business of Yorkshire Country Pork Ltd (acquired in 1992 for £3m) with other butcheries in Yorkshire and Humberside led to the formation of Cranswick Country Foods Ltd. Proceeds of a rights issue will be used to fund an expansion programme for this part of the group. Refurbishments at F.T. Sutton's Watersfoot plant, underway at the time of its acquisition are now complete, and EC licences have been obtained for both fresh and processed meat operations. Suttons is currently negotiating contracts for the supply of its speciality meat products to the main high street retailers.

D.S. Crawford Ltd

T/O £15.8m; pre-tax loss £0.24m; Emps = 1179

Crawford, which operates two bakeries – one at Leith in Edinburgh and another at Peterhead near Aberdeen – and a chain of 111 bakery and restaurant outlets in Scotland, was acquired by a management team in July 1990 from United Biscuits.

Crooks of Harleston Ltd

The company was incorporated in June 1993, and no company data is yet available. However, it recently bought Norfolk Smokehouses from the receivers.

Crown Chicken Ltd

No company data yet available.

Crown Chickens was formed as a result of a management buyout of Harvest Poultry, part of Hillsdown Holdings. Financial backing of £6m came from Barclays Bank and 3i venture capital. The company serves wholesale and retail customers with high-quality fresh whole birds and portions tray wrapped or in bulk. The deal included the EC-approved processing plant at Weybread, a hatchery at Kenninghall, Norfolk, a feed mill, and 22 broiler breeder sites in Norfolk and Suffolk. Also acquired with the deal is the Rospar van sales network based in Brighton, Sussex.

Bird throughput is likely to be about 220 000 by the end of 1994.

Crystal Drinks Ltd

T/O £14.3m; pre-tax profit £0.29m; Emps = 123 (29/2/92)

Crystal Drinks, acquired from ABF in a management buyout in 1988, concentrates on own-label carbonated drinks produced for the major retailers. It also produces Quartz, aimed at the fast-growing sports drink sector. It manufactures soft drinks for a joint venture with Gerber Foods in which Crystal has the manufacturing capacity and Gerber the sales and marketing contacts.

D

Dairyfresh Desserts Ltd

T/O £4m approx.

Dairyfresh Desserts is an independent manufacturer of frozen cakes and dairy cream desserts, supplying most major supermarkets and many national caterers. The company was formed in 1990, following a management buyout of the Emile Tissot Patisserie business of Hunter Saphir. It is currently extending its premises in Torquay by a third, with the work due for completion in mid-1994. Dairyfresh has trebled its turnover since the buyout.

Daloon Foods (UK) Ltd

T/O £6.6m; pre-tax profit £0.26m; Emps = 84 (31/12/92)

Daloon Foods claims to be Europe's largest producer of frozen crispy pancake rolls. The UK was the Danish company's largest export destination, so a production plant was set up at Newark near Nottingham in 1984 which now handles 95% of total UK requirements and is EC-approved. As well as pancake rolls Daloon also produces samosas and sweet crêpes for both retail and catering markets.

Jeffrey Davies & Davies Ltd

T/O £14.3m; pre-tax profit £0.17m; Emps = 100 (1/4/93)

Jeffrey Davies, now in its 13th year of operation, is primarily a processor of pork. It is also an important wholesaler of beef, lamb and offals. Through J.F. Edwards & Son, Davies also acts as the sole UK agent for ESA, one of Holland's premier veal producers.

Dawn Foods UK Ltd

T/O £3.7m; pre-tax loss £0.35m; Emps = 18

Dawn Foods is a premix manufacturer and wholesale distributor of bakery products. It recently acquired the dry mixes division of Evedale Foods from Albert Fisher. It has set up a new head office in Evesham. Its parent company is Dawn Food Products Inc., USA.

Dawnfresh Seafoods Ltd

T/O £19m; pre-tax profit £0.031m; Emps = 275 (27/3/92)

Dawnfresh recently moved to its new £8m seafood factory at Bothwell Park in Lanarkshire. The company claims this is the most modern seafood processing centre in Europe. Officially opened by the Queen in October 1993, the new plant employs about 170 people and has two breading lines, segregated high/low risk preparation areas, temperature controlled dispatch chamber, test kitchen and QA Laboratory. Factories at Whitehaven and Ipswich (formerly that of Starfish Ltd) have been closed with the loss of about 150 jobs. Dawnfresh is a major supplier of shellfish and other seafoods to the foodservice sector. It is planning to target export markets for value-added seafoods in France and Germany.

Dell Foods (Products) Ltd

T/O £–; pre-tax loss £0.245m; Emps = 70 (31/12/92)

Dell Foods (Products) is now a wholly owned subsidiary of Lactosan-Sanovo Holdings. Lactosan is the world's largest manufacturer of egg processing equipment and one of Europe's largest egg processors. Dell, based at Chesterfield in Derbyshire, manufactures pasteurized whole and separated egg, hard boiled eggs, and egg sandwich mixes.

Del Monte Foods (UK) Ltd

T/O £73.66m; pre-tax profit £4.19m; Emps = 182 (29/11/91)

Prior to 1988, Del Monte was owned by RJR Nabisco. In 1989, Nabisco was taken over by Kohlberg Kravis Roberts. Del Monte's US and European operations were then acquired by Merrill Lynch. To reduce debt the European operations were sold off: the fresh fruit interests went to Polly Peck International (PPI) for £557m and the canned fruits, juices and beverage business of Del Monte Foods Europe went to a management and investor buy-out team for $375m. When PPI went into receivership the Mexican investor Grupo Cabal bought Del Monte Fresh Produce for about £260m (August 1992). Del Monte Foods Europe was subsequently renamed Del Monte Foods International (DMFI) and split into Northern and Southern European divisions with Head Offices at Staines (UK) and Milan (Italy). In 1991, DMFI made a pre-tax profit of £24.73 on sales of £250m. It claims leadership in the canned fruit market with 18.4% (Libby's 4% and Dole 3.8%). In October 1992, Royal Foods and the Anglo-American Corporation of South Africa took over DMFI for £360m. Del Monte Foods (UK) has a turnover of about £73m equally split between beverages and canned fruit. It has one production unit at King's Lynn.

De Roma Ice Cream Ltd

T/O £14.25m; pre-tax profit £0.26m; Emps = 172 (28/2/93)

De Roma Ice Cream manufactures chocolate bars and filled cone products. It operates from a purpose-built plant in Wigan in Lancashire which is capable of producing a wide variety of cones in standard and larger sizes. New to its range are hazelnut, fruit and nut, and rum 'n raisin flavours. Dessert sundaes in four-packs are also available.

Derwent Valley Foods Ltd – see United Biscuits

Devon Quality Foods Ltd

T/O £2.3m; pre-tax profit £0.055m; Emps = 48 (31/7/92)

DQ Foods, based in Dawlish, make high-quality, fresh ham, cheese and egg quiches for customers, including Brake Bros, Pullman Foods, Puritan Maid and other food service companies. They have recently completed a £250 000 expansion programme which doubles manufacturing capacity and provides a new warehouse and cold store.

D.M.J. plc

T/O £13.0m; Emps = 127 (1992)

DMJ is primarily a cutter and packer of frozen meat, particularly New Zealand lamb. Formed 8 years ago by Stephen Joll and his wife, the company operates from two sites at South Kirby near Pontefract, where 8000 New Zealand lambs/week are handled. It also has a frozen pork output of about 80 000 tonnes/year and is the major producer of frozen diced and minced meat in the UK (150 tonnes/week). Customers include Asda, Iceland, CRS, Tesco, Netto and Aldi.

Recognizing the trend towards value-added lines, it has introduced two new products under its Grange brand based on New Zealand lamb flavoured with herbs and spices. It has a reciprocal arrangement with Vlees & Sate in Holland which sells DMJ products, while DMJ markets V & S kebabs in the UK.

Dove Valley Poultry Ltd

T/O £17.5m; pre-tax profit £0.6m; Emps = 210 (31/8/92)

Dove Valley, formed about 30 years ago, has recently opened a £1.5m extension to its factory complex at Ashbourne in Derbyshire. The new premises provide a further 9000 sq.ft of temperature-controlled working space and chilled storage capacity. Dove Valley offers an integrated operation from breeding through to retail packs of fresh chilled whole birds and portions. On average 250 000 birds/week pass through the new complex.

F. Duerr & Sons Ltd

T/O £18.65; pre-tax profit £0.75m; Emps = 143 (31/3/93)

In 1993, F. Duerr & Sons celebrated 112 years of successful business in preserves. About half of Duerr's business is in own-label products with the result that it produces 120 different recipes under 700 labels. In the branded market, Duerr claims market leadership in fine and medium cut marmalade and lemon cheese.

Duncan's of Scotland Ltd

(formerly Duncan's of Edinburgh)

T/O £1.3m; Emps = 42 (April 1991, prior to receivership)

In 1927 Rowntree merged with the family firm of Duncan's and by the 1950s the company employed about 2000 people. In 1967 traditional Duncan's lines were withdrawn and all efforts focused on Rowntree products. In 1987, a management buyout from Rowntree's brought some of Duncan's traditional products back to the shops, but in 1991 the company ran into difficulties and went into receivership.

Bought from the receivers by the consortium Tantallon International, the company changed its name to Duncan's of Scotland and moved to new 11 000-sq.ft premises in the Motherwell Food Park. Initially the company has concentrated on developing Duncan's well known lines – hazelnut, fruit and nut and milk chocolate bars – but a product development programme is underway to introduce new lines not only in Scotland but in England and Wales, and a new tartan chunky bar has been designed for the export market.

Dunhills (Pontefract) plc

T/O £28.83m; pre-tax profit £1.73m; Emps = 469 (31/12/92)

Dunhills, established in 1760, became part of the international Haribo organization in 1973. Following a factory extension in 1991, output increased from 300 to 450 tonnes of confectionery per week. A large part of this is own-label business with the major multiples, though the company's products are also marketed under the Haribo and Dunhill brands.

E

Elisabeth the Chef Ltd

T/O £8.78m; pre-tax profit £0.512m; Emps = 432 (25/7/92)

Elisabeth's supplies flour confectionery to the major supermarket chains. The company, based in Leamington Spa has four production sites specializing in creamed products, Danish pastries and celebration cakes.

Elsenham Quality Foods Ltd

T/O £4.3m; pre-tax loss £0.7m; Emps = 89 (1992)

Elsenham, one of the UK's leading manufacturers of jams, marmalades, condiments and chutneys was taken over in 1988 by J.M. Smucker of Ohio which bought the company from Whitbread, which had acquired Elsenham's parent company, James Burroughs. Smucker is a major player in the US jam market.

For many years Elsenham's business was largely concentrated through gift-oriented outlets, but it has recently begun selling to the major national retailers. In 1991 it won the Queen's Award for Export Achievement, and two gold medals at the SIAL exhibition in Paris for product innovation and quality.

Emley Farm Chilled Food Ltd

Catering largely for Local Authorities and Hospitals at present, Emley Farm was set up specifically to supply sandwiches, made under the stringent requirements of the Food Safety Act. Operating from a 1200-sq.ft production unit in Dewsbury, West Yorkshire, Emley produced up to 5000 filled sandwiches/hour on three shifts 24 hours/day.

Unfortunately, the company ran into financial difficulties, losing 40% of its sales when local hospitals decided to make their own sandwiches. A meeting of creditors was held in August 1993.

Eryri Food Products

T/O £3.5m; Emps = 35 (1991)

In 1991, Eryri Food Products, a traditional meat wholesaler, switched operations to the value-added sector. £700 000 was spent transforming the firm's plant at Gaerwen, Anglesey into a modern meat product plant capable of producing a range of beef, pork, lamb and ham products to customer specifications. Products include burgers, pizza toppings, ready meals and gammon steaks.

Everest Foods plc

T/O £31.9m; pre-tax profit £2.09m (31/5/93); Emps = 397 (31/5/92)

Principal activities of Everest Foods are the manufacture of frozen potato products (Everest Frozen Foods Ltd), production of eggs and egg products (Arnold Davies Ltd and Everest Egg Products Ltd) and frozen food distribution (Ashfords of Stourbridge Ltd). Everest is the UK's second largest manufacturer of frozen potato chips, and has recently spent £4m refurbishing its Wombourne plant, creating 60 new jobs. Chip sales in 1993 stood at £16.3m (£18.4 in 1992).

Ashfords acquired Caterfayre Frozen Foods of Kidderminster for £80 000 in 1992, and A La Carte of Ledbury in 1993. Both acquisitions are in line with developing the catering division of the company.

Arnold Davies, which has now severed its links with Thames Valley Eggs made a pre-tax profit of £309 000 in 1993 on a turnover of £6.7m. Margins remained tight but flock size was increased in response to increased demand for free-range eggs.

F

Farley Health Products Ltd

T/O £34.35m; pre-tax profit £10.18m; Emps = 308

Farley Health Products was acquired by The Boots Company plc, the present owner, from Glaxo in 1986. Since the acquisition, the 50-year-old Plymouth factory has been closed with the loss of nearly 300 jobs. Production of infant milk and cereal manufacture will now be centralized at the company's remaining factory in Kendal. Rusk production has been transferred to a third party: Northern Food's grocery group in Nottingham.

The product range includes Farley's rusks, Ostermilks, Breakfast, Lunch and Tea Timers and Farex baby milk cereal which has been re-named First Timers. Farley's has 90% of the infant rusk market and a 44% share of the dry baby food market sold through grocery outlets. Its baby milks are now marketed in a co-ordinated range of first milk, second milk and follow-on milk.

Farmers Boy Ltd

T/O £76.61m; pre-tax profit £2.59m; Emps = 477 (30/1/93)

Farmers Boy is owned by Wm Morrison Supermarkets plc and is the food processing operation which supplies Morrison's 46 retail superstores. It has a weekly throughput of 200 tonnes of fresh beef, pork and bacon, 60 tonnes of prepared and cooked meats, sausages and burgers, 60 tonnes of pre-packed and cut cheese, and 17 000 pizzas, quiches and pies.

Faroe Seafood Company Ltd

T/O £31.8m; pre-tax loss £0.06m; Emps = 327 (31/12/92)

Faroe manufactures a range of retail value-added fish products under the Natural Choice brand as well as products for caterers. In 1991 it extended its factory in Grimsby at a cost of £1.25m. Late in 1993 it installed a £300 000 spiral freezer which will allow greatly increased production of Faroe's fish-in-sauce products. Faroe is owned by Foroya Fiskasola of the Faroe Islands.

Fletchers Bakeries Ltd

T/O £31.43m; pre-tax profit £3.19m; Emps = 531 (30/6/92)

The company, which is a manufacturer and retailer, specializes in frozen bakery products including US-style frozen jam donuts, cake-muffins (sultana, cheese, cherry date) and scones. Northern Foods owns a 25% share in the company.

Food for Thought (U.K.) Ltd

T/O £4.4m; pre-tax profit £0.19m; Emps = 90

The company is a supplier of value-added frozen food 'components' which require short runs or complicated procedures to manufacture them. Customers include manufacturers and caterers. Food for Thought has two manufacturing units, one at Horsted Keynes in West Sussex, the other in Sunderland. The plant in Sunderland has been extensively refurbished and now has EC approval. Products include pieces of garlic butter, cooked chicken breasts, and frozen cooked omelettes in various flavours.

Forth Valley Foods Ltd – see also Highlander Snacks

Forth Valley Foods, set up in 1987 following the closure of the Golden Wonder factory at Bathgate, subsequently went into receivership in 1991. It was bought from the receivers by Unichips Finanziara of Milan, who relaunched the company using the Highlander brand name.

Fredericks Dairies Ltd

T/O £15.47m; pre-tax loss £0.85m; Emps = 186 (28/2/93)

The growth of Frederick's Dairies is closely related to the growth in sales of the premium ice cream bar and of own-label choc ices. Fredericks currently makes 10 million ice cream bars a week, which is four out of every five bars made in the UK. It developed Mars Bar ice cream, followed by Bounty ice cream. Own-label choc ices account for 88% of UK sales. Premium ice cream bars account for only 25% of sales by volume, but 48% by price.

The company is currently developing a premium ice cream on a stick and has installed new equipment supplied by Alfa Laval Hoyer in Denmark at its Skelmersdale factory. Equipment includes at tray-tunnel extrusion line, eight-line wrapping machines and automatic cartonizer.

Frederick's has recently won export orders from the United Arab Emirates and the former Soviet Republic of Gerogia to supply consignments of its Majestic brand of caramel and peanut and crispy bars.

Fresh Fruit Juice Manufacturing Co. Ltd – see Orchard House Foods

Fresh-Pak Eggs Ltd

T/O £–; pre-tax profit £32 000; Emps = 51 (1992) (30/4/92)

Hytech Foods is part of Fresh-Pak Eggs Ltd. It manufactures 30 different sandwich fillings including egg mayonnaise, Coronation chicken, apple and salad, prawn mayonnaise, chicken tikka, etc. Due to increased demand for its products in 1991 the company moved to new, purpose-built, 18 000-sq.ft premises in Loughborough. All the fillings have an 8-day shelf life and the company is an approved supplier of the British Sandwich Association. Customers range from major supermarkets to local bakeries.

Fribo Foods Ltd

T/O £11.48m; pre-tax profit £0.04m; Emps = 160 (1992)

Formerly Anglian Tendabeef, the company changed its name to Fribo Foods Ltd in August 1992 due to the strength of the Fribo brand name and the fact that the company had not been based in East Anglia since 1972. It is an independent, privately owned meat company established in 1967 which operates an EC-approved factory in Wrexham. The bulk of the company's business is in the catering sector, supplying frozen meat and further processed products under the Fribo brand (beefburgers, Range Ribs and Chopsteaks). In the last 3 years sales to the retail market have grown rapidly and now represent 9% of total turnover.

The company has recently introduced several new non-meat products including the Vegetable Super Burger.

Furniss Group Ltd

T/O £2.98m; pre-tax profit £27 000; Emps = 63 (30/9/92)

Furniss, based at Truro in Cornwall, was bought by a management team from the Argyll Group in 1987 following Argyll's decision to concentrate on retailing. The company, which employs about 50 people, continues to make its range of branded biscuits but is now owned by Artal NV in the Netherlands.

G

J.W. Galloway Ltd

T/O £35.7m; pre-tax profit £0.527m; Emps = 247 (28/2/93)
Galloway's is the parent company of Scotbeef Ltd, meat product and sausage manufacturers based in East Kilbride and of Scotbeef (Meat Packers) Ltd which operates an abattoir, cutting plant and cold store at Longleys, Bridge of Allan.

A major extension to the East Kilbride plant is planned, but to accommodate a short-term need for more production capacity, the company is now operating from the old H. Hargrave bacon plant in Polmadie, which has recently been refurbished.

Gianova Foods Ltd

Gianova Foods has a modern factory in Canterbury, Kent which manufactures a full range of dried pasta in popular shapes as well as lasagne, tagliatelle, fettuccine and spaghetti for own-label customers.

Giles Foods Ltd

Farmer Giles Foods based in Aylesbury was subject to a management buy-in valued at £1.5m. It also changed its name to Giles Foods Ltd. It operates from a £4m purpose-built factory at Broadfields in Aylesbury, producing a range of chilled foods – salads, quiches, flans and delicatessen products for the retail and catering trade. Vegetable salads and fresh fruit salads are produced by Giles Fruit Ltd, set up in 1985, while South City Foods, also part of the group, produces a range of pizzas from its base in South London.

The new Chairman and Managing Director, who were both with Kara Foods before the buy-in, are planning to develop the company's existing range and add topped or filled bread products and a range of luxury cream cheesecakes.

F.A. Gill Ltd

T/O £17.36m; pre-tax profit £0.17m; Emps = 197 (30/9/92)
This independent family business produces and distributes meat and meat products from its base in Wolverhampton.

Gillard Holdings plc

In April 1993, Peters Savoury Products (a subsidiary of GrandMet) bought the Gillards pie and sausage business at Walthamstow, the van sales operation and rights to the Gillards brand name. McLaren Foods, which was also part of the Gillards Group was bought out by its management at the same time. The other group company, Dawson & Daughters went into liquidation in March 1993, following the failure of a management buy-out. A new company, Barnaby Foods plc, is now operating from the Dawson & Daughters' premises in Hythe Road in North West London.

Glendale Frozen Foods Ltd

T/O £–; pre-tax profit £151 790; Emps = 95 (11/10/92)
Glendale Frozen Foods makes a range of meat products – grills, burgers, steaklets and ready meals – distributed through a national network of frozen food wholesalers. Recent additions to the product range include shish and doner kebabs, and Hot 'n Crispy deep fried sausages and fish and chips.

Global Group plc

T/O £75.0m; pre-tax profit £1.4m; Emps = 558

Food Division: T/O £55.6m; operating profit £1.36m (31/12/92)
Global is a diversified group with interests ranging from meat trading to shipping agency, stevedoring and fork lift truck hire. Food-related subsidiaries include: Global Meats Mid-

lands Ltd and International Global Ltd (both meat traders), Peakburgers Ltd (beefburger manufacture) Canterbury (suet product manufacture), Spacehire (Goole) Ltd which trades as Fresh Country Foods (vegetable freezing, packing and storage), Sheed Thomson International Ltd (meat importing).

A new operation has recently been established – Global Delicatessen – to source delicatessen products throughout Europe. Peakburgers had a difficult year in 1992 due to increases in raw material prices and intense price competition, while Canterbury Foods operated at full capacity throughout the year and is actively seeking large premises.

Gold Crown Foods Ltd

T/O £23.08m; pre-tax profit £2.6m; Emps = 223 (27/6/92)

In May 1993 Gold Crown was the subject of a second management buy-out funded by 3i and Henderson Venture Managers and valued at £16m. (The first buy-out was from the Argyll Group for £14m.) Based in Liverpool, the company supplies own-label tea and coffee to the major multiples as well as Mantunna brand tea. It also has a growing export trade with 40 countries. Projected turnover for 1993 is £30m.

Golden West Foods Ltd

T/O £202.38; pre-tax profit £3.10m; Emps = 485 (2/1/93)

Golden West is involved in the manufacture of bakery, syrup and ketchup supplies and the sale and distribution of supplies to McDonald's Restaurants.

Goodlife Foods Ltd

T/O £3.2m; pre-tax profit £0.1m; Emps = 68 (3/1/93)

Goodlife Foods Ltd was formed in January 1989 by the merger of Cheshire Chilled Foods and its main customer Goodlife Wholefoods. Goodlife has a factory at Warrington where it produces a range of chilled and frozen vegetarian and ethnic convenience foods including vegetable cutlets, falafel, nut cutlets, spicy bean bangers and dry mixes for nut and vegetable grills. The company uses organically grown ingredients and its products are approved by both the Soil Association and the Vegetarian Society.

Granelli McDermott Ltd

T/O £–; pre-tax profit £0.34m; Emps = 61 (1992)

Granelli ice cream was set up by Italian, Louis Granelli who arrived in Manchester in 1889 and opened a shop in Ancoats. The company, which has maintained a reputation for high-quality ice creams has been largely represented in the North of England. A management buy-in in March 1992 by former executives of Lyons-Tetley has revitalized the company's image and introduced plans to launch the Granelli brand on a national basis. The new logo features Louis Granelli and aims to fill a gap in the market between rich, luxury American-style ice creams and mass-produced products. Flavours will include cappuccino, pistachio, sorbetto al limone, cioccolatta and cassata siciliana.

H

Halal Meat Co. Ltd (Birmingham and Wales)

Birmingham: T/O £9.35m (1991, 1992 T/O not disclosed); pre-tax loss £0.24m; Emps = 93 (31/3/92)

Wales: T/O £0.5m; pre-tax profit £0.15m; Emps = 30 (5/5/93)

The company is one of the few abattoir groups in Britain specializing in halal slaughter. The Yaqoob family runs the Birmingham-based company which has abattoirs at Birmingham and Henllan in Wales. There are an estimated 2 million meat-eating Muslims in the UK.

Haldane Foods Ltd

T/O £6.3m; pre-tax loss £0.2m; Emps = 61 (31/12/93)

Haldane Foods Ltd belongs to the Haldane Foods Group which claims to be Europe's largest health food manufacturing group. Its parent company is the Archer Daniels Midlands Co. (USA) and its immediate subsidiaries are Vegetarian Feasts Ltd and Genice Foods Ltd. Other group companies include Granose Foods, Realeat, Direct Foods, Regular Tofu, Pasta Regale and Snackmasters. Its 200 vegetarian and health food products cover ambient, chilled and frozen sectors and include vegeburgers, soya milk, complete vegetarian meals, non-dairy frozen desserts, vegetarian mixes, organic baby foods, canned vegetarian protein meals, Quorn meals and Realeat vege-mince. Acquisition of Unisoy Milk 'n By-Products at Stockport in Cheshire gave the group a soy milk manufacturing unit.

In 1992, it also bought Buss Foods, a vegetarian sausage manufacturer, and the brands and goodwill of the Dietburger Co. A £1m investment in the Snackmasters plant in Abercarn, Wales has made it one of the most up to date snack-pot factories in Europe. The Saucemasters 25 000-sq.ft factory in Newport, South Wales was sold to G. Costa in 1993 and now trades as Keddie Saucemasters.

Hall & Woodhouse Ltd

T/O £66.99m; pre-tax profit £3.89m; Emps = 1440* (30/1/93)

*includes brewery staff. Hall & Woodhouse (formerly Panda Soft Drinks) is one of the largest independent manufacturers of soft drinks in the UK. (For the year ending January 1992, Panda Drinks made a pre-tax profit of £1.2m on a turnover of £33m and employed 118 people). Production is centred upon a 15-acre site in Dorset where a variety of soft drinks in cans and PET containers are produced. The Panda brand of canned carbonated soft drinks is currently worth about £24m and comes in a range of flavours plus the diet versions. Panda lemonade claims about 21% of the canned lemonade market. Independent consumer research judged Panda Cola to have a 'top brand' flavour as well as offering value for money, and better margins for retailers. Rio Riva and Rio Light, targeted at a more adult market, contain sparkling spring water blended with fruit juices in tropical and Caribbean flavours. Hall & Woodhouse distributes to the cash and carry and wholesale confectionery trade and major UK retailers.

Halo Foods Ltd

(formerly Holgates Nutritional Products)

Halo Foods has a workforce of 75 and is based at Tywyn in mid-Wales. It specializes in the development and manufacture of own-label confectionery. In April 1992, the owners Nestlé sold the company to its Managing Director, Peter Saunders. In April 1993, Holgates changed its corporate identity and name to Halo Foods. It has recently developed a new low-calorie chocolate bar containing caramel, honey, nougat and milk chocolate.

H. Hargrave & Co. (Wholesale Pork) Ltd

T/O £79m; pre-tax loss £1.14m; Emps = 564 (25/9/92)

H. Hargrave & Co was established in Spalding, Lincolnshire over 100 years ago. The company has factories at Pinchbeck in Spalding, Rutherglen in Glasgow and the former Amberley Foods plant in Middlesbrough. The Middlesbrough plant has been extensively refurbished and now forms the centre of the bacon curing operation. Its Scottish bacon plant in Polmadie has recently been refurbished and is currently being used short-term by Scotbeef who needed more production space. It also has a fleet of refrigerated delivery vehicles (Hargrave International). It is a member of the Charter Bacon scheme. It supplies butchers, caterers and retailers with fresh pork gammons, joints of vacuum packed bacon, portion-controlled bulk packs and CAP packages.

Heritage Food Group (UK) Ltd

T/O £17.34m; pre-tax profit £0.56m; Emps = 208

The group comprises Heritage Foods of Ipswich, which makes frozen pies and gateaux for caterers, and Seasons Best, a fish processor based at Cleveland. About 80% of its gateaux output goes through the cash and carry trade. Seasons Best, founded in 1987, specializes in supplying oak-smoked salmon, Yorkshire trout and mackerel.

Hey UK Ltd

T/O £27.91m; pre-tax loss £0.69m; Emps = 244 (28/12/91)

Hey UK of Pontefract, West Yorkshire derives most of its income from the manufacture of soft drinks (sold under the Citra brand), though it also distributes beers, wines, spirits and cider for major suppliers and operates a vending division. In August 1991 it was bought by the Boddington Group plc.

Highlander Snacks Ltd

T/O £5.6m; pre-tax loss £0.34m; Emps = 169 (figures relate to 1991, prior to Unichips' takeover)

Highlander Snacks is the new name given to Forth Valley Foods, which went into receivership in 1991, and was subsequently bought by Unichips Finanziara of Milan. Based in Edinburgh, it is Scotland's only crisp manufacturer.

Unichips re-launched Highlander crisps in June 1992 in nine so-called 'Scottish' flavours. The regional nature of the flavours and high crisp consumption in Scotland are factors which Unichips has successfully exploited in promotional campaigns, with sales doubling within 6 months. Through former Forth Valley Foods contacts, the company already has markets established in Germany, Italy, Spain and France. Unichips Holdings (UK) Ltd is a sales operation for crisps and snacks with sales of £4m and 145 employees. (See also Tucker Foods.)

Hitchen Foods plc

T/O £7.5m; pre-tax profit £0.17m; Emps = 311 (1992)

The business was founded in 1939 by Gerald Hitchen who retired in 1989; the company is now run by his three sons: Bernard, Harry and Anthony. In 1989, £3m was invested in relocating to a 12-acre site in Wigan; the factory itself covers 4 acres. The company is a vegetable and salad processor handling 25 000 tonnes of vegetables a year, primarily potatoes, carrots and onions which are distributed nationally in a fleet of 25 temperature-controlled vehicles. Customers include ready meal manufacturers, e.g. Danbys Foods.

Peter Hollingsworth Ltd

T/O £4.0m; pre-tax profit £82 000; Emps = 135 (31/7/92)

The company has been involved in a series of takeovers. In 1964 it was acquired by Imperial Foods and then sold to Associated British Foods in 1969. Finally in 1972 the company recovered its independent status when it was bought by the current chairman, Peter Hollingsworth. The company makes pies, sausages and other meat products distributed through van sales, major supermarkets and cash and carry outlets.

Horizon Farms

T/O £16.3m; pre-tax profit £0.5m; Emps = 172 (27/9/92 – figures for Clifford Kent Ltd)

Horizon Farms is owned by the Clifford Kent Ltd company of Liskeard. It produces over 4 million eggs/week, supplying most of the major multiples. It is the fifth-largest packer in the UK. It holds a 50% stake in Horrells Dairies of Peterborough. Under the terms of the joint venture, Horizon controlled the egg production side leaving Horrells to concentrate on its dairy and other interests. However, in 1992 Horrells discontinued its production of fruit juices and mineral waters.

Horlicks Farms & Dairies Ltd

T/O £20.3m; pre-tax profit £3.2m; Emps = 154 (31/3/93)

Horlicks Farms was bought by the management from SmithKline Beecham following a decision by the pharmaceuticals and health care products company to concentrate on its mainstream businesses. The management team, led by current managing director, Stephen Curtis, had financial backing from Citicorp, National Westminster Bank and venture capital group, 3i.

The Ilminster-based business is engaged primarily in making mature cheddar cheese for a number of major retailers: Kwik Save, Nurdin & Peacock, CWS and the East Anglia & Greater Nottingham Co-operative Society. It also operates a licensed cattle breeding centre.

Hyman Foods Ltd

T/O £4.4m; pre-tax profit £0.07m; Emps = 86 (30/4/92)

Hyman Foods, established in 1981, produces a range of burgers, grillsteaks, doner kebabs, hot dogs, and chicken sandwiches. Customers include Iceland, Kwik Save, Morrisons, Asda, and Capital Foods. The company trades as Steak Express. In 1991, Hyman was bought by the giant American corporation ConAgra. Considerable investment has been made by the new parent company in both plants at Ashton under Lyne and Stockport, and a brighter, American-style look has been given to packaging. Brands include Hickory Farm and Clover Valley, with the recent addition of pure beef or vegetarian Sloppy Joes.

Hytech Foods – see Fresh-Pak Eggs Ltd

I

Icelandic Freezing Plants Ltd

T/O £50.59m; pre-tax profit £0.65m; Emps = 505

This Grimsby-based company is the UK subsidiary of Icelandic Freezing Plants Co. of Iceland. Originally the UK company concentrated on the wholesaling of bulk fish but diversified into further processed fish products in 1985 through the acquisition of Hull-based Brekkes Foods.

Ilchester Cheese Co. Ltd

T/O £4.36m; pre-tax profit £0.065m; Emps = 41

Ilchester produces a wide range of speciality and waxed cheeses, cheese party packs, and a new range for children called Cheesey Feet and Cheddly Bears. The latter are foot-shaped and bear-shaped. Its Kings Choice is Stilton infused with red wine; Applewood is a smoked cheddar and Abbeydale is a Double Gloucester with chives and onions.

International Cuisine Ltd

T/O £2.6m; pre-tax loss £86 000; Emps = 72 (1992)

International Cuisine, launched in 1989 with a £1.2m investment at Consett in County Durham, is to spend a further £1m expanding its facilities for high-quality chilled sauces and ready meals. The expansion programme should double capacity from 150 000 to 300 000 meals/week and increase staffing from 70 to 100. The company which has created a range of 'around the world' chilled dishes under both branded and private labels plans to increase foodservice sales to pubs and contract caterers.

Inverurie Scotch Meat Co. Ltd

Inverurie Scotch Meat is one of about 15 major meat companies located in Scotland's Grampian region. All beef is sourced within a 30-mile radius of Inverurie, most direct from farms. This reduces transport costs and animal stress and ensures meat quality. The company supplies retail butchers, catering butchers, meat manufacturers, and supermarket chains with bone-in, boned or vacuum packed beef, lamb and pork from its EC-approved abattoir and boning hall in Inverurie. The company was formed in 1969. During the latter half of 1993, it joined forces with Aberdeen Meat Marketing and the Edinburgh Meat Plant to form a new meat company – Scotch Premier Meat Ltd. Scotch Premier Meat is a member of the UK's leading farming co-operative, the ANM Group Ltd (Aberdeen & Northern Marts).

J

Jacksons Foods Ltd

T/O £2.9m; pre-tax profit £0.226m; Emps = 35

Jacksons is a private manufacturing company supplying the wholesale, local authority, bakery and industrial catering trade with cake mixes, starch-based products, diabetic and jelly products and tea blends.

W. Jordan (Cereals) Ltd

T/O £17.2m; pre-tax profit £0.846m; Emps = 170 (28/2/93)

W. Jordan, which has been involved in milling since 1855, now has two production facilities, one at Bedford and the other at Biggleswade where it has recently invested a total of £0.5m on new equipment. Originally the company concentrated on the health food trade and in 1980 it launched the Original Crunchy Bar. It currently claims to hold about 22% of the crunchy bar market. It also has a strong position in the ready-to-eat cereals market with a wide range of fruit and nut mueslis and crunchy cereals. Its most recent product is Country Crisp which has been supported by a £3m TV campaign. About 80% of its business is now with major multiples.

J.W. Holdings Ltd – see Allan & Dey Ltd

Jugg Foods plc

T/O £3.94m; pre-tax profit £8000; Emps = 66 (31/12/92)

Jugg produces gourmet ready meals (fish, game, poultry and ethnic) for hotels, restaurants and pub chains.

K

Kavli Ltd

T/O £13m; pre-tax profit £0.71m; Emps = 152 (31/12/92)

Kavli, based at Gateshead in County Durham, is the UK subsidiary of O. Kavli of Norway which celebrates its centenary in 1993. Its principal product, Primula, holds 15% of a static cheese spread market. It has introduced new flavours and added two 'light' products to its range, but regards these developments as somewhat cosmetic and is looking for more adventurous ways to take the company forward – possibly by moving into dips and pâtés, or by acquisition.

In 1991 Kavli acquired the Lambourn Food Co. which makes Scooples and offers synergy with the existing Kavli range of dips, snack and crispbreads.

Kays Cooked Meats Ltd

T/O £6.4m; pre-tax profit £0.029m; Emps = 56

'Mrs Kays' is a manufacturer of high-quality cooked meats with manufacturing sites in London (Walthamstow Head Office), Essex and Lancashire.

Kibun Co. (UK) Ltd

T/O £15.6m; pre-tax loss £3.35m; Emps = 198 (31/12/92)

Kibun produces a range of frozen surimi products – mainly crab sticks – at its purpose-built factory on the Motherwell Food Park. It is the brand leader in the £20m crab stick market with a 70% share. The plant produces approximately 6 tonnes per week, 20% of which is exported. The company is currently diversifying into the ready meals market with its Oriental Parcels.

Kinnerton (Confectionery) Co. Ltd

T/O £13.9m; pre-tax profit £1.03m; Emps = 407

Kinnerton (Confectionery) of London has acquired Gilchris Confectionery of Fakenham, Norfolk which was formed via a £1m management buyout from Renshaw & Co. in 1988. Gilchris makes marzipan fruits and petits fours under own-label for Marks & Spencer, Boots and Sainsbury. Since the acquisition, £0.3m has been invested on new production equipment for chocolate novelties at its 45 000-sq.ft factory and office complex in Fakenham, Norfolk.

Kitchen Range Foods Ltd

T/O £15.25m; pre-tax profit £0.725m; Emps = 171 (31/1/93)

Kitchen Range Foods was set up in 1971 as a manufacturer of frozen gourmet foods. In 1977 Kitchen Range began to supply McDonalds – who at the time had only five restaurants in the UK – with pies. Business with McDonald's expanded and in 1990 a factory at Peterborough dedicated to McDonald's products was opened. During the 1980s a range of value-added products mainly for catering markets were developed. These included potato waffles, coated vegetables, vegetable grills and nuggets. The company now makes a number of own-labels for the main supermarkets as well as selling its own branded goods, Netto and Morrisons.

In July 1993, the Cambridge facilities were refurbished and expanded to house an enlarged product development team, a laboratory and a test kitchen.

L

Lagan Meat Co. Ltd

T/O £–; Emps n/a

Lagan Meat, a Northern Ireland-based meat company formed in 1961, now handles about 750 cattle and 1000 sheep per week. Most of its output goes to the major UK supermarkets. A new company, Hilton Meat Products, was set up in 1985, supplying all the major UK supermarkets with vacuum pre-packed offals (liver, heart and kidney). Its output has now reached about 100 tonnes/week.

Lancashire Dairies Ltd

T/O £38.5m; pre-tax profit £1.06m; Emps = 444 (31/10/92)

Lancashire Dairies, a family-owned milk processing company established during the last century, is now a major supplier of flavoured, skimmed and sterilized milk. It operates three depots, one at its head office site in Manchester, another at Bradford and the third in Brentford. The acquisition of Bradford Model Dairies from Express Dairy in 1988 was made to meet demand for flavoured milk products which could no longer be met by the Manchester production facility alone. The purchase doubled capacity.

The Brentford-based operations of Superlife Foods were also strengthened in 1988 by the purchase of Natural Food Products. Superlife makes flavoured milks, long-life fruit juices and UHT milk. The plastics-bottled sterilized and UHT milk business of Daisy Dairy (part of Waterford Foods) was recently acquired.

The group also packs for Waitrose, Gateway, Spar and others, and exports to the Far East and the West Indies.

Lark Valley Foods Ltd

Lark Valley Foods, set up in January 1986, operates from a custom-built plant in Bury St Edmunds. It processes about 60 tonnes of bacon per week as rashers, joints or steaks sold under own-label or the Suffolk Crown brand, 90% of which is modified atmosphere (MA) packed. Approximately 40 tonnes of its bacon is British, of which half is bought-in ready cured and the rest is cured on site. The company is a member of the British Charter Quality Bacon scheme.

Laxgate Ltd

T/O £71.97m; pre-tax profit £0.75m; Emps = 231 (28/2/93)

Laxgate was formed out of a management buyout from Hunter Saphir in 1989. The £10.7m buyout included HS Foods (formerly Berisford Foods) and Haigh Castle & Co., whose combined turnover was £45m. Following the buyout Meadowbrook Foods was formed as the beverage division; Belvoir Bacon acquired in 1990 was renamed Belvoir Foods.

Haigh Castle & Co. imports and distributes canned goods and grocery products. Belvoir Foods is a manufacturer of hams, cooked meats, shaved meats, pies and sausages, which employs about 200 people at two factories in Mantle Lane and Baker Street in Leicester. The white meat division at Baker Street in Coalville has recently been refurbished at a cost of £500 000. Both plants now meet EC hygiene standards.

Mr Lazenby's Traditional Sausages

Set up in 1982 by Richard and Rosemary Lazenby, the company produces 65 000lbs of high-quality traditional and exotic sausages/week from its purpose-built factory in Stokesley, North Yorkshire. Some are produced for own-label, but the majority of sales are through its branded premium products. Products include Cumberland rings, pork burgers, tandoori and tikka, barbecue, pork and mustard, American premium, etc. Customers include Safeway, Presto, Asda and the Co-op.

Leaf United Kingdom Ltd

T/O £50m; Emps = 500 (1993)

Leaf is a UK subsidiary of the Finnish foods, pharmaceutical and packaging group Huhtamaki which derives 59% of its £750m turnover from confectionery and has operations in the USA, Canada, Holland, Germany, Belgium, Sweden, UK, Ireland, Poland and China. In 1989, Leaf acquired Elizabeth Shaw which has factories in Bristol and Southport. As well as manufacturing the Elizabeth Shaw luxury range, Chewits and new Fizzy Chewits, Leaf also distributes various confectionery items made by other Leaf companies including the Hellas range, Sport Life chewing gum and LO bars.

Huhtamaki also has a subsidiary in Eire, established in 1959, which makes bubble gum (Jawbreaker, Bubble King), candy products, cool pops (Mr Freeze) and liquorice products, 90% of which are exported.

John J. Lees plc

T/O £10.1m; pre-tax profit £0.4m; Emps = >300 (31/3/91)

In 1991 Lees, a manufacturer of biscuit and confectionery items, was taken over by Northumbrian Fine Foods for £5m. However, adverse market conditions and a demoralized sales team meant that by February 1993, it was losing £100 000/month. Northumbrian subsequently sold Lees for £2.1m to a newly formed company – the Claymore Group.

Leopardi Ice Cream Ltd – see New England Food Company

Le Pain Croustillant

T/O £9.4m; pre-tax profit £0.364m; Emps = 110 (1992)

Manufactures frozen and part-baked French bread products. It uses a genuine French recipe incorporating soft-milled flour and a sour sponge dough known as 'le vain'. Le Pain is owned by London Superstores Ltd.

Lion Mark Holdings Ltd

T/O £60.32m; pre-tax profit £3.51m; Emps = 392 (31/12/91)

Lion Mark Holdings has been involved in the importing and milling of spices and herbs in the UK for many years through its chief subsidiary, Lion Foods. Since 1989 it has diversified into other areas through various acquisitions: Richardson Foods, a supplier of mayonnaise and salad dressings, bought from the receivers; Hockenhull & Hayes, a producer of cooked turkey and chicken breasts and rolls; Blackpool-based Meal Maker, which produced potato waffles, croquettes and fritters for the catering and retail market but was subsequently closed; the sauces, toppings and mustard business of Croda International; Hereford Herbs, a supplier of fresh frozen herbs; Tarbert Fyne Foods, Scottish salmon smokers, forming the basis of Lion Speciality Foods. Other group companies include a French subsidiary Lion Foods S.A.R.L., a pizza producer Stateside Foods, set up in 1989 and a commodity trader Victoria Merchants Group (cocoa, pepper, spices, fruit concentrates, etc.).

New products include its Exotics range of dips and salad dressing and fresh herbs pasteurized and preserved in lemon juice, salt and oil with a 2-year shelf life.

Chinney Holdings, a Hong Kong-based private investment company has a controlling interest in the group and has backed investment in both acquisitions and upgrading of the two factories at Runcorn (retail and industrial) and one at Wrexham (foodservice).

Livera Foods plc

T/O £ >10m; Emps = 600 (1993 estimated)

In March 1993, Livera Foods opened a new £8m frozen desserts factory in Stockton on Tees. The plant can produce 6000 gateaux/hour. Within the first 6 months orders exceeded £14m and the predicted annual turnover of £10m was reached in the first half year. The workforce rapidly expanded from the initial 150 to 600. The company has particularly sought staff from among the long-term unemployed. The parent company is Mr Chris Liveras Ltd (Cyprus).

Despite optimism about the company's future in 1993, press reports early in 1994 suggest that Livera Foods has gone into receivership.

Lofthouse of Fleetwood Ltd

T/O £18.09m; pre-tax profit £5.05m; Emps = 255 (31/12/92)

The company's main product, Fisherman's Friend lozenges, has been made since 1865. The lozenges command a 9% share, by volume, of the UK £80m medicated confectionery market though a massive 80% of the company's turnover is derived from exports.

Long Clawson Dairy Ltd

No up-to-date accounts filed

The company traces its beginnings to 1911 when a co-operative was formed by 12 local dairy farmers. Today there are 57 farmers involved with Long Clawson which has cheese production units at Bottesford, Hose, Harby and Long Clawson. The unit at Bottesford has recently been the subject of a £200 000 extension equipped with new dairy plant. The company packs a range of speciality cheeses though Long Clawson is best known for its Stilton which accounts for more than half the turnover. It has recently introduced a new low-fat, soft cheese in plain, herb and garlic flavours to complement the successful low-fat Blue Vinney cheese. Long Clawson's total annual production amounts to about 11 000 tonnes.

As well as cheese, the co-operative also operates retail milk rounds covering an area of 400 sq. miles.

Lord Rayleigh's Dairies Ltd

T/O £60.37; pre-tax profit £0.36m; Emps = 519 (31/3/93)

Lord Rayleigh's is a major independent regional producer of milk products – cream, yoghurt, cottage cheese, flavoured milks and butter – for the major high street retailers, as well as own-label. The Chelmsford-based company has six regional depots covering the whole of East Anglia, London and the south east.

Lovell's Confectionery Ltd

T/O £5.98m; pre-tax profit £0.392m; Emps = 113 (31/3/93)

In 1989, G.F. Lovell plc, a loss-making confectionery company was bought for around £1.5m by two ex-Tarmac managers who planned to bring it back to profitability and increase turnover from £4m to £100m within 3 years. Other companies acquired as part of the plan were in the construction industry. The new group was renamed Albrighton plc.

Lovells continued to be unprofitable and in June 1991, a management team bought it back from Albrighton. The new team reduced the workforce, and used agents rather than an in-house sales team to sell the goods.

Customers now include Sainsbury's, NISA, Spar and Woolworth's. A range of branded bagged goods and countlines are produced for the cheaper end of the market (e.g. Big Deal and Top Ten at 10p). Goods for grocery chains are packed by Portfolio Food's Sweetmate division.

The company operates from purpose-built premises in Newport built 6 years ago to which £500 000's worth of new equipment has been added by the current management.

M

Macfish Ltd

Emps = 500 approx.

Associated Fisheries plc and Geest plc held 50:50 shares in Macfish until September 1992 when it was sold to Foodmark A/S of Denmark.

Macfish, whose business includes that of D.A. Macrae, produces smoked salmon, fish products, scampi and marinades. Extension of its main plant at Fraserburgh with the £2.5m installation of up-to-date fish filleting and coating equipment has been in response to increased demand for many of its value-added lines. With the closure of the Aberdeen smokery, all smoking activities are now located at Fraserburgh. New products include Hat Tricks for children and Light and Healthy breaded cod and haddock fillets for health-conscious consumers. These two products achieved sales of £4m in their first year and took over 5% of the frozen breaded fish market. Success of the low-fat lines brought Macfish additional own-label business from three major supermarket chains.

Products under the Macrae label include marinaded herrings, cockles, mussels and lumpfish roe as well as traditional smoked fish products like smoked haddock.

M & W Mack Ltd

T/O £148.88m; pre-tax profit £3.21m; Emps = 524 (30/4/93)

Mack is an importer, packer and distributor of fresh produce. The company has six wholesale branches operating as Mack Markets in Birmingham, Bristol, Cardiff, Gateshead, Sheffield and Southampton. Mack Multiples Division has a major prepacking depot for fruit and vegetables at Paddock Wood in Kent and a new banana ripening facility at Bishop's Stortford in Hertfordshire.

The prepacking depot has been extended from 78 000 to 127 000 sq.ft. It has five independently operated cold storage areas handling up to 100 000 tonnes of fruit and vegetables a year. On the same site banana rooms built at a cost of £1.5m will increase fruit handling capacity by 20 000 tonnes p.a. and the Bishop's Stortford facility by a further 15 000 tonnes.

The company has managed to increase its profits in the last year by 10% despite pressure on margins and over-supply in European markets for fresh produce.

Mackie's Aberdeen Dairy Co. Ltd

Emps = 300 approx.

Mackie's is an independent farming and dairy company, handling over 12 million litres of milk in glass, cartons and plastic non-returnable containers. Trading as Mackie's Quality Farm Foods, it makes cream and luxury ice creams, and produces potatoes, grain, pork, bacon and some eggs on its Aberdeenshire farm.

Macphie of Glenbervie Ltd

T/O £14m; Emps = 120 (1992)

Macphie was set up in 1973 primarily to produce innovative ingredients for bakers, caterers and manufacturers. Products include non-dairy cream, UHT glazes and savoury sauces, and a range of complex blends and concentrates for the bakery trade. The company's biggest selling product over the last 9 years has been Glen Delight, which has 35% of the catering market for cream alternatives and is handled by the CPC (UK) Ltd sales team. In 1992, it introduced its Highland Kitchen range of savoury, hot pudding, dessert and 'tea time' mixes for caterers.

Marlow Foods Ltd

T/O £3.9m; pre-tax loss £4.05m; Emps = 55 (31/12/92)

Marlow Foods, based in High Wycombe, manufactures Quorn mycoprotein from the Fusarium microfungus. Quorn can be used fresh in chunks or minced in vegetarian foods and ready meals. Demand for Quorn is outstripping supply and the company is rapidly expanding its production facilities in Cleveland (a new fermenter) and North Yorkshire (further processing plant) at a cost of £26.5m. These developments should increase Quorn production from 7000 to 14 000 tonnes/year, most of which will be sold in the Netherlands, Germany, France and Italy.

In March 1993 Marlow Foods was acquired by the Zeneca Group plc.

Mathers (Inverurie) Ltd

T/O £26.35m; pre-tax profit £0.11m; Emps = 96 (31/12/92)

Mathers entered the meat trade about 30 years ago. Today it operates an EC-approved abattoir and cutting plant in Inverurie supplying beef, lamb and pork to a range of customers in the UK and overseas. Within the UK the company operates its own fleet of refrigerated vehicles; J. Bain & Son is responsible for the company's export trade.

E.C. Matheson & Son Ltd

Matheson's is a specialist supplier of Scotch beef to the retail meat trade from its abattoir and cutting plant at Brechin in Angus. Boneless vacuum packs of beef as well as lamb and pork are distributed to most parts of the UK and EC.

Matthews Foods plc

T/O £9.58m; pre-tax profit £0.11m; Emps = 80

Matthews Foods is a major independent company supplying branded and own-label margarines, spreads and lards, as well a comprehensive range of margarines and shortenings for bakers. It operates from an up-to-date factory in Yorkshire and has a contract with Wincanton Distribution for refrigerated delivery of its products.

Maxway Food Products Ltd

T/O £4.4m; pre-tax profit £0.138m; Emps = 128 (28/9/92)

Maxway takes in fresh chilled chicken, then cooks, strips and packs it for supply to food manufacturers. Strict hygiene controls are operated at the company's new Northumbria plant, and it also has BS 5750 accreditation.

McIntosh Donald Ltd

T/O £67m (1991)

McIntosh Donald was formed 7 years ago when James McIntosh, a meat wholesaler bought William Donald & Son. The company operates an abattoir, cutting and meat wholesaling operation at Porthleven, Aberdeen. In 1993, it opened a new plant in Porthleven costing £3.5m and comprising a boning and packing hall and effluent treatment unit. It handles Scotch beef and lamb. (See also McIntosh of Dyce.)

McIntosh of Dyce Ltd

T/O £11.47m; pre-tax loss £0.06m; Emps = 340 (26/2/93)

In 1988 James McIntosh acquired Lawson's of Dyce from Mattessons Wall's and renamed the company McIntosh of Dyce.

In 1992, a senior management team bought the company from James McIntosh for an undisclosed sum, leaving James McIntosh to concentrate on its core slaughtering and meat wholesaling activities.

It now produces more than 50 volume lines under the McIntosh name including sausages, cooked meats, pies, savoury rolls and vegetable and cheese pies. Many own-label products are also manufactured. It has capitalized on the growth of the children's market with Wee Ones – mini pork and beef skinless sausages. It is a member of the Chilled Foods Association.

McKey Food Services Ltd

T/O £53.5m; pre-tax profit £1.35m; Emps = 177

McKey Food Services, owned by the McDonalds Corporation, USA, is the manufacturing company for McDonalds Restaurants (UK). It has recently opened a new 5400-sq.ft factory in Scunthorpe to handle 1 million lb of boneless beef/week. The Scunthorpe plant joins McKey's existing facility at Milton Keynes which, with a similar capacity, turns out 0.5 million burgers/day. McKey, which sources its meat (8000 cattle/week) from abattoirs across Europe, is easily the largest burger producing operation in the UK. The company also produces bacon and sausages for McDonalds and supplies the Ministry of Defence.

McLaren Foods Ltd

T/O £10; pre-tax profit £39 000; Emp = 107 (30/6/92)

McLaren Foods was set up in 1987 supplying retailers with meat, high-quality chilled burgers and other value-added meat products. Marks & Spencer was one of its clients. This side of the business was sold to Hillsdown with Gillards taking over the remainder which was in the catering sphere.

In May 1993, a management team who already had a 20% stake in McLaren's bought out Gillards' 80% share. McLaren offers a daily delivery service to caterers from depots at Ashford in Kent (also its head office) and Chalgrove in Oxford. It is a member of the National Association of Catering Butchers. Its parent company is Barnaby Holdings Ltd. (See also Gillard Holdings.)

McQuillan Meats Ltd

After recently completing a new £2m factory in 1987, McQuillan Meats of Northern Ireland has been looking to the mainland to increase turnover. It produces a value-for-money frozen range under its Crofter's Kitchen brand which includes beef and lamb steaklettes, burgers, chicken nuggets and chips, and sliced meats in gravy. It also has premium products under its Goodstile brand – Chicken and Beef Ranchers, Beef Wellington – as well as marinades, crumbed products and fresh burgers.

Meadow Vale Foods Ltd

T/O £3.5m; pre-tax profit £0.12m; Emps = 100 approx.

Meadow Vale Foods is part of the relatively new Meadow Vale Food Group involved in food processing, trading and distribution which operates from Oswestry. (It is part of Lloyds group of companies, a major egg producer.) Group companies include Arthur Barras Foods (wholesaling of frozen poultry and other foods), Coachman Foods Ltd and Meadow Vale Trading. Meadow Vale also has a controlling interest in Seafrost, a company in North Shields which produces a range of chicken and fish dishes for retailers and caterers.

Country Style Foods, with poultry processing operations at Wrexham (further processed products) and Mold (raw poultry meat) has been merged with Arthur Barras Foods. The parent company has recently spent £300 000 on the Wrexham plant with a view to obtaining BS 5750 and EC certification.

Meadow Vale Trading specializes in the re-sale and distribution of poultry and frozen and chilled foods which result from over-production, out-of-season, short-coded and discontinued lines.

Meltis Confectionery Ltd

T/O £21.2m; pre-tax profit £0.62m; Emps = 369 (30/4/93)

In January 1991, Meltis Confectionery was bought from Jacobs Suchard (part of Phillip Morris) by two Suchard directors. At that time the Bedford-based company had a low profile and was turning over about £3m a year. The new management focused on seasonal and novelty chocolate lines, using the Thunderbirds and Fireman Sam themes on promotional material. Another innovation has been the development of sales through charity items like the Variety Club's Golden Heart Egg. About 80% of Meltis's production is in chocolate goods. Sales of Newberry Fruits have fallen from £10m in the early 1980s to about £3m at the present time. Contract packing and own-label supplies account for half of total production. Meltis is also importing confectionery from the Continent through its subsidiary Cofisell.

Demand for Meltis goods has increased so rapidly that about 200 extra staff are likely to be recruited in 1994.

Meridian Foods Ltd

T/O not disclosed; pre-tax profit £0.125m; Emps = 66 (31/1/93)

Meridian produces or distributes a wide range of organic and health foods, speciality unrefined oils, fruit spreads, Horizon peanut butter and Questo cook-in sauces. Most of its business has traditionally been with the health food trade. In 1992 a new national manager was appointed with the specific aim of gaining access to mainstream grocery markets.

Merrydown plc

(formerly Merrydown Wine plc)

T/O £19.2m; pre-tax profit £1.71m; Emps = 193 (31/3/93)

In January 1993 Merrydown Wine changed its name to Merrydown plc and in February moved from the unlisted securities market to a full stock exchange listing. Despite the recession the company recorded rising profits and sales in the 6 months to September 1993: pre-tax profits up 23% to £930 000 and sales up 31% to £11.3m.

Merrydown manufactures a range of ciders and its subsidiaries produce speciality vinegars (Martlett Natural Foods), pack honey (West Country Honey Farms) and manufacture premium jams, preserves, chutneys and sauces (Dorothy Carter). Group companies produce both branded and own-label goods. A new operating division, formed in April 1993, combines Martlett Vinegars and West Country Honey Farms and Martlett Natural Foods.

Soft drinks now account for 27% of Merrydown's sales following acquisitions of Piermont from Taunton Cider, Sorelle from Premier Teas and PLJ and Schloer from

SmithKline Beecham. (Schloer has sales of £6m and PLJ of £0.7m.) The latter deal cost £8.5m. It has also recently signed a distribution agreement with the Gaymer Group for Merrydown Draught Cider. Merrydown exports to 25 countries, the largest markets being in Germany and Greece.

John Millar & Sons (1844) Ltd

T/O £3.72m; pre-tax profit £0.07m; Emps = 67 (31/12/92)

After a century at its Edinburgh premises, this well-established Scottish confectioner has moved to the former Golden Wonder factory at Broxburn. The move will result in a 50% increase in production. The staff of 74 will be retained at the new site. The primary outlets for the company's products – mints, barley sugars, fruit drops and liquorice – are in Scotland, north-east England and Ireland.

Millway Foods Ltd

T/O £9.6m; pre-tax loss £1.3m; Emps = 151 (31/12/92)

Millway Foods was created in March 1988 by a management buy-out of the Stilton factory at Harby in Leicestershire previously operated by St Ivel. Located on an 8-acre site, the factory produces about 3000 tonnes of cheese per year, about half of which is Stilton. The buy-out was funded with help from the venture capital group, 3i, and the Bank of Scotland.

Millway is a significant manufacturer of Stilton, producing 1200 tonnes/year in mild, mature and vegetarian versions. In addition, because Stilton is now frequently used by manufacturers in soups, dips, etc. it also supplies Blue Stilton crumb. Speciality cheeses include White Stilton with Apricot or Mushroom and Garlic, Double Gloucester with Chives and Onions and Cambridge Blue, a full-fat soft blue cheese.

Moray Seafoods Ltd

From its base in the fishing port of Buckie, Moray Seafoods produces a range of prepared seafood products and ready meals (using prawns, scampi, shellfish and other seafoods). The company supplies caterers, chain stores and national wholesalers in the UK and overseas, and operates a direct service via its own fleet of temperature-controlled vehicles. It also has its own microbiology laboratory staff capable of carrying out quality audits on request.

Morning Foods Ltd

T/O £29.46m; pre-tax profit £0.41m; Emps = 125 (31/10/92)

Morning Foods, based at Crewe in Cheshire, is one of the UK's largest independent cereal millers and packers, producing a range of breakfast cereals under the Mornflake brand (oats, oatmeal, oatbran Hawaiian Crunch).

Müller Dairy (UK) Ltd

T/O £76.78m; pre-tax profit £1.05m; Emps = 57 (31/12/92)

Since Müller launched its innovative Fruit Corner yoghurt in the UK in 1987, the brand has grown rapidly and now claims 23% of the £520m yoghurt market. In the first year of production (1992) from a new £30m factory in Market Drayton in Shropshire it produced 360 million pots worth £86m. By July 1993, the value of the brand had risen to £125m. In addition to the much copied split pot Fruit Corner, Müller also introduced the 'thick and creamy' brand (worth £6m) and the virtually fat-free yoghurt, Müllerlight. Müllerice is also produced at Market Drayton. At present Müller UK is resisting requests to produce for own-labels.

Initially all products were imported from parent company Molkerie Alois Müller GmbH at Aretsreid in Germany, but growth of the brand meant that UK production was the most cost effective option. The new plant was completed just in time to avoid the worst effects on prices of the sterling's devaluation.

N

Needlers plc

T/O £23.13m; pre-tax loss £0.9m; Emps = 427 (31/12/92)

Nidar Bergene AS, a member of the Norwegian Nora Group, (UHC Orkla A/S) acquired Needlers from Hillsdown in 1990. The deal allowed Needlers to market the Nidar Bergene chocolate range in the UK and provides it with an opportunity to market Needler products in Scandinavia. Needler brands include Fruit, Mint and Butterscotch Sensations and Extra Strong Mints sold in jars, boxes, bags and multi-packs. Despite a 11% decline in the boiled sweet market since 1989, Needlers Sensations have increased sales volumes by 38% and doubled market share.

Henry Newbould Ltd

T/O £8m; Emps = 300 (1993)

Established in 1856 by Wilson Newbould, the company now operates from a £1.7m purpose-built factory in the Business Enterprise Zone of Cleveland. Newbould makes a range of pies, sausages, black puddings, flans, scotch eggs, sausage rolls and cooked meats for customers which include CWS, Dewhurst, Morrison, Presto and Safeway. Some of the output still goes to its 15 retail shops, where newly developed products are test marketed. Although having a reputation for traditional products, Newbould won an MLC Enterprise Award for its Bolognese Mizza (a meaty bolognese pizza).

New Covent Garden Soup Company

T/O £7.2m; pre-tax profit £0.784m; Emps = 81 (1992)

The New Covent Garden Soup Company was set up 1988 by a former stockbroker Andrew Palmer. Following 2 years of product and process development, the company produces fresh chilled soups with a 14-day shelf life in foil lined gable top, Elopak cartons. All varieties are made with vegetables from New Covent Garden or direct from the grower and are priced at the premium end of the market. Special 'Soup of the Month' promotions are planned. Initially, they were available through independent retailers in London and the south and in certain Tesco stores, but distribution has now been extended to the North and Scotland.

In March 1992, the company added fresh chilled sauces to its product range – Provencal, Florentine, Mediterranean, and Parmigiana.

New England Food Company

T/O £3.6m; pre-tax loss £1.2m; Emps = 56 (31/1/92)

New England produces a range of luxury American-style ice creams at a purpose-built factory and cold store in Newton Abbot in Devon. Following acquisition by the Cygnus Venture Capital Fund, New England took over both the Leopardi and Cricket St Thomas regional brands and manufactured a super-premium brand, Ultimate, in eight flavours.

In July 1993, the company ran into financial difficulties and went into receivership. Three directors of New England subsequently acquired some of its assets and goodwill and formed Newton Ice Cream which now trades as New England Ice Cream.

J.N. Nichols (Vimto) plc

T/O £47.0m; pre-tax profit £8.4m; Emps = 254 (31/12/93)

The Nichols group of companies, set up in 1908 as a supplier of fruit essences, pills and potions, is engaged in the soft drinks, drinks vending and catering supply industries. Trading subsidiaries include: Solent Canners, Cabana Soft Drinks and Nichols Foods Ltd.

The principal product of the company, Vimto, is sold as a cordial, aerated in bottles and cans, in ready-to-drink and diet versions. It holds 2% of the UK soft drinks market. The company's head office and factory is at Wythenshaw in Manchester, where a substantial investment has recently been made in new plant and equipment. Solent Canners, Southampton, who joined the group in 1980 provide in-house canning facilities for Vimto. In 1986 Independent Vending Supplies of Skelmersdale (re-named Nichols Foods in 1990) and Cabana Soft Drinks of Preston were acquired. Cabana supplies draught soft drinks to pubs and catering outlets. It recently launched Cola Gold. Nichols Foods already a leading vending ingredient manufacturer, successfully joined the can vending market in 1992, with its own Vimto branded machines.

Nichols exports Vimto products to 30 countries, with markets recently opened up in Canada's eastern provinces, Guyana in South America, South Africa, Egypt, Tanzania and Russia. Of its £47m turnover, about £4.3m come from export sales.

Nimmo (Holdings) Ltd

T/O £15.36; pre-tax loss £0.04m; Emps = 130 (30/9/92)

This is a holding company for a group of companies, run by the Nimmo family, and is engaged in the manufacture and retailing of cooked meats, bacon, sausages and burgers.

Noon Products plc

T/O £10.9m; pre-tax profit £0.6m; Emps = 154 (31/12/92)

Noon Foods, a family business, manufactures chilled and frozen Indian ready meals from two modern, purpose-built factories in Southall, London. Recipes are authentic and prepared from high-quality ingredients and the factories place strong emphasis on hygiene standards with designated high and low risk areas. 60% of the company's sales are to J. Sainsbury.

North Downs Dairy Co. Ltd

T/O £22.75m; pre-tax profit £0.51m; Emps = 34 (30/6/92)

The company, formed in 1983, is a packing and marketing organization for Pilgrims Choice mature cheddar. The cheddar is manufactured by Tower Farms of Taunton in Somerset. It has a custom-built warehouse and cheese packing facility at Wincanton in Somerset. The company claims to be the UK's largest independent factor of matured cheddar. While the market for mature cheddars has grown, major retailers now demand retail packs which focus on a price point. This has led North Dairy to offer a 400g pack for medium mature and fully mature cheddar, rather than 500g, which allows them to be sold within the £2 and £2.50 price barriers respectively.

North of Scotland Milk Marketing Board

Milk Fund: Sale of milk £13m; surplus before tax £0.6m
Commercial Division: T/O £14.7m; trading surplus £0.1m;
Emps = 241 (total of Board and Commercial Division) (31/3/93)

The Board is one of five MMBs responsible for the marketing of milk in the UK. In common with the other four MMBs its future after the abolition of the statutory milk marketing scheme in 1994 is somewhat uncertain.

In the year to March 1993, total milk production was 56.1 million litres, a decrease of 3% on 1992 figures. Milk quality improved dramatically following introduction of the Somatic Cell Count Scheme. 32.2 million litres was sold as liquid milk, the remainder as cheese manufacturing milk. A £2m investment at the Nairn Creamery is expected to boost profits in the coming year. Profits from the Kirkwall Creamery suffered from a one-off payment to the Milk Fund in relation to cheese pricing systems. Orkney Cheese won first prize at the International Food Exhibition in London.

Northern Co-operative Society Ltd (NORCO)

Faced with debts of £7.4m, NORCO embarked on a drastic disposal programme involving most of its superstores. When this failed to overcome the problem it entered discussion with CWS on a possible merger. However, CWS declined, leaving NORCO no option but to call in liquidators, Ernst & Young, in June 1993. Since then a number of the smaller neighbourhood stores have been bought by C.J. Laing.

Northumbrian Fine Foods plc

T/O £15.9m; pre-tax profit £0.15m; Emps = 200 approx (31/3/92)

Northumbrian Fine Foods, listed on the stock exchange, has delivered a rather mixed performance in the last 2 years, recording a pre-tax loss of £5.6m in the year to March 1993 but regaining profitability in the half-year to 30/9/93 following the sale of the loss-making Lees business (see John J. Lees, Section [2.3]). There was a pre-tax profit of £104 000 on sales of £8.09m for the latter period.

The company has interests in biscuits, cakes, confectionery, and healthy snack foods. Subsidiaries include Heather Cameron, Prewitts, Country Fitness Foods and Sunwheel Natural Foods. In 1991 it sold its loss-making Danish subsidiary Danish Natural Foods for £1.3m and bought John J. Lees cake and biscuit business for £5m and Biscuits for the Connoiseur of Huddersfield for £1.75m. However, Lees proved to be unprofitable and was re-sold only 16 months later to the Claymore Group for £2.1m.

In August 1993, Northumbrian raised cash to promote its new Dunkers biscuit through a rights issue. The issue was 71.2% subscribed. Dunkers are a novel oval shape and remain crisp when dunked in hot liquid. In the half-year to 30/9/93 Northumbrian made a pre-tax profit of £0.1m on a turnover of £8.09m (1992 figures; pre-tax loss of £1.06; sales of £11.1m).

O

Oak Forest Foods Ltd

Emps = 20 (1993)

Oak Forest Foods is a new company set up in 1993 in Tipton, West Midlands following the closure of Amberley Foods in February 1993. Three senior managers from Amberley invested their redundancy money in the venture which was also supported by the DTI, the Black Country Development Corporation and Sandwell Council. The new company makes ten different sausages, hams, ox tongues, chawl, haslett, brawn and black puddings. Products are chilled not frozen and made with high-quality ingredients and meat from EC-approved abattoirs. All the employees were formerly with Amberley Foods.

Oak Lodge Salmon Ltd

(formerly Hines Bros)

T/O £2.2m; pre-tax profit £0.023m; Emps = 27 (31/12/91)

Oak Lodge is a leading salmon smoking company, processing over 600 tonnes of salmon each year. It also produces hot-smoked tuna cutlets, marinaded halibut, gravadlax, fish hors d'oevre, smoked halibut and trout. At its East London factory, it has a modern, environmentally acceptable, 7500-sq.ft smokehouse.

Orchard House Foods Ltd

(formerly Fresh Fruit Juice Manufacturing Co. Ltd)

T/O £17.68; pre-tax profit £0.67m; Emps = 260 (31/10/92)

Fresh Fruit Juice, set up in 1985, now supplies most of the multiples with freshly squeezed juice. Its Orchard House range includes orange, grapefruit and apple juice. With its sights firmly set on Europe, the company has set up a production unit in Monchengladbach, West Germany and acquired Arcadia, a Belgian producer of fresh fruit juice, along with its French subsidiary, Arcadia-Vivagrum.

In 1991, the company changed its name to Orchard House Foods with a view to preparing whole fruit products and desserts.

P

Panda Drinks Ltd – see Hall & Woodhouse Ltd

Paradise Cuisine

Emps = 50

Paradise Cuisine, set up in 1973 started out selling frozen dishes to Harrods. It now operates from two West Country sites. At Warminster, gourmet entrées, pâtés, mousses and sauces are produced. The Wincanton site produces luxury desserts and gateaux for major retailers and caterers. Its founder Jenny Flew won Barclay's Businesswoman of the Year Award in 1991.

Parkham Food Group

T/O £17m, £23m predicted for 1993; Emps = 180 (1992)

Parkham is a turkey and chicken processor operating from two plants in Liversedge, West Yorkshire. It manufactures roast, smoked or honey roast turkey breast joints, stuffed turkey breast, chicken burgers, goujons, and battered or pepared chicken breasts. Products are primarily targeted at the food service sector through its Poultry Chef division. Total manufacturing space is 57 000 sq.ft. Following rapid growth of the company in the UK in the last 8 years, its is now actively seeking export markets with an order already won from Switzerland.

Pasta Reale Ltd

T/O £5.2; pre-tax profit £0.093m; Emps not diclosed (30/9/92)

Pasta Reale is one of the leading manufacturers of fresh pasta. In the last 15 years the market for fresh pasta has grown from nothing to £24m in 1992, with Pasta Reale claiming a 42% share, about the same as the Pasta Company, with Dolmio and Buitoni taking the remainder. Customers include Waitrose, Asda, Safeway, Presto, Morrisons, Wm Low and Bookers.

The company has moved for the fourth time to new premises in Crawley which are equipped with machinery capable of turning out 90 000 kg of fresh pasta/week. Over 20 plain and filled pasta types are produced, with three new seafood dishes recently added to the list.

Patak's (Spices) Ltd

T/O £–; pre-tax loss £0.162m; Emps = 150 (30/4/92)

The company began trading in 1958 from a basement kitchen, but now operates from a 4-acre site in Wigan, producing over 240 varieties of pickles, pastes, chutneys, canned curry sauces, poppadoms and ready-to-eat savoury snacks. It claims to hold 35% of the market for Indian food ingredients. There are now seven canned sauces and six in glass jars. Patak's entire range has recently been re-packaged with nutritional information, recipe suggestions and a recipe book offer. As well as being available through all the major retailers, Patak's also supplies the cash and carry sector and the majority of UK Indian restaurants.

Paulig UK Ltd

T/O £5.3m; pre-tax loss £1.7m; Emps = 47 (1992)

Paulig UK is the UK subsidiary of Finnish coffee company, Oy Guystav Paulig, which has a 42% share of the Finnish coffee market and worldwide sales of £125m. Paulig UK initially ran two plants: a high-capacity coffee production plant at Wimbledon in London and the James Ashby tea and coffee company in Milton Keynes, acquired in 1987. In September 1990, the company acquired the rights to manufacture and market Melitta roast and ground coffee in the UK. Since this acquisition, Paulig UK has decided to concentrate on developing the Melitta brand and own-label coffee, and has withdrawn from the tea market, closing and selling its Milton Keynes plant.

The company also markets a coffee whitener Melitta Compleat and dominates the world coffee filter paper markets with a share of 36%.

In February 1994 it acquired Allied-Lyons roast and ground coffee business.

Pershore Poultry Ltd

T/O £8m (approx); pre-tax profit £0.3m; Emps = 108 (26/3/93)

Pershore Poultry specializes in air-chilled, fresh whole chicken and portions, processing about 80 000 birds/week. It also operates three farms, has a turnover of £8m and works closely with Brandons Poultry (poultry farmers) and Foston Farm Produce, which supplies fresh and frozen chicken and turkey, whole, portioned or pre-packed through its refrigerated national distribution service.

Perimax Meat Co. Ltd

T/O £4.0m; pre-tax profit £0.324m; Emps = 70 (31/12/92)

Perimax is one of the UK's leading suppliers of mechanically recovered meat. The company was founded in 1978 and now has factories in Shepton Mallet in Somerset and Arbroath in Scotland. Until recently, the company was part of Hillsdown which sold the Somerset-based company in 1990.

Phoenix Foods Manufacturers Ltd

T/O £5.11m; pre-tax loss £0.092m; Emps = 80 (31/3/93)

Owned by Howard Foundation Holdings, Phoenix formulates, manufactures and packs dry grocery products for manufacturers and retailers. Products range from hot beverage to dessert mixes and can be matched to customer requirements for products with reduced fat and additive levels.

Pilgrim Food Group Ltd

Chaucer Foods (renamed Pilgrim Food Group) was recently the subject of a £1.7m management buyout from parent company, Hazlewood, which retains a 25% stake. Under Hazlewood, which acquired the company in 1986, Chaucer Foods outgrew its former factory in Ashford, Kent and moved to new premises in Hull, where turnover is currently about £2.5m. The company exports 60% of its output and claims to be the largest independent manufacturer of croutons in the UK.

Pinegain Ltd

T/O £43.77m; pre-tax profit £0.41m; Emps = 389 (31/3/93)

Pinegain was formed in 1989 to carry the management buyout of Marr Frozen Foods of Hull from Andrew Marr International (AMI). AMI has retained a minority interest in the business which employs about 350 people. Since the buyout, Marr has acquired Merson & Gerry which operates a 400-sq.m. fish processing facility in Aberdeen, and the fish products division of William Jackson & Son (now trading as Junella Seafoods in the wholesale sector). In August 1993 it bought Lewis Frozen Foods of Warrington. Lewis supplies frozen fish portions and steaks to supermarkets. The modern processing and packing equipment at Lewis's plant will be transferred to Marr's Hull factory.

In December 1993, Pinegain restructured the group into two divisions: Retail Services responsible for the Hull factory and frozen seafood and prawns and chilled seafood products; Catering Services handling nine branches and the whole business of Junella.

Playtime Foods Ltd – now part of the JLI Group

Plumrose Ltd

T/O £36.8m; pre-tax loss £0.15m; Emps = 407 (26/12/92)

The UK and US activities of canned and chilled meat manufacturer Plumrose were bought from the parent company, East Asiatic Company of Copenhagen, in July 1993 by Vestjyske Slagtier which is Denmark's second-largest slaughterhouse chain. Collaboration arrangements with Plumrose in Germany may also be sold to Vestjyske at a later date. East Asiatic followed up the deal by selling off Plumrose's interests in Australia and New Zealand to Pacific Dunlop for A$225m.

In the UK Plumrose has factories at King's Lynn and Glenbrook in Northern Ireland. Its head offfice is at Willerby near Hull. Following a major re-launch of its brands in July 1991, Plumrose has captured 25% of the branded chilled meats sector with its American-style wafer-thin meats and Continental specialities. Its new range of Delifresh savoury rolls, joints and steaks is claimed to combine the freshness of delicatessen products with the convenience of long shelf-life.

Plumtree Farms Ltd

T/O £5.9m; pre-tax profit £0.43m; Emps = 99 (1/5/92)

Plumtree Farms produces fresh and frozen sausages and burgers for the catering and fast food markets from its custom-built premises in Wallis Road, London. In August 1992 Plumtree bought the cooked meats division of Chivers Farm Foods including its Ponders End factory. Products will be targeted towards the catering sector, with the ham products additionally marketed to the burgeoning sandwich sector.

Early in 1993, the company invested in a new burger machine capable of producing 26 000 burgers/hour.

Pollard Confectionery Ltd

Pollard's family-owned ice cream and confectionery business, founded in 1916, went into receivership early in 1993. Bristow's of Devon subsequently bought the confectionery interests and a management team the ice cream business.

Princes Ltd

T/O £260m; pre-tax profit £5.66m; Emps = 579 (31/12/92)

This long-established importing and manufacturing company was acquired from Nestlé in 1989 by the Japanese trading business, Mitsubishi, along with the edible fats supplier Trex, giving the Japanese industrial group a foothold in the UK.

The Princes brand currently encompasses over 150 retail lines, with extensive ranges in canned fish, meats, fruit, vegetables and fruit juices. The company claims a 25% share in the £210m canned meats sector and 25% of the £354m canned fruit sector. The company's main factory is in Liverpool with a second plant at Southport, where £2m has recently been invested in new machinery.

In 1991, Princes bought G. Barraclough, a soft drinks manufacturer heavily involved in the own-label market, for an undisclosed sum. At the time of the takeover Barraclough's had a turnover of £3m. Since then Princes has invested £6m in plant and equipment at Barraclough's factory and developed the Gee-Bee Twist 'n Squeeze and Princes Jucee brands. More recently Cardiff-based Cima Foods was bought. Cima turns over £30m through its Cima, Fruit-jus and Spartan Spring Water brands.

L.E. Pritchitt & Co. Ltd

T/O £36.8m; pre-tax profit £1.68m; Emps = 300 (28/5/93)

Pritchitt Foods with its UK headquarters in Bromley, Kent and its main production unit at Newtownards in Northern Ireland, is a privately owned dairy product manufacturer. In the year to May 1993, pre-tax profit increased by 123% on the 1992 figure largely due to cost cutting rather than increased turnover.

Pritchitt sells long-life dairy-based ingredients to catering, bakery and ingredient industries. Brands include Millac fat-filled milk powder, Roselle UHT alternative cream products and Comelle ice cream mixes. The company exports to 60 countries, the most important markets being in the EC, Middle East, Mediterranean area, Far East, West Indies and Africa. Exports sales now account for nearly half of turnover. Pritchitt also offers technology transfer consultancy services.

Pukka Pies Ltd

T/O £9.82m; pre-tax profit £1.4m; Emps = 196 (30/5/92)

Pukka Pies was formed in 1962 by Trevor Storer, the present chairman. Today the company produces about 100 000 pies and pasties/day, and a small quantity of sausages, sold primarily to the fish frying trade. The company operates its own van sales fleet, distributing in a 50-mile radius of Leicester, while national distribution is handled through a network of 14 distributors.

Puredrive Ltd

T/O £2.6m; pre-tax loss £0.89m; Emps = 54 (31/12/92)

Set up in 1989 by three former Mattessons Wall's executives, Puredrive operates from a 20 000-sq.ft purpose-built plant on a greenfield site in Corby, Northants. They were backed by the resources and European meat processing expertise of the Van de Laan group of companies, which includes Rea Valley, the UK's leading canner of tongues and other meat products. Puredrive's strategy was not to create a new brand but to supply own-label consumer packs as well as bulk packs for slicing by retailers or packers. Examples of products include garlic sausage, black pudding, chicken roll, chopped ham and pork, etc. A microwaveable sausage, very similar to a Mattessons Wall's product caused a dispute between Puredrive's directors and their former employer Unilever.

Th. S. Van der Laan has recently acquired full control of the company, whose ultimate parent company is Meatpoint BV in the Netherlands.

Puritan Maid Ltd

T/O £126.9m; pre-tax loss £1.9m; Emps = 825 (total food and non-food supplies) (31/1/93)

Puritan Maid is the supplies division of Forte plc. Until recently it employed about 300 people in food manufacturing, but financial strains within the group caused by the recession have prompted the sale of the burger manufacturing business to the Sims Food Group and the closure of the bakery and ice cream units. A small operation producing frozen entrées and airline meals is based at Poynters Road, Dunstable.

Q

J.G. Quicke & Partners

The Quicke family have farmed at Newton St Cyres in Devon for over 400 years. The present partnership produces various traditional farmhouse foods including speciality cheeses, meat products, luxury Devonshire dairy ice cream in sophisticated flavours and flavoured yoghurts. Cheeses available include mature and smoked Cheddar, Cheddar with herbs, vegetarian Cheddar, Double and Single Gloucester and Red Leicester. Quicke also supplies major supermarkets with own-labels but plans to increase distribution of Quicke's branded products through independent delicatessens in 1994.

To meet increased demand Quicke has recently completed a new packing area and opened a farm shop in October 1993.

R

Rahbek-Food Ltd

T/O £18.6m; pre-tax profit £0.602m; Emps = 225 (31/12/92)

Rahbek-Food is the UK division of Rahbekfisk A/S company of Denmark. At its factory in Redditch, it produces a wide range of chilled and frozen value-added fish products for leading retailers in the UK and Europe. The ultimate holding company is Foodmark A/S of Denmark, which also owns Macfish Ltd.

Rayner & Co. Ltd

T/O £18.36; pre-tax profit £0.71m; Emps = 241 (31/12/92)

In June 1990, Hero Conserven, the Swiss manufacturer of jams and other foods, acquired a controlling interest in Rayner Burgess, one of the the UK's oldest food manufacturers. Hero has a worldwide turnover of about £280m.

Rayner's main site in London produces food colours, flavourings and essences under the Rayner brand, and many traditional condiments and sauces (horseradish, mint, etc.) under the Burgess brand. Rayner's subsidiary, Cauldron Foods, produces plain or flavoured tofu and other vegetarian foods from a new factory at Portishead near Bristol. Rayners also handle Hero's range of jams and Swiss-style potato products (Rosti, Fines Herbes, and Gratin Dauphinoise).

In January 1993, the assets, liabilities and trade of Rayner Burgess Ltd were transferred to Rayner & Co. Ltd. For the year ending December 1992, Rayner Burgess had a turnover of £15.5m and a pre-tax profit of £36 000.

Rea Valley (Canned Meats) Ltd

T/O £11.1m; pre-tax profit £1.12m; Emps = 75 (31/12/91)

The company was formed in 1925 producing cured frozen and canned meat. Today it specializes in canned ox and pork lunch tongues in slicing sizes for the catering and delicatessen trade. Th. S. Van der Laan, the principal shareholder in Rea is one of Holland's leading exporters of canned meat products and bacon to the UK.

J.F. Renshaw Ltd

Renshaw is a supplier of ingredients to bakers and food manufacturers. Following poor performances for several years the company now employs only about half the workforce it did in 1989, and is owned by the German company Schwartauer Werke GmbH & Co. Amongst a string of disposals, its Cremanaze Salads subsidiary was sold to Leiseur in 1991.

J.F. Renshaw has a 50% holding in R. & W. Scott of Carluke (Scotbloc manufacturers). Renshaw products are now being distributed to cash and carries through Scott's.

In July 1993, Renshaw announced that it would be consolidating all UK operations in a new building at its Liverpool site as part of a £12m development programme.

Ringtons Ltd

T/O £22.17m; pre-tax profit £2.53m; Emps = 556 (31/12/92)

The company is engaged primarily in the blending and marketing of tea and coffee, though Ringtons also has a subsidiary involved in electric vehicle manufacture.

Roach Foods Ltd

T/O £72.1m; pre-tax profit £0.62m; Emps = 578 (31/12/92)

Roach is an independent supplier of bacon and ham, operating in three areas: retail distribution; catering distribution; bacon and ham production, slicing and packing. Its original factory and head office are at Ebenezer, Bugle near St Austell in Cornwall. In 1991, it was awarded a DTI grant of £9m to extend its Redruth bacon slicing factory and create 180 jobs in the area. Additional cold storage facilities have been added to the Bodmin distribution centre.

Robertson & Baxter Ltd

T/O £90.55; pre-tax profit £27.35; Emps = 472 (31/12/92)

Though Robertson & Baxter is better known for its whisky, the Scottish company is also involved in food manufacture through Arran Provisions and Paterson-Bronte. Arran Provisions, set up in 1981 on the Isle of Arran, moved to custom-built premises in 1989 and has doubled its output of mustard, jams and preserves. Approximately 70% of its output is sold by leading high street and specialist food stores under their own label. The company employs 60 people and has a turnover of about £3m.

Paterson Bronte was formed in 1984 by the merger of Paterson with a Yorkshire-based company, Bronte, the combined company being acquired by the Argyll Group before being bought by Roberston & Baxter in 1987. It produces a range of quality biscuits from premises at Livingstone in Scotland where it employs a workforce of about 280, supplying major multiples, cash and carries and the catering trade. Approximately 35% of its business is in private label products.

Rubicon Products Ltd

T/O £2.73; pre-tax profit £0.13m (30/4/91)

T/O £–; pre-tax profit £0.23m (30/4/92)

Rubicon was set up in 1982 as a manufacturer of still and sparkling ethnic soft drinks. Its first product was called Rubicon Passionade. It was followed by Mango Crush made from imported mango pulp. Now Rubicon's Exotic canned carbonated drinks come in passion fruit, guava and mango flavours and in the diet versions. Similar flavours are available in 1-litre Tetra Pak cartons, as well as Coconut Water, a new product launched at the Chilled Food Fair in April 1993.

S

S & A Foods Ltd

T/O £10.5m; pre-tax profit £1.57; Emps = 115 (31/3/93)

S & A Foods started life in the kitchen of Purween Warsi. Subsequently the company became part of the Hughes Food Group. When Hughes went into liquidation in 1991 Sahib and Purween Warsi bought the company back in a deal funded through 3i for £2.3m. The company manufactures Indian and Chinese ready meals in chilled and frozen forms under the Shahi amd Laziz brands. It also produces many own-labels. Demand for S & A products means that the company is expanding into a new factory and creating 150 new jobs.

Saxby Bros. Ltd

T/O £13.06m; pre-tax profit £0.13m; Emps = 370 (4/4/93)

The business was founded in 1904 by two brothers, Ted and Herbert Saxby, and the company still occupies the same site on which the brothers built their original factory in 1920. Today, Saxby is best known for its Melton Mowbray pork pies of which annual production is about 10 million. The company processes 600 pig carcasses per week in the production of pork pies, sausages, quiches, pizzas and scotch eggs. Following a £4m extension in 1988, the company has moved into vegetable- and fruit-filled pies and pastries. As well as producing for the larger multiples, Saxby sells through its own retail outlets. They have recently developed a new catering range under the Irchester Grange Fine Foods brand. After 7 months trading sales have reached £1m.

Schwan's Europe Ltd

A subsidiary of Schwan's USA, the company is experiencing rapid growth in Europe with its Great Texas Creations and Chicago Town pizzas. It took over Take to Bake International, a pizza base manufacturer, which gave it a base in Leyland, Lancashire and was incorporated in May 1993. It already sells its products through Asda, Tesco, Safeway, CRS and Iceland and plans to add garage forecourts and convenience stores to this list.

Sco-Fro Foods Ltd

T/O £8.6m (1991); T/O £–; pre-tax profit £0.39m; Emps = 12 (30/4/92)
While marketing a wide range of canned vegetables through its grocery division, Sco-Fro Foods also claims to be Scotland's leading seafood specialist. Much of the company's turnover comes from the export market, to the extent that it won a Queen's Award for Export Achievement in 1988. Sco-Fro's biggest customers are the French, Italian and Spanish. While scampi and scallops formed the original base of the business, it is now developing markets for surimi and warm-water prawns. It has also opened a new salmon smoking unit at the Motherwell Food Park.

Scotch Premier Meat Ltd

During the latter half of 1993, Aberdeen Meat Marketing, the Inverurie Scotch Meat Co. and the Edinburgh Meat Plant were merged to form Scotch Premier Meat Ltd. The new company is a member of the UK's leading farming co-operative the ANM Group Ltd (Aberdeen and Northern Marts).

Scotfresh Ltd – see Top Hat Holdings

R. & W. Scott Ltd

T/O £–; pre-tax profit £1.6m; Emps = 89 (31/12/92)
R. & W. Scott was founded in 1880 by two brothers, Robert and William, whose fruit growing farm is still owned by the Scott family. Post-war expansion resulted in the building of a chocolate factory next to the original preserves site. One of the company's biggest selling lines is the Scotbloc range of chocolate flavour cake coverings, the UK brand leader. Other products include: ice cream coating chocolate, chocolate drops, chocolate and hazelnut spreads, jams, marmalades, lemon cheese and canned fruit.

A 50% interest in the company has been acquired by the German company Schwartauer Werke GmbH & Co. through J.F. Renshaw. Scott's recently entered an agreement by which it will distribute Renshaw's ready-to-roll Regal Ice cake icing, and several other home baking ingredients, to cash and carries.

Benjamin Shaw & Son Ltd

T/O £66.25m; pre-tax loss £1.09m; Emps = 383 (31/1/93)
Benjamin Shaw & Son, formed in 1871 and one of the UK's largest soft drink manufacturers, ran into financial difficulties after investing £10m in a new high-speed canning facility at its Pontefract factory in 1991. Mounting interest charges coupled with problems with the new plant led to a pre-tax loss of £1.09m in 1992. In a complex refinancing package, through a new holding company – Orderpart – Rutland Trust now owns 84% of Shaws, and senior management the remaining 16%. The purchase price was £5.69m.

Ben Shaw makes own-label soft drinks for Tesco, Sainsbury and Safeway as well as Shaws, Suncharm, Fountain Head and Pennine Spring brands. Recent additions include a range of children's mini cans called Can-d-Pops. It also runs a retail delivery service and a draught dispenser drinks supply company. It has manufacturing sites in Huddersfield and Pontefract.

Shearway Foods Ltd

T/O £12.13m; pre-tax profit £0.218m; Emps = 108 (30/9/92)

Shearway packs and supplies frozen vegetables and fruits to multiple retailers and wholesalers. It has recently completed a £3m investment programme at its Orchard Services Ltd operation near Maidstone in Kent with a new cold store, packing facility, pallet store, QA laboratory and development kitchen. The factory has four packing lines capable of handling 16 000 tones of frozen fruit/year. The new factory underlines Shearway's commitment to move into more value-added areas of production.

In February 1994, Shearway became a partly owned subsidiary of the Belgian vegetable processor, Digrom. Another Digrom subsidiary, Ardovries, will merge its UK sales operation with Shearway.

Sidoli Ice Cream Ltd

Sidoli makes frozen confectionery and ice cream at a 50 000-sq.ft factory and 15 000-sq.ft cold store in Welshpool. It is a family business which has grown from 10 people in 1982 to around 160 today. The company recently won a Rural Enterprise Award from the Development Board for Rural Wales. It has successfully expanded outside local markets and now exports apple pies and gateaux to France and Spain. It specializes in products for hotels and restaurants.

The Silver Spring Mineral Water Company Ltd

T/O £26.12m; pre-tax profit £0.61m; Emps = 167 (31/1/92)

Improvements to the Silver Spring production plant in the late 1980s doubled capacity and sales and allowed it to expand its base as a leading own-label soft drinks supplier to supermarkets and cash and carries. Its brands include Bing (first made in 1856), Rola Cola and Spring Up. Sales of its 250 ml Fizz Kids increased by about 150% during 1993 summer months. The company now plans to launch the six brightly coloured flavours in 2-litre PET bottles.

Simpson Ready Foods Ltd

T/O £–; pre-tax profit £0.351m; Emps = 130 (31/1/92)

Originally founded as a business making mincemeat, lemon curd and jam, the company moved into canning meat just before the second world war. The Manchester-based company currently produces 35 branded meat lines and 24 for own-label. Products include steak and kidney pudding, beef and chicken casseroles, Irish Stew, minced beef, frankfurters, coq au vin and many others. It now markets a completely meat-free canned hamburger.

Ruddin Shelley, the Manchester-based import and trading company, was acquired in 1989.

A.C. Skelton & Sons Ltd

T/O £11.7m; pre-tax profit £0.052m; Emps = 741 (30/9/92)

Skelton is a bakery and butchery concern which began operating as a baker in 1931 but later diversified into meat in 1963. It now has a chain of 43 shops in the Humberside area selling a range of bread and flour-based confectionery: it also sells bacon, ham, scotch eggs and meat pies through its delicatessen counters.

F. Smales & Son (Fish Merchants) Ltd

T/O £45.42m; pre-tax profit £0.11m; Emps = 270 (30/8/92)

Smales claims to be the UK's largest independent fish and seafood company. Located close to one of the UK's major fishing ports in Hull, it has now opened its new £4m fish processing factory and cold store, funded in part by MAFF and EC grants.

The Billy Boy brand, established in 1975, is currently used for the company's catering products which accounts for most of Smales' production. The retail sector, however, is growing; products include value-added lines such as breaded fillets and glazed prawns.

Walter Smith (Birmingham) Ltd

T/O £18.52m; pre-tax profit £0.2m; Emps = 349 (3/4/93)

Birmingham-based Walter Smith operates a chain of multiple butchers outlets in and around Birmingham. Its factory at Ward End has been the subject of much recent investment culminating in a £250 000 upgrade with a view to obtaining EC approval. The decision to concentrate on its red meat activities means that pie production has ceased and the bakery been sold. Through its manufacturing arm – Walter Smith Fresh Foods – the company manufactures traditional sausages, cooked meats, burgers, and many minced and diced meats packed in controlled atmosphere packs under the Chase Farm label.

Walter Smith is a substantial supplier of regional multiples, Midlands Co-ops and wholesalers.

The group also has a chain of 40 specialist butchers shops in the Midlands and two farms, one at Sutton Coldfield and another at Ashbourne, Derbyshire.

The Snack Factory Ltd

T/O £20.8m; pre-tax profit £2.0m; Emps = 330 (31/12/92)

The company operates from a greenfield site at Skelmersdale in Lancashire, where it has invested £22m in improving facilities to allow it to respond quickly to market opportunities. It produces both own-label and branded snacks. In addition to its Quintins brand it now markets the Really Ruthless range of adult snacks which come in hot and spicy flavours. At the other end of the spectrum Noddy Toyland snacks are intended for children under seven. It also produces a low-fat range under the Jack Sprat label.

Solway Foods Ltd

T/O £13.29m; pre-tax profit £0.47m; Emps = 258 (31/3/93)

The company, formed in 1989, specializes in chilled salads and sandwiches for major supermarkets and foodservice outlets. 260 staff process 500 tonnes of salads and 750 000 rounds of sandwiches each week at its 50 000-sq.ft EC- and USDA-approved premises at Corby in Northants.

Turnover has increased by 50% each year since the start and this led to the company winning the 1993 Grocer/IFE Business Award. Of the £13m turnover about £2.4m is derived from own-label sandwiches – 16.6% of the market by volume. In salads, 73% of production is for own-label, the remainder being sold under the Salad Company brand. Predicted sales for 1993 should exceed £20m.

Speciality Foods Ltd

T/O £6.3m; pre-tax loss £0.173m; Emps = 171 (2/7/93)

Speciality Foods is a holding company set up to encompass the interests of Leeds-based kosher food manufacturer Rakusens, Kippax Biscuits of Colne in Lancashire, and Quintessential Catering. Other immediate subsidiaries include Vegetarian World Ltd and SF2 Ltd.

R. & R. Spink & Sons (Arbroath) Ltd

This long-established, family-owned fish processing company has recently invested £0.2m in its premises at Arbroath which produces a range of products from smoked salmon to peppered mackerel. The company supplies Wm. Low & Co., Scotland's leading retail chain, with all its fish products, as well as catering outlets and department stores.

Stella Products Ltd

T/O £–; pre-tax profit £0.35m; Emps =102 (30/9/92)

Stella Products, incorporated in 1985, operates from a £5m factory in Liverpool. The company supplies ice cream and ice lollies to cash and carries, freezer centres, and the retail multiples.

A.G. Stewart & Co. (Catering)

Emps = >100

Trading as Stewart of Buckie, the company is a catering and wholesale butcher supplying the Scottish catering trade with vacuum-packed fresh or frozen beef, lamb and pork and meat products.

St Merryn Meat Ltd

T/O £48m; pre-tax profit £0.8m; Emps = 271 (1992)

St Merryn Meat operates meat plants at Bodmin, Probus and Falmouth and is a major supplier of retail beef packs to Tesco. It has invested a total of £4.5m in its packing facilities at Falmouth and Bodmin to offer retailers a fully integrated service from slaughter, through primal cuts to retail packages.

It has recently bought Hillsdown's Torrington abattoir in North Devon, in a two-stage deal involving Anglo Beef Processors, who bought the abattoir and sold it on to St Merryn. Acquisition of Torrington will increase the company's output of primal cuts destined for retail packing, and will enable a move into pork processing in addition to beef and lamb. Torrington has EC approval to handle 5000 lambs, 3000 pigs and 800 cattle/week.

Its projected turnover for 1993 is around £90m.

Stonegate Farmers Ltd

T/O £56.6m; pre-tax profit £1.1m; Emps = 442 (3/10/92)

Stonegate is a major egg producer and packer, based at Hailsham in Sussex. Subsidiaries include Primrose Valley Eggs Ltd, Stonegate Winchester Ltd and Stonegate Breeders Ltd.

Strathmore Foods Ltd

T/O £4.6m; pre-tax profit £0.33m; Emps = 153 (30/6/92)

This is the new name of Strathmore Meats Ltd from July 1993 and it reflects the company's widening range of products under the Countryside and Tay Valley Patisserie brands. Countryside includes Scotch pies, black puddings and other savoury pies; Tay Valley covers frozen gateaux, cheesecakes and other desserts. Demand for the company's products has led to the extension of both its factories in Forfar, including £150 000 spent on a new production line for cheesecakes. Recent product launches include 16-portion Turkish Delight Gateaux for the catering sector and Tangy Macaroni and Cheese Pies. Countryside also has a new range of bake-off products for independent bakeries.

Surrey Hills Foods Ltd

T/O £9.7m; pre-tax profit £0.139m; Emps = 117 (31/3/93)

This London-based, privately owned company supplies pizza toppings and component meats to the ready meal, sandwich-making and fast food sectors.

Sweet 'n Savoury Frozen Foods Ltd

T/O £22.84m; pre-tax profit £0.46m; Emps = 585 (26/9/92)

Sweet 'n Savoury is an independent frozen food manufacturer, formed in 1990 as a result of a management buy-out of part of Campbell's Foods, with a factory in Hartlepool producing ready meals, burgers, breaded and fried foods, gateaux and dairy cream sponges. Equipment for dessert and gateaux making was bought from Allied Frozen Foods' Calne factory in 1992. The company makes many own-label products and claims a 16% supply share of the frozen gateaux market and a 30% share of the frozen cream sponge market. Its brand names are Sweet Carolines and Prime Chef.

William Sword Ltd

T/O £–; pre-tax profit £0.178m; Emps = 148 (1992)

William Sword was founded in the last century, supplying bread rolls, pies and cakes to the Lanarkshire mining town of Airdrie. The company moved to a new factory in Cumbernauld which now employs a staff of 148 operating two fully automated pastry-producing plants and a small butchery department producing scotch pies, Forfar bridies, sausage rolls and steak pies sold to the wholesale and retail sectors. An area from Thurso to Newcastle is served by the company's fleet of sales vans.

T

Taveners plc

T/O £12.3m; pre-tax profit £0.21m; Emps = 252 (31/12/92)

Liverpool-based Taveners is one of the UK's oldest manufacturers of sugar confectionery. The product range includes various children's lines, Princess marshmallows and the newly-launched chocolate covered marshmallow 'Trello'. Since the closure of the Glasgow factory in 1989, production has been concentrated at the Liverpool site.

In 1989 Oy Karl Fazer AB, a Finnish manufacturer of chocolate and sugar confectionery, acquired the 30% shareholding previously held by A.G. Barr. The £400m turnover Finnish company has exclusive rights to sell Taveners' products in Scandinavia. Exports currently account for about 30% of the company's turnover.

The ultimate parent company is Toms Fabrikker A/S.

Tendafrost Frozen Foods Ltd

T/O £36.31m; pre-tax profit £–; Emps, not disclosed (30/6/92)

This Lincolnshire-based company, set up in 1978, claims to be the UK's largest independent processor of frozen vegetables with 24% of the market, a position which was consolidated by the acquisition of Scotfresh (Eyemouth) in 1989 which added 25 000 tonnes to Tendafrost's existing annual production of 65 000 tonnes of fruit and vegetables. Tendafrost controls about 150 000 acres of land.

In 1991, working with Intermerk in Germany, Tendafrost's Scotfresh operation has developed a range of 12 different family-sized pizzas and now claims 13% of the market.

Conservation Grade vegetables have also been added to Tendafrost's standard range (Scottish Harvest label). These are produced in an environmentally friendly manner with the absolute minimum use of pesticides. Tendafrost is also planning to diversify into canned vegetables.

Thames Valley Eggs Ltd

T/O £60m; pre-tax profit £4.11m; Emps = 445 (26/9/92)

Thames Valley Eggs (TVE) is Europe's largest shell egg packing and marketing company. In 1992 it bought the Vestey Group's 50% interest in South Western Egg Products.

In April 1993, it announced that it was restructuring in order to become a private limited company. The restructuring was needed to keep pace with changes in egg marketing and increased demand for free-range and barn-produced eggs.

Thayers Ice Cream Ltd

T/O £5m approx; Emps = 40 (1992)

Thayers Ice Cream was formed in Cardiff in 1937 as a grocery shop by a Mr Thayer, who later met an American serviceman who sent him a book on ice creams. The business grew, moving to its present site in 1968. In 1988 GrandMet-Eden Vale bought the plant and invested £2.5m in equipment to produce Ski frozen yoghurt. GrandMet subsequently sold Express/Eden Vale to Northern Foods, who in turn sold Thayers to its management team.

Thayers makes mainly premium ice cream, a few lollies and a small amount of non-dairy ice cream. Most of its business is through Londis, Spar Welsh wholesaler, Capper & Co., CTNs and delicatessens in Wales and the West Country. Although the plant installed by Express provides the company with the opportunity of developing the frozen yoghurt business, the age and design of the plant means the new management will probably have to invest in upgrading facilities to seriously expand the business to major retailers.

Thornley & Son Ltd

T/O £6.6m; pre-tax loss £0.01m; Emps = 160 (4/4/92)

This established, family-run meat company was founded in 1921. It operates a slaughterhouse, processing plant, four retail outlets and a wholesale distribution operation covering the north of England. The EC-approved slaughtering and boning hall has a throughput of about 5000 pigs/week. In addition to its best known product – black pudding – the company produces pork cuts, ham, bacon, pies and sausages.

Tillery Valley Foods Ltd

T/O £4.44m; pre-tax profit £0.428m; Emps = 142 (31/3/93)

Tillery Valley Foods makes a range of Asian vegetarian chilled and frozen meals under the Annapurna brand name. There are 24 different recipes, 12 based on pulses and 12 on vegetables. The dishes are manufactured particularly for hospitals, schools and social services. Tillery Vallery is a major cook–chill supplier to the NHS and a supplier of recipe dishes to major retailers.

Production capacity at the 3000-sq.m. plant in Abertillery is 60 000 frozen meals/day. Strict high- and low-risk food separation practices are carried out to ensure optimum hygiene standards are maintained.

Tinsley Foods Ltd

T/O £49.3m; pre-tax profit £2.98m; Emps = 1147* (31/3/92)

* includes farming and growing operations

This Spalding-based company is involved in farming, vegetable growing and pre-packing, and food manufacture. Most of its production is own-label for a major high street retailer. The prepared foods department employs about 400 people producing a range of salads and ready meals.

Top Hat Holdings Ltd

T/O £6.87m; pre-tax profit £0.37m; Emps 166 (26/6/92)

Top Hat Holdings is a collection of companies formerly held by the farmer's co-operative Scotfresh, which became a limited company in 1986. Since then it has sold its plant at Eyemouth to Tendafrost Frozen Foods. The remaining businesses – Top Hat Foods, Scotfresh Chilled Foods and Scotfruit – now form Top Hat Holdings. In 1992 Scotts Kitchen joined the group, trading as Cake & Co. and producing hand-made cakes.

Top Hat Foods claims to be Scotland's largest producer of ready meals under its Freezer Chef brand (worth £5m at rsp) and own-labels (which account for 40% of turnover). Its modern premises near Dundee have EC approval and have been extended to increase production capacity by 30%. Products cover a range of meat and vegetable-based products.

Scotfresh Chilled Foods has sandwich factories at Cupar in Fife and Denton in Manchester. It has recently acquired the goodwill of Britannia Fresh Foods, and transferred its sandwich making operations from Bathgate to Cupar.

In November 1993, Scotfruit bought the Rosemount Fruit business from Howgate Foods Ltd of Dundee for an undisclosed sum. Howgate plans to concentrate on its core cheese business. Scotfruit is a major supplier of raspberries, strawberries and blackcurrants.

Town & Country Petfoods Ltd

The company was set up in 1974 by current managing director, John Parkinson, a former chief buyer at Pedigree Petfoods. The company markets a range of canned cat and dog food, and also semi-moist pet foods. The Hi-life brand is now sold through a number of major retailers.

Treats Ice Cream Ltd

T/O £15.63m; pre-tax profit £1.33m; Emps = 270 (31/12/92)

Birds Eye Wall's sold Treats to its management in March 1991 for an undisclosed sum. Since then the new independent company has invested £7m in a new factory in Leeds (2200 sq.m.) and purchased land and buildings adjacent to the Crossgates site. Since the buyout it has continued to produce established products but has also capitalized on independent retailers' dissatisfaction over supplier reliability and dictation of terms and conditions by the major manufacturers.

Brand names include Strika, Festival, Juicy Lucy, Split and Refresher.

Tucker Foods Ltd

In 1991 Tucker's became part of the giant Italian snack food company, Unichips SpA of Milan. Its operations were merged with Highlander Snacks (formerly Forth Valley Foods) and its equipment sold to other group companies.

Tudor Dairies Ltd

T/O £848 890; pre-tax profit £22 740; Emps = 80 (31/3/91)

T/O and pre-tax profit not disclosed for 1992.

Tudor Dairies is a small ice cream business which belonged to UB Ross Young's until 1991 when it was bought by a team headed by Neil Smith. Since then, Tudor has developed into the UK's fourth-largest ice cream manufacturer. More recently Smith acquired Guanaria frozen desserts from Albert Fisher for £2m and this has been integrated into Tudor's operations at Henley-in-Arden.

The most recent development (Jan 1994) has been the purchase of Loseley ice cream from Booker plc.

U

ULN (UK) Ltd

T/O £14.03m; pre-tax loss £0.43m; Emps = 105 (31/12/92)

In 1991, ULN, a subsidiary of the French Dairy company Union Laitiere Normande acquired the former St Ivel cheese plant at Westbury in Wiltshire with the ultimate objective of supplying 30 000 tonnes of cheese/year. It has moved its cheese processing operations from Hemel Hempstead to Westbury.

Unique Cuisine Ltd

T/O £7.5m

Unique Cuisine is part of Steel Aviation Services, which is an airline caterer with branches at many airports. The factory at Prestwick Airport was originally a flight kitchen, but 7 years ago it started making frozen entrees for other companies, e.g. Sharwoods and realized the potential of the new markets open to it.

Eventually the flight catering began to assume less importance (now just 7% of turnover) and Steel changed its name to Unique Cuisine. Products and customers are very varied and will expand further on completion of its upgraded premises. They include multi-portion packs of ready meals (paella, curries, etc.) cooked bacon for sandwich manufacturers and frozen concentrated soups supplied to restaurants at Littlewoods, Asda, Safeway, Chef & Brewer and Pizzaland.

W

Waldens Wiltshire Foods Ltd

T/O £16.27m; pre-tax profit £0.13m; Emps = 519 (30/6/92)

This Trowbridge-based group operates in three main areas: packing lard, producing frozen ready meals and operating a chain of retail shops. The company moved out of butter packing in 1987 due to intense competition in a declining market.

Waldens is a major supplier of 'meals on wheels' dinners, producing 20 million frozen meals/year, not only for meals on wheels but also for schools and hospitals. It also has a range of retail frozen ready meals sold under the Waldens Wiltshire Farm label.

Its immediate sudsidiaries are Frigfreight Ltd, Ladydown Motors Ltd and Wiltshire Foods Ltd.

Walkers Nonsuch Ltd

T/O £4.34m; pre-tax profit £0.309m; Emps = 68 (31/3/93)

Walkers Nonsuch claims to be Britain's oldest independent toffee manufacturer. Its long-standing Andy Pack range has recently been updated with new colour co-ordinated packaging. It also manufactures a dairy fudge bar. Walkers' products are aimed at the premium end of the market.

Walkers Shortbread Ltd

T/O £28.86m; pre-tax profit £3.82m; Emps = 572 (31/12/92)

Founded at Aberlour-on-Spey in 1898 by Joseph Walker, the company is still independent and managed by the Walker family. Walkers is the largest employer on Speyside and the world's leading producer of shortbread. As well as shortcake, the company also produces oatcakes, rich fruit cakes, biscuits and meringue. It has also introduced chocolate-covered shortbread in triangles and rings. These are sold predominantly to independent outlets, distributed by its own sales force and delivery vehicles in Scotland, while distribution in England and Wales is by Petty Wood. Over 25% of its production is exported to markets in the EC, North America, Middle East and Far East.

Wander Ltd

T/O £48.77m; pre-tax profit £2.29m; Emps = 379 (27/12/92)

The company is a subsidiary of Swiss pharmaceutical company Sandoz Nutrition, which has invested £20m over the last 10 years or so, making the Kings Langley factory the biggest Wander factory in the world. Currently it processes 20 000 tonnes per annum of malt extract, approximately half of which is sold as an ingredient to other food manufacturers – for inclusion into various cereals, confectionery, bread and baby food products – with a 50/50 split between UK sales and export. The company has added other products – the Ovaltine Chocolate Collection and flavoured milk – to its best-known consumer product, Ovaltine.

Webbs Poultry & Meat Group (Holdings) Ltd

T/O £21.0m; pre-tax profit £1.12m; Emps = 360 (27/3/93)

Formed 34 years ago as a spent hen processing company, Webbs was the subject of a management buyout by present managing director, Mike Audin, in 1984.

The company handles 220 000 birds/week and operates through two divisions: Webbs Poultrymeat Division – frozen oven ready chicken, poultry meat for food manufacturers and oven-ready boiling fowl; Webbs Cooked Meats Division – further processed poultry products, chicken and turkey roll, cooked, diced and stripped poultry meat.

It has recently spent £0.5m on its EC-approved processing plant at Cullingworth, West Yorkshire. Newly installed equipment includes a 6000 birds/hour evisceration machine. It also has a further processed factory at Haworth.

Wells Soft Drinks Ltd

T/O £47.7m; Emps = 350 (31/3/92)

A member of the Robinson Group of companies, Wells is a major supplier of own-label soft drinks in the UK. From its modern plant at Tenbury Wells in Worcestershire it is able to respond particluarly rapidly to demand for new products and provides one-stop shopping to soft drinks buyers with a full range of soft drinks options from mineral waters and carbonates through to ready-to-drink products. The company claims to have one of the world's largest in-line bottle producing and filling facilities, and is one of the biggest UK manufacturers of PET containers.

It has launched a new range of sugar-free drinks in ready to drink and dilutable forms and also produces one of the best selling mineral waters, Stretton Hills.

Westlake Foods Ltd

Westlake Foods Ltd of Nottingham bought the Sheffield plant in 1988. Since then it has built a reputation based on the supply of high specification, seam boned meat for use in value-added foods. In 1992, Aberdeen Meat Marketing purchased of a 50% share in the Sheffield plant from Westlake Foods and more recently acquired a majority shareholding in Westlake Sheffield. The refurbished Sheffield plant is EC-approved and employs 70 staff.

The remaining part of Westlake Foods (now renamed Westlake International) is to concentrate on its core businesses of meat marketing and technical consultancy, worldwide. In 1992, Westlake International Ltd employed 87 people, and made a pre-tax loss of £0.13m on a turnover of £16.47m.

Westler Foods Ltd

T/O £26.01m; pre-tax profit £0.83m; Emps = 389 (31/3/93)

Westler Foods was formed over 30 years ago, supplying canned meat to the catering trade, which remains its main line of business. The Yorkshire-based company is perhaps best known for its range of hamburgers, hot dog sausages, meatballs and minces. Mexican foods are increasing in popularity and the company's canned chili con carne and 453-g bags of Nacho Chips under the Casa Fiesta brand are selling well. It also supplies vacuum-packed prime beef to supermarket chains and other meat processors.

Wilkin & Sons Ltd

T/O £8.4m; pre-tax profit £0.44m; Emps = 183 (31/12/92)

Wilkin & Son produces a range of about 60 products under the familiar Tiptree brand including conserves, jellies, lemon curd, marmalade, honey, chutney, pickles and relishes, with much of the fruit being grown on its own 1100-acre estate. The company supplies about 200 accounts directly via its own distribution system.

R. & K. Wise Ltd

T/O £14.7m; pre-tax profit £1.4m; Emps = 473

Wholesale baker and confectioner, Wise has recently bought a 102 000-sq.ft building on the Groundwell industrial estate in Swindon. The new site will provide an additional manufacturing unit near to its 87 000-sq.ft factory at Headlands Grove in Swindon.

Witwood Food Products Ltd

T/O £5.4m; pre-tax profit £1.12m; Emps = 29 (31/3/93)

Witwood is a small company making crumb coatings, batter mixes, non-pre-fry coatings, toppings for recipe dishes and microwave coatings.

John Wyeth & Brother Ltd

T/O £124.1m; pre-tax profit £18.52m; Emps = 1334 (30/11/92)

John Wyeth is involved in the manufacture and sale of pharmaceuticals, fine chemicals and infant foods. It currently holds about 40% of the £102.8m baby milks market with its range of SMA products. Its closest competitor in this market is Cow & Gate Nutricia which also holds 40%. Despite falling birth rates and the recession the baby foods market grew by 5% in 1992, mainly due to the development of 'follow-on' milks for babies over 6 months old. Since the Department of Health's recommendation that the use of cow's milk should be delayed well beyond 6 months if possible, the market for milks suitable for infants aged from 6 months to 2 years has grown rapidly.

Because the Wyeth name was not well known to consumers the baby foods operation has been re-named SMA Nutrition but is still based at the same address in Taplow, Berkshire.

The ultimate parent company is the American Home Products Corporation (USA).

X,Y,Z

Yeo Valley Farms (Production) Ltd

T/O £21.8m; pre-tax profit £1.02m; Emps = 270

Yeo Valley supply fresh-chilled yoghurts and desserts to the major high street retailers from their farms and production facility at Blagdon, south of Bristol.

Zest Foods Ltd

Having built a reputation for its Pesto sauces, Zest has now opened a new 5000-sq.m. factory at Newtown in mid-Wales to produce fresh herb sauces in sachets for Tesco and Safeway.

Part Three

COMPANY ADDRESSES

A

Abacus Frozen Foods Ltd, Station Road, Mintlaw, Peterhead, AB42 8EE; (0771) 22963; Fax (0771) 23279

Abel Eastern Foods Ltd, Unit 10, Lindsay Square, Deans South Industrial Estate, Deans, Livingston, Scotland EH54 8RL; (0506) 442020; Fax (0506) 441120

Abel Eastern Foods Ltd, Grange Road, Honstoun Industrial Estate, Livingston, Scotland EH54 5DD; (0506) 442020; Fax (0506) 441120

Abbey Cakes Ltd, Bradwell Abbey, Milton Keynes, Buckinghamshire; (0908) 311622; Telex 826932

Abercroft Cakes Ltd, Bentley Moor Lane, Adwick-le-Street, Doncaster, DN6 7BD; (0302) 721706; Fax (0302) 330627

Aberdeen & District Milk Marketing Board, PO Box 117, Twin Spires, Bucksburn, Aberdeen AB9 8AH; (0224) 696371; Fax (0224) 697749

Aberdeen Meat Marketing Ltd, Tillybraik Road, Banchory, Grampian AB3 3UN (03302) 3333; Telex 73685; Fax (03302) 3659

Abergavenney Meat Packers, Lancaster Road, Carnaby, near Bridlington, East Yorkshire YO15 3QX

AB Ingredients, Barn Way, Lodge Farm Industrial Estate, Northampton NN5 7UW

ABP – see Anglo Beef Processors

ABR Foods Ltd, 22 Sallow Road, Weldon Industrial Estate, Corby, Northamptonshire NN17 1JX; (0536) 65291; Telex 34695; Fax (0536) 402228

AB Technology Ltd, Salthouse Road, Bracknalls, Northamptonshire NN4 0BD; (0604) 768999; Telex 311848; Fax (0604) 701503

Acatos & Hutcheson plc, Orchard Place, London E14 0JH; 071-987 2066; Fax 071-538 1796 (registered office)

Ackermans Chocolates Ltd, Unit 9, St Pancras Commercial Centre, 63 Pratt Street, London NW1 0DP; 071-482 3731; Fax 071-482 4651

Acorn Confectionery Ltd, Brookside Road, Breadsall, Derbyshire DE7 6AF

Acorn Smokeries plc, Unit 5–7, New Rock Industrial Estate, Chilcompton, Bath BA3 4JE; (0671) 233294

Adams (Dalkeith) Ltd, Sherwood Industrial Estate, Bonnyrigg, Midlothian EH19 3LW; 031-663 1081

Adams Foods Ltd, Sunnyhills Road, Barnfields Estate, Leek, Staffs ST13 5SP; (0538) 399111; Fax (0538) 399918

Adams Pork Products Ltd, Crescent House, Fulney Lane, Spalding, Lincolnshire PE12 6EZ; (0775) 766161; Telex 32858; Fax (0775) 710717

Adams Pork Products Ltd, West End Road, Frampton, near Boston, Lincolnshire; (0205) 722488

Adams Pork Products Ltd, Sleaford Road, Ruskington, Lincolnshire; (0526) 832216

Addeys Confectionery (Derby) Ltd, Chequers Lane, Derby; (0332) 44094; Fax (0332) 382160

W.J. Adkins (Cambridge) Ltd, Church End, Cherry Hinton, Cambridge CB1 3LD; (0223) 244141; Telex 81574

Adpac – see CPC (UK Ltd), Adpac Division

AFB Bakeries – see Family Bakery

Agrimark Ltd, Agriculture House, 55 Bucks Road, Douglas, Isle of Man; (0624) 73731; Telex 629459

Ainsley's of Leeds Ltd, 20 Roundhay Road, Leeds LS7 1PZ; (0532) 451561

Allan & Dey Ltd, 45 Sinclair Road, Aberdeen AB1 3PL; (0224) 896000; Fax (0224) 872920

Allied Bakeries Ltd, Sunblest House, Fairfield Avenue, Staines TW18 4BA; (0784) 451366

Allied Bakeries (N. Ireland) Ltd, 58 Greenville Road, Belfast BT5 5JZ; (0232) 70545

Allied Breweries Ltd, Burton on Trent, Staffordshire

Allied Frozen Foods, Henwood Industrial Estate, Ashford, Kent TN24 8DU; (0233) 627061; Telex 965395; Fax (0233) 30040

Allied Frozen Foods, Kingshill Industrial Estate, Bude, Cornwall EX23 8QN; (0288) 353101; Fax (0288) 353103

Allied Frozen Foods, Lee Mill Industrial Estate, Lee Mill, Ivybridge, Devon PL21 9PE; (0752) 895050; Fax (0752) 896764

Allied Frozen Foods, Nursteed Industrial Estate, Nursteed Road, Devizes, Wiltshire, SN10 3DY; (0380) 725161; Fax (0380) 726871

Allied Frozen Foods, Porte Marsh Industrial Estate, Porte Marsh Road, Calne, Wiltshire SN11 9BW; (0249) 813708; Fax (0249) 821620

Allied-Lyons plc, 24 Portland Place, London W1N 4BB (registered office)

Allied Mills Ltd, Kingsgate, 1 King Edward Road, Brentford, Essex CM14 4HG; (0277) 262525; Telex 996500; Fax (0277) 200320

Allied Oils and Fats Ltd, Hoghton House, Hoghton Street, Southport, Merseyside PR9 0PP; (0704) 530777; Fax (0704) 544737

Alpine Soft Drinks Ltd, Richmond Way, Chelmsley Wood, Birmingham, West Midlands B37 7TT; 021-770 6816

Alpine Soft Drinks Ltd, Cleggs Lane, Worsley, Salford, Manchester

Alveston Kitchens, Warwick House, Birmingham Road, Stratford-upon-Avon, Warwickshire CV37 6SH; (0789) 68411; Telex 311282

Alvis Bros. Ltd, Lye Cross Farm, Redhill, Bristol BS18 7RH; (0934) 862320; Fax (0934) 862213

Ambrosia Creamery – see CPC (United Kingdom)

R.H. Amar & Co. Ltd, Unit 7, Corinium Estate, Raans Road, Amersham, Buckinghamshire HP6 6JJ; (0494) 721351; Telex 83387; Fax (0494) 431453

Leonard Ames (Ampthill) Ltd, Doo Little Mill Farm, Steppingley Road, Ampthill, Bedford MK45 2NX; (0525) 402256

Amin Enterprises, Woden Road, Wolverhampton WV10 0UB; (0902) 351104

AMS Turkeys Ltd, Sandlow Green Farm, Holmes Chapel, Cheshire CW4 8AS; (0477) 32329; Fax (0477) 537598

An Bord Bainne, Gratton House, Mount Street Lower, Dublin 2; 010 353 1 661 9599; Fax 010 353 1 661 2776/8

Anchor Foods Ltd, PO Box 82, Frankland Road, Blagrove, Swindon, Wiltshire SN5 8YZ; (0793) 32181; Telex 444463; Fax (0793) 616084

Anchor Seafoods Ltd, Devonshire House, Handcross, West Sussex RH17 6BJ; (0444) 400363; Fax (0444) 400949

Anglia Oils Ltd, King George Docks, Hull HU9 5PX; (0482) 701271

Anglian Tendabeef Co. Ltd – see Fribo Foods Ltd

Anglo-Austrian Patisserie Ltd, 16 Ravensbury Terrace, London SW18 4RL; 081-947 8607 or 081-947 7370 (sales)

Anglo Beef Processors Ltd, Northampton Road, Blisworth, Northampton NN7 3DR; (0604) 858001; Telex 311023; Fax (0604) 858581

Anglo Beef Processors Ltd, Irthlingborough Road, Wellingborough, Northants NN8 1RA

Anglo Beef Processors Ltd, Murton Lane, Murton, York YO1 3OS

Anglo Beef Processors Ltd, Battlefield Road, Harlescott, Shrewsbury SY1 4AH

Anglo Beef Processors Ltd (Wessex Foods), Common Lane North, Beccles, Suffolk NR34 9BL

Anglo Beef Processors (ABP Nuneaton) Ltd, St Mary's Road, Nuneaton, Warwickshire CV11 5AS; (0682) 383401; Telex 311155

Anglo Beef Processors (ABP Ellesmere) Ltd, Hordley, near Ellesmere, Shropshire SY12 9BL; (0939) 22333; Telex 35413

Anglo Beef Processors (Fermanagh), Drumclay, Enniskillen, County Fermanagh, Northern Ireland BT74 6BN; (0365) 23711; Telex 747384; Fax (0365) 23006

Anglo Beef Processors (Newry), Warrenpoint Road, Newry, County Down, Northern Ireland BT34 2PD; (0693) 63211; Telex 74375; Fax (0693) 63211

Anglo Beef Processors (ABP Scotland) Ltd, Birniehill Meat Complex, Whitburn Road, Bathgate, West Lothian, Scotland EH42 2HP; (0506) 632722; Fax (0506) 632802

Anglo Irish Beef Processors, 87 Charterhouse Street, London EC7M 6HJ

Anglo-Irish Meat Packers (Newry) Ltd, Greenbank Industrial Estate, Warren Road, Newry, County Down BT34 2PD; (0693) 3211

Antonelli Brothers Ltd, The Bakery, Weymouth Road, Winton, Eccles, Manchester M30 8WW; 061-789 4485 or 061-789 5592

Antrim Creameries Ltd, Church Road, Ballynure, Ballyclare, County Antrim, Northern Ireland (0960) 312276

Arcadian Confections Ltd – see Ernest Jackson & Co.

Armagh Down Creameries, Greenbank Industrial Estate, Warrenpoint Road, Newry, County Down BT34 2SJ; (0693) 62224

Geo. H. Armitage & Son, 29 Sandford Road, Balby, Doncaster, Yorkshire (0302) 853506

Armitage Bros. plc, Colwick, Nottingham NG4 2BA; (0602) 614984

Armour Foods (UK) Ltd, 193 St John Street, London EC1P 1EN; 071-253 8400; Telex 25261

Asda Group plc, Asda House, Southbank, Great Wilson Street, Leeds LS11 5AD; (0532) 435435; Telex 556623; Fax (0532) 418666 (registered office)

Ashbury Confectionery Ltd, Darwin Road, Willowbrook Industrial Estate, Corby, Northants NN17 5XZ; (0536) 401229; Fax (0536) 401176

Ashby Dairy Company, Smisby Road, Ashby de la Zouch, Leicester LE6 5UF; (0530) 412858

Ashdown Foods, 4 Canteloupe Mews, Canteloupe Road, East Grinstead, West Sussex RH19 3BG; (0342) 318888

Ashford Creameries Ltd, PO Box 2, Lowndes Road, Stourbridge, West Midlands DY8 3SR; (0384) 371551; Telex 336703

Ashford's of Stourbridge, Bradley Road, Stourbridge, West Midlands DY8 1UH

Ashmount Foods Ltd, Thornton Road, Thornton, Bradford, West Yorkshire BD13 3QG; (0274) 832965

Ashmount Foods Ltd, Armytage Industrial Estate, Brighouse, West Yorkshire HD6 1QF

Associated British Foods plc, Weston Centre, Bowater House, 68 Knightsbridge, London SW1X 7LR; 071-589 6363; Telex 263255; Fax 071-584 8560 (registered office)

Associated Canners Ltd, West House, Bixteth Street, Liverpool L3 9SR; 051-236 0363

Associated Co-operative Creameries Ltd, Whiteleys Road, Whitby, North Yorkshire YO21 3PB

Associated Co-operative Creameries Ltd, Chainbridge Road, Blaydon-on-Tyne NE21 5SZ

Associated Family Bakers Ltd, Wycombe Lane, Woodburn Green, Buckinghamshire HP10 0HH; (0628) 522061; Fax (0628) 523483

Associated Family Bakers (Surrey) Ltd, Chertsey Road, Byfleet, Surrey KT14 7AX; (0932) 353251; Fax (0923) 342100

Associated Family Bakeries, Second Way, Avonmouth, Bristol BS11 8DQ; (0272) 825145; Fax (0272) 825635

Associated Fisheries plc, The Gatehouse, 16 Arlington Street, London SW1A 1RD (registered office)

Associated Fresh Foods Ltd – see MD Foods plc

Attwell Farms Ltd, Seafield Lane, Beoley, Redditch, Worcestershire B98 9DB; (0527) 66191; Fax (0527) 62975

Audhali Farm Gate Fresh Poultry Ltd, Five Ways Road, Hatton, Warwick CV35 7HZ; (0926) 484552; Fax (0926) 484243

Thomas Auld & Son (Bakery), 5–9 Brisbane Street, Greenock PA16 8LS; (0475) 25288

Aulds (Foods) Ltd, Bothwell Road, Hamilton, Lanarkshire ML3 0DW; (0698) 283959; Fax (0698) 891609

Aulds (Foods) Ltd, 1 Bamford Avenue, Inchinnan Business Park, Inchinnan, Renfrewshire, PA4 9RG; 041-812 1126; Fax 041-812 1127

Authentic Indian Foods, Unit 1A, Stockport Trading Estate, Yew Street, Stockport SK4 2JZ; 061-480 8651; Fax 061-480 6249

Avana Bakeries Ltd, Wern Trading Estate, Rogerstone, Newport, Gwent; (0633) 892441

Avana Group, Avana Buildings, Pendyris Street, Cardiff, South Glamorgan CF1 7YH; (0222) 25521; Telex 497537; Fax (0222) 83263

Avebe UK Ltd, Thornton Hall, Thornton Curtis, South Humberside DN39 6XD; (0469) 32222; Fax (0469) 31488

Avonmore Country Ladle Soups, 55 Burscough Street, Ormskirk, Lancs L39 3ND; (0695) 571852; Fax (0695) 571853

Avonmore Foods plc, Avonmore House, Kilkenny, Eire; (010 353) 56 22907; Telex 80704; Fax (010 353) 56 21909 (registered office)

Ayrshire Foods Ltd, Eglinton Park, Kilwinning, Ayrshire KA13 7QB; (0294) 52141; Telex 778021

B

B & B Meat Products, 9 Second Avenue, Bletchley, Milton Keynes, Buckinghamshire ML1 1ED; (0908) 79776 or 76370

John C. Bain & Son, Duthie Road, Tarves, Ellon, Grampian (06515) 675; Telex 739669; Fax (06515) 794

A.J. Baker (Taunton) Ltd, Comeytrowe Manor, Taunton, Somerset TA4 1EE; (0823) 87891

Edward Baker Ltd, Cornard Mills, Sudbury, Suffolk CO10 0JA; (0787) 72353; Telex 987111

F.W. Baker Ltd, 1 Regent Place, Rugby, Warwickshire CV21 2PJ (registered office)

F.W. Baker Ltd, Crick, Northampton NN6 7TZ; (0788) 823711; Fax (0788) 822960

Banham Poultry Produce Ltd, Bunns Bank Industrial Estate, Old Buckenham Road, Attleborough, Norfolk NR17 1QB; (0953) 455001

A.J. & R.G. Barber Ltd, Maryland Farm, Ditcheat, Shepton Mallet, Somerset; (0749) 8666; Fax (0749) 86678

A.J. & R.G. Barber Ltd, (factory), Sandy's Moor, Taunton Road, Wivelsicombe, Somerset TA4 2JU; (0984) 24642; Fax (0984) 24353

F.E. Barber Ltd, Hanover Street, Manchester M60 4ES; 061-834 9089; Telex 667046; Fax 061-839 0037

Bard Bros., Romford Road, Forest Gate, London E7 8AH; 081-472 3338; Fax 081-552 9261

Barnaby Foods Ltd, 28–31 Hythe Road, London NW10 6RX; 081-969 0006

Barnett Confectioners Ltd, Salesbury Street, Nottingham NG7 2AE; (0602) 784642

Barnett & Foster Ltd, Denington Estate, Wellingborough, Northants NN8 2QJ; (0933) 226337; Telex 311342

Baron Fresh Food Services, Waleswood Food Factory, Kiveton, Sheffield, S31 8PF; (0909) 770861; Fax (0909) 515868

Baron Meats Ltd, Weasenham Lane, Wisbech, Cambridgeshire PE13 2RD; (0945) 61441; Fax (0945) 475105

A.G. Barr plc, 1306 Gallowgate, Glasgow G31 4DS; 041–554 1899; Telex 778807

A.G. Barr plc, North Road, Atherton, Manchester M29 0R; (0942) 882691

G. Barraclough (Soft Drinks) Ltd, Swaledale House, West Yorkshire Industrial Estate, Toftshaw Lane, Bradford BD4 6SX; (0274) 688002; Telex 517676; Fax (0274) 651088

Arthur Barras Foods Ltd, Unit 26, Abenbury Way, Wrexham, Clwyd LL13 9UZ; (0978) 660491; Telex 265871

Arthur Barras Foods Ltd, Unit 6, Antelope Industrial Estate, Rhydymwyn, near Mold, Clwyd, North Wales CH7 5HS; (0352) 83441; Telex 617091

Barratt & Co. Ltd, PO Box 80, Livesey Street, Owlerton, Sheffield, South Yorkshire S6 2AP; (0742) 349391; Telex 54343; Fax (0742) 334706

Barretts & Baird (Wholesale) Ltd, PO Box 7, The Abattoir, Harvills Hawthorn, West Bromwich, West Midlands B70 0TG; 021-556 1277; Telex 338427; Fax 021-556 1985 or 021-556 1989

Bartlett Foods (UK) Ltd, 13 West Street, Cromer, Norfolk NR27 9HZ; (0263) 51125

Bart Spices Ltd, York Road, Bedminster, Bristol BS3 4AD; (0272) 773474; Fax (0272) 720216

Basildon Dairy Foods Ltd, Oxford Road, The Grove, Wantage, Oxon OX12 7QD; (0235) 766221

Basildon Dairy Foods Ltd, Honeywood Road, Off Crane Farm Road, Basildon, Essex SS14 3HT; (0268) 520591; Fax (0268) 288502

Basildon Dairy Foods Ltd, Copse Road, Lufton Trading Estate, Yeovil BA22 8RN; (0935) 766666; Fax (0935) 31838

Bass plc, 66 Chiltern Street, London W1M 1PR; 071-486 4440; Fax 071-486 7186 (registered office)

Geo. Bassett & Co. Ltd, Livesey Street, Sheffield S6 2AP; (0742) 349391; Telex 54343; Fax (0742) 334706

A.J. Batten Ltd, Spindle Way, Crawley, Sussex (0293) 22241

Baughan's Foods Ltd, East Street, Braintree, Essex CM7 6JJ; (0376) 21141; Telex 987164

G.G. Baxter Ltd, Sarre, Birchington, Kent CT7 0LD; (0843) 46361; Telex 96408; Fax (0843) 47032

Baxters of Speyside Ltd, Fochabers, near Elgin, Grampian IV32 7LD; (0343) 820393; Telex 73327; Fax (0343) 820286

Bay Tree Candies, Unit 1F, Barbican Rise, Looe, Cornwall PL13 1QQ; (050 36) 2413

Beaverlac (Fine Foods) Ltd, Beverley Manor, 250 Park Avenue, Hull HU5 4DA; (0482) 493171

Beck Food Group Ltd, New Hammond Beck Road, Wyberton, Boston, Lincolnshire PE21 7TA; (0205) 352020; Fax (0205) 364048

The Beddington Fruit Co. Ltd, Highlands Hill, Swanley, Kent BR8 7NH; (0322) 62306; Telex 897109

The Beddington Fruit Co. Ltd, Liphook Way, Maidstone, Kent ME16 0lQ; (0622) 679245

Bee & Cee Foods Ltd, Valley Road, Birkenhead, Merseyside, L41 7ED; 051-653 7377; Telex 629651

Beech's Chocolates Ltd, Fletcher Road, Preston, Lancashire PR1 5AD; (0772) 792808; Telex 677130

Belgravia Salads Ltd, 54 A & B Minerva Road, Park Royal, London NW10; 071-968 3368

Ken Bell (International) Ltd, 22–40 Brentwood Avenue, West Jesmond, Newcastle-upon-Tyne NE2 3DH; 091-281 5311; Fax 091-281 0611

David Bell & Sons (Bakers) Ltd, 152 Shottskirk Road, Shotts, Lancashire (0501) 20222

Bellamy Seafoods Ltd, 4–6 Windmill Street, Peterhead, Grampian AB4 6LB; (0779) 75491; Telex 73697; Fax (0779) 70419

Belmor Confectionery Ltd, Wantz Road, Dagenham, Essex RM10 8PR; 081-595 0031; Telex 8951012

Belvoir Foods Ltd, Mantle Lane, Coalville, Leicestershire LE67 3DW; (0530) 36501; Fax (0530) 510708

Bendicks (Mayfair) Ltd, Moorside Road, Winnall, Winchester, Hampshire SO23 7SA; (0962) 844800; Telex 477037; Fax (0962) 841547

Beni Foods Ltd, Michigan Drive, Tongwell, Milton Keynes MK15 8HQ; (0908) 210410

Bennett's of Doncaster Ltd, Gunhills Lane, Armthorpe, Doncaster, South Yorkshire DN3 3EB; (0302) 832335; Fax (0302) 835338

Wm Bennett Frozen Foods Ltd, 28 Orgreave Close, Dore House Industrial Estate, Handsworth, Sheffield S13 9NP; (0742) 697791

Bensons Crisps plc, Marquis Street, Kirkham, Preston, Lancashire PR4 2HY; (0772) 683383 (registered office)

Bensons Crisps plc, Tregwilym Road, Rogerstone, Newport, Gwent NP1 9XP

Bensons Crisps plc, Freckleton Road, Kirkham, Preston, Lancashire PR4 3RB; (0772) 68383

Berisford International plc, Berisford Wing, 1 Prescot Street, London E1 8AY (registered office)

Berrydales Ltd, 4th Floor, Centre Heights, 137 Finchley Road, Swiss Cottage, London NW3 6JG (registered office)

Berrydales Ltd, Berrydale House, 5 Lawn Road, London NW3 2XS; 071-722 2866; Fax 071-722 7685

Berwick Salmon Fisheries Co. Ltd, 1 Main Street, Spittal, Berwick-on-Tweed TD15 1QY; (0289) 7474

Beswicks Ltd, New Mill Street, Littleborough, Lancashire OL15 8YL; (0706) 74331; Telex 63102

Betabake (Ipswich), Whitehouse Road, Ipswich, Suffolk IP1 5LZ; (0473) 240240

Betabake (Norwich), Diamond Road, Norwich, Norfolk NR6 6AN; (0603) 400620

B.G. Foods Ltd, Caxton House, 22 St Johns Hill, Sevenoaks, Kent TN13 3NP. National Sales Dept: (0732) 450701; Fax (0732) 460433 or 458127

Billy Boy Frozen Foods Ltd, West Dock Street, Hull, North Humberside HU3 4HL

Binghams (Cooked Meats) Ltd, 148–154 Western Road, Sheffield S10 1LD; (0742) 685390

Birchtrent Ltd, 290b Ealing Road, Alperton, Wembley, Middlesex; 081-903 4471

Birds Eye Wall's Ltd, Station Avenue, Walton-on-Thames, Surrey KT12 1NT; (0932) 228888; Telex 261255

Birds Eye Wall's Ltd, Martins Score, Lowestoft, Suffolk NR32 1JG

Birds Eye Wall's Ltd, Rant Score, Lowestoft, Suffolk NR32 1UN; (0502) 3131

Birds Eye Wall's Ltd, Ladysmith Road, Grimsby, South Humberside DN32 9SF; (0472) 359191; Fax (0472) 242719

Birds Eye Wall's Ltd, Hessle High Road, Hull, North Humberside HU4 6SH; (0482) 506081

Birds Eye Wall's Ltd, Walls House, Barnwood, Gloucester Gl4 7HB; (0452) 418588; Fax (0452) 392344

Birmingham Dairies, Island Road, Handsworth, Birmingham; 021-523 9211

Blackmore Vale Farm Cream Ltd, Wincombe Lane, Shaftesbury, Dorset SP7 8PJ; (0747) 51855

Blackwood & Son (Products) Ltd, 1 Cowley Road, Blackpool, Lancashire FY4 4NE; (0253) 63619; Fax (0253) 63940

Blue Bird Confectionery Ltd, Hunnington, Halesowen, West Midlands B62 0EN; 021-550 8011; Telex 337102; Fax 021-501 1521

Bluecrest Convenience Foods Ltd, Castle Park, Flint, Clwyd, North Wales CH6 5XA

Bluecrest Foods Ltd, Ladysmith Road, Grimsby, South Humberside DN32 9TN; (0472) 44781; Telex 527768

Blue Chilled Fish Ltd, Bluecrest House, Fish Dock Road, Grimsby, South Humberside DN31 3PF

Bluecrest Foods Ltd, Humberstone Road, Grimsby DN32 8HP; 0472 242242

Blue Prince Mushrooms (Orchidwood Mushrooms Ltd), Great Bellhurst, Hobbs Lane, Beckley, East Sussex TN21 6TU; (0797) 260411; Fax (0797) 260603

Boboli UK Ltd, 38 Raleigh Street, Walsall, West Midlands WS2 8RB; (0922) 612555

Bonds Confectionery Ltd, 800 Lea Bridge Road, Leyton, London E17 9DJ; 081-556 1311; Fax 081-539 6254

Bonnet & Sons, 38 Huntriss Row, Scarborough, North Yorkshire; (0723) 361033

Booker plc, Portland House, Stag Place, London SW1E 5AY; 071-828 9850; Telex 888169; Fax 071-630 8029 (registered office)

The Boots Co. plc, 1 Thane Road West, Nottingham NG2 3AA; (0602) 506111; Telex 377811; (0602) 687151 (registered office)

Borthwicks plc, Priory House, St John's Lane, London EC1M 4BX; 071-253 8661; Telex 23716; Fax 071-253 8417

La Boulangerie Francaise – see Falconis

Bourne Salads Ltd, Church Farm, Swinderby, Lincoln LN6 9LU

Bourne Salads Ltd, Tunnel Bank, Cherry Bank Road, Bourne, Lincolnshire PE10 0DJ

Bovril (CPC United Kingdom), PO Box 18, Wellington Road, Burton-on-Trent, Staffordshire DE14 2AB; (0283) 63781

Bow Tie Confectionery Products, 457 London Road, Westcliff-on-Sea, Essex SS0 9LG; (0702) 341159

G.D. Bowes & Sons Ltd, The Abattoir, Brandon Road, Watton, Thetford, Norfolk IP25 6LW; (0953) 881555; Telex 975309; Fax (0953) 882455

Jas. Bowman & Sons Ltd, Ickleford Mill, Hitchin, Hertfordshire SG5 3UN; (0462) 34401; Telex 826043

Bowyers (Wiltshire) Ltd, 55 Stallard Street, Trowbridge, Wiltshire BA14 8HH; (0225) 777367; Telex 444347; Fax (0225) 751076

Boxford Suffolk (Holdings) Ltd, Hill Farm, Boxford, Sudbury, Suffolk CO10 5NY; (0787) 210348; Fax (0787) 211396

BP Nutrition (UK) Ltd, Wincham, Northwich, Cheshire CW9 6DF; (0606) 41133; Telex 668994; Fax (0606) 41963

W.D. Brabin Ltd, Shaw Road, Speke, Liverpool L24 9JT; 051-486 4990; Fax 051-486 5938

Bradfords Bakeries, Birmingham Road, West Bromwich B71 4JH; 021-553 2988

Brains Frozen Foods, St George's House, Bayshill Road, Cheltenham, Glos. GL50 3AE; (0242) 236101; Fax (0242) 512084

Brake Bros. (Frozen Foods) Ltd, Enterprise House, Godinton Road, Ashford, Kent TN23 1EU; (0233) 637730; Fax (0233) 648200

Brandon's Poultry Ltd, Holly Bank Farm, Scropton, Derby DE6 5PS; (0283) 812396

Brandway Group Ltd, Bishopdyke Road, Sherburn-in-Elmet, Leeds, W. Yorkshire LS25 6JA; (0977) 684937; Fax (0977) 683654

Brash Bros Ltd, 6 Church Street, Waltham Abbey, Essex EN9 1DX; (0992) 788388; Fax (0992) 788404

Brays (Cardiff) Ltd, Units 2/3 Ty-Verlon Industrial Estate, Cardiff Road, Barry, South Glamorgan CF6 3BE; (0446) 736069

Brekkes Foods Ltd, Wm. Wright Dock, Hull, North Humberside HU3 4PB; (0482) 27396; Telex 52109 (registered office)

Bridge Farm Dairies Ltd, Worlington Road, Mildenhall, Bury St Edmunds, Suffolk IP28 7DU; (0638) 714212; Telex 81597 (registered office)

Bridgewater Country Foods Ltd, Bridgewater House, Bridge Street, Lisburn, Co. Antrim, Northern Ireland BT28 1XY; (0846) 677070; Fax (0846) 660008

Bristow's of Devon Ltd, Marsh Lane, Crediton, Devon EX17 1ET; (0363) 774631; Telex 42746; Fax (0363) 772792

Briter Beef Products Ltd, 17 Dolly Lane, Leeds LS9 7TU; (0532) 436078

British Bakeries Ltd, RHM Centre, PO Box 178, Alma Road, Windsor, Berkshire SL4 3ST; (0753) 857123; Fax (0753) 857123

British Bakeries Ltd, Mothers Pride Bakery, Main Road, Watnall, Nottingham NG16 1HB; (0602) 383391; Fax (0602) 458274

British Bakeries Ltd, Mothers Pride Bakery, Gain Lane, Thornbury, Bradford, West Yorkshire BD3 7DX; (0274) 665211

British Bakeries Ltd, Mothers Pride Bakery, Wheatfield Road, Westerhope, Newcastle-upon-Tyne NE5 5LB; 091-286 9831; Fax 091-214 5246

British Bakeries Ltd, Mothers Pride Bakery, Cale Lane, Aspull, Wigan, Lancashire WN2 1HD; (0942) 42371; Fax (0942) 498066

British Bakeries Ltd, Mothers Pride Bakery, Stephenson Way, Three Bridges, Crawley, West Sussex RH10 1UB; (0293) 519244; Fax (0293) 565523

British Bakeries Ltd, Mothers Pride Bakery, Toynbee Road, Eastleigh, Hampshire SO5 4YU; (0703) 644577; Fax (0703) 641612

British Bakeries Ltd, Mothers Pride Bakery, Belmont Road, Erith, Kent DA8 1JZ; (0322) 341144; Fax (0322) 350249

British Bakeries Ltd, Mothers Pride Bakery, Chaucer Road, Forest Gate, London E7 9NA; 081-472 9881; Fax 081-548 0842

British Bakeries Ltd, Mothers Pride Bakery, Oldfield Lane North, Greenford, Middlesex UB6 0AU; 081-578 6771; Fax 081-813 2596

British Bakeries Ltd, Mothers Pride Bakery, Granby Avenue, Garretts Green, Birmingham B33 0TL; 021-784 8211; Fax 021-624 2036

British Bakeries Ltd, Wimbush Bakery, Green Lane, Small Heath, Birmingham B9 5DA; 021-772 2151; Fax 021-766 6971

British Bakeries Ltd, Valley Road, Exwick, Exeter, Devon EX4 2AY; (0392) 79161; Fax (0392) 493404 (depot only)

British Bakeries Ltd, Mothers Pride Bakery, Burrington Road, Plymouth PL5 3LX; (0752) 775511; Fax (0752) 706519

British Bakeries Ltd, Mothers Pride Bakery, Apollo Road, Belfast BT12 6LP; (0232) 381131

British Bakeries Ltd, Glen Road, Londonderry BT48 0DA; (0504) 264122 (depot only)

British Bakeries Ltd, Mothers Pride Bakery, Seaforth Road, Ayr KA8 9JA; (0292) 268481; Fax (0292) 266919

British Bakeries Ltd, 338 Clepington Road, Dundee, Angus DD3 8SL; (0382) 84111 (depot only)

British Bakeries Ltd, Mothers Pride Bakery, 783 Duke Street, Glasgow G31 1LL; 041-556 5211; Fax 041-554 3508

British Bakeries Ltd, Mothers Pride Bakery, 1 Burnett Road, Inverness IV1 1RZ; (0463) 234511

British Beef Co. Ltd, Newport Pagnell Road, Hardingstone, Northampton NN4 0BN; (0604) 64911; Telex 311759

British Beef Co. Ltd, The Old Rectory, Cold Higham, Towcester, Northampton NN12 8LR

British Beef Co. Ltd, Hollow Road, Bury St Edmunds, Suffolk; (0284) 1212; Telex 888857

British Beef Co. Ltd, Padholme Road, Peterborough, Cambridgeshire; (0733) 68341

British Beef Co. Ltd, Church Road, Bamber Bridge, Preston, Lancashire; (0772) 39011

British Beef Co. Ltd, Flanshaw Way, Wakefield, West Yorkshire

British Cocoa Mills (Hull) Ltd, 145 Cleveland Street, Hull HU8 8AD; (0482) 24305

British Creameries Ltd, West Marton, North Yorkshire BD23 3UE; (0282) 843463; Fax (0282) 844850

British Edible Oils Ltd, Sussex House, Burgess Hill, West Sussex RH15 9AW

British Fish Canners (Fraserburgh) Ltd, Bath Street, Fraserburgh, Grampian AB4 5DY; (0346) 23191; Telex 73209; Fax (0346) 29247

British Pepper & Spice Co. Ltd, Rhosili Road, Brackmills Industrial Estate, Northants NN4 0LD

British Sugar plc, PO Box 26, Oundle Road, Peterborough PE2 9QU; (0733) 63171; Telex 32149 or 32273; Fax (0733) 63171 (registered office)

British United Turkeys Ltd, Warren Hall, Broughton, Chester, Cheshire CH4 0EW; (0244) 661111; Telex 61134; Fax (0244) 661105

British Vinegars Ltd, 87 South Lambeth Road, London SW8 1RE; 071-735 8100

B. & A. Britton Ltd, Tartan House, Lea Road, Waltham Abbey, Essex EN9 1AS

B. & A. Britton Ltd, Lammas Road, Leyton, London E10 7QU; 081-539 0531; Telex 896828

Britvic Soft Drinks Ltd, Britvic House, Broomfield Road, Chelmsford, Essex CM1 1TU; (0245) 261871; Telex 99173; Fax (0245) 267147

Britvic Soft Drinks Ltd, Swinnow Lane Industrial Estate, Bramley, Leeds LS13 4HT

Broadhurst Bakeries, Cake & Biscuit Factory, Gadbrook, Northwich, Cheshire (0606) 3321

Broadland Foods Ltd, 111 Green Barr St, Birmingham, West Midlands B9 4BB; 021-773 5955; Telex 333604; Fax 021-771 1207

William Brock & Son Ltd, Spiersbridge Avenue, Spiersbridge Industrial Park, Thornliebank, Glasgow G46 8NL; 041-638 6475; Fax 041-638 9146

Brooke Bond Foods Ltd, Leon House, High Street, Croydon, Surrey CR9 1JQ; 081-686 8899; Telex 264609; Fax 081-688 8652

Brooke Bond Foods Ltd, Redbourn, St Albans, Hertfordshire; (0582) 853111

Brooke Bond Foods Ltd, Kennington Lane, Ashford, Kent TN24 0LU; (0233) 25733

Brooke Bond Foods Ltd, Waterden Road, Hackney, London E15 2EE

R.F. Brookes Ltd, The Wern Industrial Estate, Rogerstone, Gwent NP1 9YE; (0633) 895931

R.F. Brookes Ltd, Magna Road, South Wigston, Leicester LE8 2ZH

Brookfield Foods Ltd, Unit 105, 10th Avenue, Deeside Industrial Park, Deeside, Clwyd CH5 2UA; (0244) 281481; Fax (0244) 281430

B. Brooks (Norwich) Ltd, Little Melton, Norwich, Norfolk NR9 3NP; (0603) 810137; Telex 975103; Fax (0603) 812272

B. Brooks (Norwich) Ltd, Somersham Road, St Ives, Cambridgeshire PE17 4LR; (0480) 300333; Fax (0480) 300718

F. Broomfield, 100 Blackheath Road, Greenwich, London SE10 8DD; 081-692 8181

Brossard UK, 3 Brunel Drive, Newark Industrial Estate, Newark, Notts NG24 2EG; (0636) 73831; Fax (0636) 705729

Brossard Memory Lane Cakes Ltd, Maes-y-Coed Road, Cardiff CF4 4XR; (0222) 623351; Fax (0222) 624327

Brown Brothers Manufacturing Ltd, Unit 1 Greystone Avenue, Kelloholm Industrial Estate, Kirkconnel, Dumfries & Galloway DG4 6RD; (0659) 66406; Fax (0659) 66414

Brown & Polson – see CPC (United Kingdom) Ltd

W.S. Brown & Sons Ltd, 27 Broad Street, Hanley, Stoke-on-Trent ST1 4HU; (0782) 24584

Henry Brown (Bakers) Ltd, Abbeyfield Road, Nottingham NG7 2SZ; (0602) 866473

Sarah Brownridge, Whole Country Foods, Little Catterton Lane, Tadcaster, North Yorkshire LS24 8EA; (0937) 835599; Fax (0937) 835897

Buchan Meat Producers Ltd, Markethill Road, Turriff, Grampian AB5 7HW; (0888) 63751; Telex 73328; Fax (0888) 63251

Buckingham Foods Ltd, Wimblington Drive, Redmoor, Milton Keynes MK6 4AH; Fax (0908) 699655

James Budgett Sugars Ltd, Beacon House, Rainsford Road, Chelmsford, Essex CM1 2PY; (0245) 261261; Fax (0245) 490824

Buitelaar (Kilmarnock) Ltd, Cambuskeith, Dundonald Road, Kilmarnock, KA2 0AZ; (0563) 27730; Fax (0563) 27460

Buitoni Foods, Wilson Road, Huyton, Liverpool L36 6AE; 051-480 7333

Burger King (UK) Ltd, 10 Windmill Road, Chiswick, London W4 1SD; 081-994 6454

Burlton (Farms) Ltd, Valley Farm, Meldreth, Royston, Hertfordshire SG8 6JP; (0763) 60445; Fax (0763) 262994

Burton, Son & Saunders Ltd, College Street, Ipswich, Suffolk IP4 1DE; (0473) 56234; Telex 98224

Burtons Gold Medal Biscuits Ltd, Columbia Centre, Market Street, Bracknell, Berkshire RG12 1LF; (0344) 412121; Telex 849437; Fax (0344) 487862

Burtons Gold Medal Biscuits Ltd, Quality House, Sighthill Industrial Estate, Edinburgh EH11 4HN; 031-453 6622; Telex 72508; Fax 031-458 5036

Burtons Gold Medal Biscuits Ltd, Llantarnam, Cwmbran, Gwent NP44 3YL; (0633) 482271; Telex 497723; Fax (0633) 871539

Buss Foods Ltd, Drury Lane, Ponswood Industrial Estate, Hastings, East Sussex TN34 1XL; (0424) 432321; Fax (0424) 426483

Buss Foods Ltd, 31 Queensway Meadow Industrial Estate, Meadows Road, Newport, Gwent

Butcher & Baker Foods Ltd, Crackley Way, Peartree Lane, Dudley, West Midlands DY2 0VW; (0384) 254389; Fax (0384) 456334

A.E. Button & Sons Ltd, Grange Farm, Redgrave, Diss, Norfolk IP22 1SA; (0379) 898345; Fax (0379) 890039

Buxted Chicken Ltd, Flixton, Bungay, Suffolk NR35 1NT; (0986) 892792; Telex 97447; Fax (0986) 895632

Buxted Fresh Foods Ltd, Dallamires Lane, Ripon, Yorkshire HG4 1TU; (0765) 2481; Telex 57743; Fax (0765) 4938

Bysel Ltd, Selby House, Batley Road, Heckmondwike, Yorkshire WF16 9ND; (0924) 403857; Telex 55293; Fax (0924) 405368

C

Cabana Soft Drinks Ltd, Cann Bridge Street, Higher Walton, Preston, Lancashire PR5 4DL; (0772) 34045; Telex 67214

Cadbury Ltd, PO Box 12, Bournville Lane, Bournville, Birmingham B30 2LU; 021-458 2000; Telex 338011; Fax 021-451 4139

Cadbury Ltd, Station Road, Somerdale, Keynsham, Bristol BS18 2BH; (0272) 861789

Cadbury Beverages, 25 Berkeley Square, London W1X 6HZ; 071-830 5027

Cadbury Beverages Northern Europe, 28 Clarendon Road, Watford, WD1 1JJ; (0923) 210500

Cadbury Schweppes plc, 1–4 Connaught Place, London W2 2EX; 071-262 1212; Telex 334413; Fax 071-262 1212 (registered office)

Cadec Ltd, Station Road, Rowley Regis, Warley, West Midlands B65 0LJ; 021-561 3366; Telex 335166; Fax 021-559 2429

Caledonian Seafoods Ltd, Station Road, Mintlaw Station, Grampian AB4 8EB; (07712) 23668; Fax (07712) 23072

Callard & Bowser Group, Waterton Industrial Estate, Bridgend, Mid Glamorgan CF31 3DJ; (0656) 58011; Telex 497585; Fax (0656) 60215

Calthwaite Dairy Products Ltd, Brunel House, Brunel Way, Carlisle, Cumbria CA1 3NQ; (0228) 518239

Calypso Soft Drinks Ltd, Tattenhall, Chester, Cheshire CH3 9PX; (0829) 70315; Fax (0829) 70166

Cambrian Soft Drinks Ltd, P O Box No. 5, Derby Street, Bolton BL3 6JF; (0204) 23251; Fax (0204) 23395

Campbell Grocery Products Ltd, Hardwick Road, King's Lynn, Norfolk PE30 4HS; (0533) 692266; Telex 81136; Fax (0553) 692177

Campbell's UK Ltd, Heathrow Boulevard, 284 Bath Road, West Drayton, Middlesex UB7 0DT; 081-564 8686; Fax 081-564 8620

Campbell Frozen Foods Ltd, Campbell House, 101 Broadway, Salford M5 2UQ; 061-953 1212; Telex 668228; Fax 061-848 7519

Canadian Pizza Crust UK Ltd, 6–8 Missouri Avenue, Weaste, Salford, Manchester M5 2NP; 061-736 7660; Telex 665888; Fax 061-745 7114

Canvin International Ltd, Canvin House, Meadow Lane, Cardington, Bedfordshire MK44 3SB; (0234) 838322; Telex 825909

Canvin International Ltd, The Mount, Lamberhurst, Kent TN3 8ES

Cargill UK Ltd, 3 Shortlands, London W6 8RT; 081-741 9090; Telex 298965

Carousel Meat Packers Ltd, Sidings Lane, Simondswood, near Kirkby, Liverpool L33 4YR; 051-548 9793; Telex 627968

Carousel Food Services Ltd, 6 Seddon Place, Skelmersdale, Lancashire WN8 8EB; (0695) 50500; Fax (0695) 29677

Carr's Milling Industries plc, Old Croft, Stanwix, Carlisle, Cumbria CA3 9BA; (0228) 28291; Telex 64347; Fax (0228) 511293 (registered office)

Carricks (Caterers), Ponteland Road, Cowgate, Newcastle-upon-Tyne NE99 1DD; 091-0286 9044

Carter Bond Ltd, Attercliffe Road, Sheffield, South Yorkshire S4 7XG; (0742) 21172

Carter's Drink Group Ltd, 366 Tamworth Road, Sawley, Long Eaton, Nottingham NG10 3AW; (0602) 735161; Telex 37731; Fax (0602) 736017

Carters Gold Medal Soft Drinks Ltd, Citrus Grove, Side Ley, Kegworth, Derby DE7 2FJ; (0509) 674915; Telex 341177; Fax (0509) 673461

Carylon Foods Ltd, Carylon House, Holmbush Industrial Estate, St Austell, Cornwall PL25 3JL

Case & Sons Ltd, Motcombe, Shaftesbury, Dorset SP7 9HJ; (0747) 53011 or 52314; Fax (0747) 51050

Castle Dairies Ltd, Pontygwindy, Caerphilly, Mid-Glamorgan CF8 3HU

Cauldron Foods Ltd, Units 1–2, Portishead Business Park, Portishead, Bristol BS20 9BF; (0275) 818448; Fax (0275) 818353

Cavaghan & Gray Ltd, Harraby Factory, London Road, Carlisle, Cumbria CA1 3EN; (0228) 24451; Telex 64420; Fax (0228) 23171

Cavaghan & Gray Ltd, Brunel House, Brunel Way, Carlisle, Cumbria CA1 3NQ; (0228) 518200; Fax (0228) 518215

Cawston Group Ltd, The Winery, Cawston, Norwich, Norfolk NR10 4BQ; (0603) 871444

Caxton Chocolate Co., Denmark Street, High Wycombe, Buckinghamshire HP11 2ES; (0494) 25291; Telex 837562; Fax (0494) 448321

C & D Continental Patisserie Ltd, Eton Hill Road, Radcliffe, Manchester M26 9XQ; 061-724 4000; Fax 061-724 7950

CDB Meats Ltd, Concord House, 41 Overy Street, Dartford, Kent DA1 1UP; (0322) 77844; Telex 886286; Fax (0322) 70260

Central Edible Oils, Britannia House, Dock Road, Birkenhead, Merseyside LH1 1HX; 051-630 4636; Telex 629603

Central Midlands Co-operative Society Ltd, Central House, Hermes Road, Lichfield, Staffordshire WS13 6RH; (0543) 414140; Fax (0543) 415813 (registered office)

Channel Foods Ltd, Lower Quay, Newham Industrial Estate, Newham, near Truro, Cornwall TR1 2DT; (0872) 71591

Charbonnel et Walker Ltd, 28 Old Bond Street, London W1X 4BT; 071-629 4396

Chaucer Foods Ltd, Unit 26, Brighton Street Industrial Estate, Freight Liner Road, Hull, North Humberside HU3 4UN; (0482) 588088; Fax (0482) 588082

Cheale Meats Ltd, Orchard Farm, Little Warley, Brentwood, Essex CM13 3EN; (0277) 811631; Telex 995121; Fax (0277) 810836

Cherry Valley Farms Ltd, Rothwell, Lincoln LN7 6BR; (0472) 371371; Fax (0472) 371545

Cherry Valley Farms Ltd, North Kelsey Moor, Caistor, Lincolnshire LN7 6HH; (0472) 371371

Cherry Valley Farms Ltd, Pelham Road, Cleethorpes, South Humberside DN35 7JZ; (0472) 692283

Cheshire Wholefoods Ltd, Stadium Industrial Estate, Sealand Road, Chester CH1 4LU; (0244) 375330; Telex 61168; Fax (0244) 390855

Chibnalls Bakeries, Springfield Road, Hayes, Middlesex UB4 0LH; 081-573 3601

Chic-O-Roll Co. Ltd, Unit 2, Lyon Industrial Estate, Lyon Way, St Albans, Hertfordshire AL4 0LL; (0727) 35251; Telex 8812296; Fax (0727) 33837

Childhay Manor, Blacknell Lane, Crewkerne, Somerset TA18 7HJ; (0460) 77422

Chiltern Hills, Bourne End Mills, Hemel Hempstead, Hertfordshire HP1 2RW; (0442) 865522; Telex 825075; Fax (0442) 865147

Chitty Group Ltd, The Abattoir, Slyfield Industrial Estate, Moorfield Road, Guildford, Surrey GU1 1RP

Chivers Hartley Ltd, The Orchard, Chivers Way, Histon, Cambridge CB4 4NR; (0223) 233333; Fax (0223) 234008

Church Farm Turkeys (Chichester) Ltd, Highleigh Road, Sidlesham, near Chichester, West Sussex PO20 7NRT; (0243) 56580; Fax (0243) 56758

Church's (Pork Butchers) Ltd, 224 High Street, Epping, Essex; (0378) 73231

Churchfields Dairies Ltd, Festival Avenue, Salisbury, Wiltshire SP2 9HF

Cima Foods Ltd, Portmanmoor Road, East Moors, Cardiff CF2 2HB; (0222) 465335

Citra Soft Drinks Ltd, Chantry Bridge Works, Wakefield, West Yorkshire WF1 5PL; (0924) 373651

City Bakeries, 180 Glentanar Road, Glasgow, Aberdeen G22 7UP; 041-336 5345

Classic Meats Ltd, 52 Doogary Road, Omagh, County Tyrone BT79 0BQ; (0662) 3201; Telex 747107; Fax (0662) 3201

Clearwater Products, Ludbridge Mill, East Hendred, near Wantage, Oxfordshire OX12 8LN; (0235) 833732; Fax (0235) 835586

Clifford Kent Ltd, Horizon Farm, Tremar, Liskeard, Cornwall PL14 6EA; (0579) 45938; Telex 45131; Fax (0579) 48581

Clifford Foods (Unigate), Western Road, Bracknell, Berkshire RG12 1QA; (0344) 425741; Telex 847567; Fax (0344) 48700

T.W. Clothier & Sons Ltd, Whitehouse Farm, Wyke Champflower, Bruton, Somerset; (0749) 812340

Coca-Cola & Schweppes Beverages Ltd, Charter Place, Vine Street, Uxbridge UB8 1EZ; (0895) 231313

Coca-Cola & Schweppes Beverages Ltd, Grosvenor Road, St Albans, Hertfordshire AL1 3AD; (0727) 59242; Telex 338011

Coca-Cola & Schweppes Beverages Ltd, Cray Road, Sidcup, Kent DA14 5DF; 081-302 8982; Fax 081-300 3320

Coca-Cola & Schweppes Beverages Ltd, Canvermoor Division, Don Pedro Avenue, Beckbridge Industrial Estate, Normanton, West Yorkshire WF6 1TD

Coca-Cola Bottlers (Ulster) Ltd, The Green, Lambeg, Lisburn, County Antrim, Northern Ireland BT27 5SS; (0846) 674231; Telex 748131; Fax (0846) 671049

Coles Traditional Bakery, 19 High Street, Saffron Walden, Essex CB10 1AT; (0799) 22126

H.H. Collins (Broadway) Ltd, 28 North Street, Broadway, Worcestershire WR12 7DT; (0386) 852374

Colman's of Norwich, Carrow, Norwich NR1 2DD; (0603) 660166; Telex 975225

Connors Seafoods Ltd, The Harbour, Boddam, Peterhead, Grampian AB4 7AU; (0779) 78991; Telex 739645; Fax (0779) 71014

Conrad & Essenhigh Ltd, 2 Commerce Way, Walrow, Highbridge, Somerset TA9 4AG; (0278) 783600

E. Cookson & Sons, Exhibition Bakery, Lytham, Lancashire FY8 5AD; (0253) 739010

Cooley Fresh Foods Ltd, Guy's Farm, Washway Road, Moulton Marsh, Spalding, Lincolnshire PE12 6LQ

Coombe Bakery, Gould Road, Twickenham, Middlesex TW2 6RT; 081-894 2121

Frank Cooper Ltd, Claygate House, Esher, Surrey; (0372) 62181

Co-operative Retail Society Ltd, 29 Dantzic Street, Manchester M4 4BA; 061-832 8152; Fax 061-832 7355 (registered office)

Co-operative Wholesale Society Ltd, PO Box 53, New Century House, Manchester M60 4ES; 061-834 1212 (registered office)

Co-operative Wholesale Society Ltd, Biscuit Factory, PO Box 7, The Pinnacles, Harlow, Essex CM19 5AU; (0279) 442721; Telex 817531; Fax (0279) 641314

Co-operative Wholesale Society Ltd, Milk Group, 334/338 Ravenhill Road, Belfast BT6 8GL; (0232) 458481; Fax (0232) 456755

Co-operative Wholesale Society Ltd, Milk Group, Lygon House, 50 London Road, Bromley, Kent BR1 3RA

Copella Fruit Juices – see Boxford Farms (Holdings) Ltd

Cornish Quality Bacon Ltd, Chapel Street, Redruth, Cornwall TR15 2BZ; (0209) 215134

Coronation Rock Co. Ltd, 11 Cherry Tree Road North, Blackpool, Lancashire FY4 4NY; (0253) 62366; Telex 677334

G. Costa & Co Ltd, Staffordshire Street, London SE15 5TL; 071-639 3456

Cotteswold Dairy Ltd, Northway Lane, Tewkesbury, Glos. GL20 8JE

Country Banquet Foods Ltd, Westlegate House, Westlegate, Norwich, Norfolk NR1 3LR; (0603) 760123; Telex 975426; Fax (0603) 760124

Country Choice Foods (Group) Ltd, The Mill House, Main Road, St Paul's Cray, Orpington, Kent; (0800) 521366

Country Fitness Foods, Whaley Bridge, Stockport SK12 7LY; (066 33) 2051

Country Maid Bakeries, Sunblest Bakery, Chester Road, Saltney Ferry, Chester CH4 8RN; (0244) 680200

Country Maid – Hot Plate Bakery, Dunnings Bridge Road, Netherton, Liverpool L30 6TG; 051-525 5959

County Dairies Ltd (Unigate), Langford Lane, Kidlington, Oxford OX5 1JB; (08675) 78111; Telex 837392; Fax (08675) 78334

C.D. Coveney & Sons Ltd, 209 Upper Fant Road, Maidstone, Kent ME16 8BY; (0622) 26214

Cow & Gate Nutricia Ltd, Cow & Gate House, Manvers Street, Trowbridge, Wiltshire BA14 8YX; (0225) 768381; Telex 449328; Fax (0225) 768847

Cow & Gate Nutricia Ltd, Glastonbury Road, Wells, Somerset BA5 1TP; (0749) 74949

Cow & Gate Nutricia Ltd, Whitehorse Business Park, Trowbridge, Wiltshire BA14 0XQ

Cox Seafoods Ltd, 2 March Road, Buckie, Grampian AB5 2BY; (0542) 831736 or 831220; Telex 739266; Fax (0542) 835113

CPC (United Kingdom) Ltd, Falside Road, Paisley PA2 2JX; 041-884 5111

CPC (United Kingdom) Ltd, Claygate House, Littleworth Road, Esher, Surrey KT10 9PN; (0372) 62181; Telex 27106; Fax (0372) 62181

CPC (United Kingdom) Ltd, Adpac Division, Heming Road, Washford Industrial Estate, Redditch, Worcestershire B98 0TS; (0527) 510801

CPC (United Kingdom) Ltd, Wellington Road, Burton-on-Trent DE14 2AB; (0283) 511111; Fax (0283) 510194

CPC (United Kingdom) Ltd, Ambrosia Creamery, Lifton, Devon PL16 0BB; (0566) 784361; Fax (0566) 784776

Craigmillar, Sussex House, Civic Way, Burgess Hill, West Sussex RH15 9AW; (04446) 46300; Telex 877293.

Craigmillar, Craigmillar House, Stadium Road, Bromborough, Wirral, Merseyside L62 3NU

Cranswick plc, The Airfield, Cranswick, Driffield, North Humberside YO25 9PF; (0377) 70649; Fax (0377) 70994

Cranswick Country Foods Ltd, Staithe Road, Preston, Hull, North Humberside HU12 8TD

M.A. Craven & Son Ltd, Candyland House, Poppleton Lane, York, North Yorkshire YO2 6BA; (0904) 792121; Telex 57748l; Fax (0904) 790613

Cravendale Foods – see MD Foods plc

D.S. Crawford Ltd, Wardieburn House, 496 Ferry Road, Edinburgh EH5 2DL; 031-551 1101; Telex 727369

R. Crawford & Co. Ltd, Bert Road, Bensham Lane, Thornton Heath, Surrey CR4 7EP; 081-684 5114

Creamery Fare Continental Ice Cream Ltd, 305 Elveden Place, London NW10 7SY; 081-965 9044

Creamola Food Products Ltd, 114 Cornwall Street, Glasgow 041-427 1241

Cregagh Foods Ltd, 90–92 Grosvenor Road, Belfast, Northern Ireland BT12 5DR; (0232) 222001; Telex 74203; Fax (0232) 241009

Cremanaze Salads Ltd, 56–60 New Cross Road, London SE14 5BB; 071-639 6671

Crispa Produce Ltd, East Preston, West Sussex BN16 3AD

Cromwells Chocolatiers Ltd, PO Box 60, Malvern, Worcestershire WR13 6RZ; (0684) 310910; Fax (0684) 310379

Crooks of Harleston Ltd, 39 Redenhall Road, Harleston, Norfolk IP20 9HE; (0379) 854262; Fax (0379) 854100

Crosse & Blackwell Ltd, St George's House, Croydon, Surrey CR9 1NR; 081-686 3333; Telex 23117

Crosse & Blackwell Ltd, Kirk Square, Peterhead, Aberdeenshire AB4 6RQ; (0779) 72071

Crossfield Foods Ltd, The Bakery, Crossfield Street, Blackburn, Lancashire BB2 3AD; (0254) 55495; Fax (0254) 663602

Crown Chicken Ltd, The Norwich Business Park, Whiting Road, Norwich NR4 6OJ

Crown Chicken Ltd, Green Farm, Edge Green, Kenninghall, Norwich, Norfolk NR16 2DR; (0953) 887736; Fax (0953) 888376

Crown Foods Ltd, Brock House, Grigg Lane, Brockenhurst, Hampshire SO42 7RE; (0590) 22773

Crystal Drinks Ltd, Wakefield Road, Featherstone, West Yorkshire WF7 5HH; (0977) 797171; Fax (0977) 791735

Cuisine Sans Frontieres, Weybourne House, Lenton Street, Alton, Hants GU34 1HH; (0420) 543279; Fax (0420) 542209

A.W. Curtis & Sons Ltd, Long Leys Road, Lincoln LN1 1DX; (0522) 27212

Curwin Poultry Ltd, Church Lane Farm, Three Mile Cross, near Reading, Berkshire RG7 1HB; (0734) 883493; Telex 847250

D

Daintee Chocolate Confectionery Co. (Blackpool) Ltd, Clifton Road, Marton, Blackpool, Lancashire FY4 4QB; (0253) 61201; Telex 67679; Fax (0253) 792006

Dairyborn Foods Ltd, Dairyborn House, Crescent Road, Luton, Bedfordshire LU2 0AL; (0582) 457979; Fax (0582) 400957

Dairy Crest Ltd, Portsmouth Road, Thames Ditton, Surrey KT7 0EH; 081-398 4155; Telex 8956671; Fax 081-398 9603

Dairy Crest Foods, Dairy Crest House, Portsmouth Road, Surbiton, Surrey, KT6 5QL; 081-398 4155; Telex 927604; Fax 081-398 0705

Dairy Crest Ingredients, Philpot House, Rayleigh, Essex SS6 7HH; (0268) 775522

Dairy Crest Ingredients, Crudgington, Telford, Shropshire TF6 6HY; (0952) 541373

Dairyfresh Desserts Ltd, Barton Hill Way, Torquay, Devon TQ2 8JG; (0803) 323966

Dairy Produce Packers Ltd, Milburn Road, Coleraine, County Londonderry, Northern Ireland BT52 1QZ; (0265) 56231; Telex 74632; Fax (0265) 56412

Dairy Products, Church Square, Bures, Suffolk CO8 5ED; (0787) 227451

Dairyvale International, 1 Somers Road, Reigate, Surrey RH2 9DU; (0737) 222261

Daisy Dairy Foods Ltd, Blackburn Road, West End, Oswaldtwistle, near Accrington, Lancashire BB5 4LT; (0254) 391511; Fax (0254) 394732

Dale Farm Dairies Ltd, 169 Upper Newtownards Road, Belfast, Northern Ireland BT4 3JA; (0232) 471341; Telex 747724; Fax (0232) 655461

Dale Farm Foods Ltd, Station Lane, Holme-on-Spalding Moor, York YO4 4AN; (0696) 60377

Dalehead Foods Ltd, Eastern Way, Bury St Edmunds, Suffolk IP32 7AB; (0284) 767126; Telex 818334; Fax (0284) 750537

Dalehead Foods Ltd, Cambridge Road, Linton, Cambridge CB1 6JD; (0223) 891000; Telex 818334; Fax (0223) 891117

Dalehead Foods Ltd, Coldhams Lane, Cherry Hinton, Cambridgeshire CB1 3JJ; (0223) 248004; Telex 818334; Fax (0223) 212296

Dalehead Foods Ltd, Hallam Way, Old Mill Lane Industrial Estate, Mansfield Woodhouse, Mansfield, Nottinghamshire NG19 9BG; (0623) 656065; Telex 818334; Fax (0623) 422955

Dalehead Foods Ltd, Oakleigh Green, Westerleigh, Bristol BS17 4QZ; (0272) 571919; Fax (0272) 573188

Dalepak Foods plc, Dale House, Leeming Bar, Northallerton, North Yorkshire DL7 9DQ; (0677) 24111 or 24222; Telex 587582; Fax (0677) 2443 (registered and head office)

Dalgety plc, 100 George Street, London W1H 5RH; 071-486 0200

Dalgety Foods Ltd, Mill Reef House, Cheap Street, Newbury, Berkshire RG14 5DD; (0635) 33001; Telex 849797

Dalgety Fresh Produce, Floods Ferry Road, Doddington, March, Cambridgeshire PE15 0UW; (0354) 77361

Daloon Foods (UK) Ltd, 162 Ewell Road, Surbiton, Surrey KT6 6HG; 081-390 5889; Telex 8952573

Daloon Foods (UK) Ltd, Brunel Drive, Newark, Nottinghamshire NG24 2EG; (0636) 701000; Telex 377775; Fax (0636) 72581

James Dalton (Seasonings & Spices) Ltd, Camval Road, Harrogate, North Yorkshire HG1 4PY; (0423) 885255

Percy Daltons Famous Peanut Co. Ltd, Old Ford Works, Dace Road, London E3 2PE; 081-985 9241; Telex 896993

Danby's Foods Ltd, The Old Cheese Factory, Masham, Ripon, North Yorkshire HG4 4EH

Danepak Ltd, Caxton Way, Thetford, Norfolk IP24 3SB; (0842) 4521; Telex 81288; Fax (0842) 65804

H.E. Daniel Ltd, Longfield Road, Tunbridge Wells, Kent TN2 3EY; (0892) 511444; Telex 957103

S. Daniels plc, South Quay Plaza, 183 Marsh Wall, London E14 9SH

S. Daniels plc, Swan House, Farndale Road, Staveley, Chesterfield, Derbyshire S43 3YN; (0246) 280065; Fax (0246) 280136

Danish Bacon plc, Parkway, Welwyn Garden City, Hertfordshire AL8 6NN; (0707) 323421; Telex 28485; Fax (0707) 371765

Danone Branch (UK), Eastman Way, Hemel Hempstead, Herts HP2 7DU; Fax (0442) 236856

Dansco Dairy Products Ltd, The Creamery, Aberad, Newcastle Emlyn, Dyfed

Dansco-Tolona Pizza Products, 7/8 Priorswood Place, East Pimbo, Skelmersdale, Lancashire WN8 9QB; (0695) 20662; Telex 627770; Fax (0695) 28806

Dartington Foods (UK) Ltd, Lee Mill, Ivybridge, Devon PL21 9PE; (0752) 896172; Telex 42551; Fax (0752) 896172

Davies Brook & Co. Ltd, Coningsby Walks, Catherine Street, Hereford HR1 2DZ; (0432) 272437

Jeffrey Davies & Davies Ltd, Unit C2, Arctic House, Rye Lane, Dunton Green, Sevenoaks, Kent TN14 5HW; (0732) 450948; Telex 957552

Dawn Foods (UK) Ltd, Worcester Road, Evesham, Worcs WR11 4QU; (0386) 41241; Fax (0386) 443608

Dawnfresh Seafoods Ltd, Bothwell Park Industrial Estate, Uddingston, Lanarkshire G71 6LS

Dawn Pac Ltd (Dawn Meats), Cross Hands Business Park, Cross Hands, Dyfed SA14 6RB; (0269) 831100; Fax (0269 831912

Dayla Ltd, 80–100, High Street, Aylesbury, Bucks HP20 1QZ; (0296) 20261

Daylay Eggs Ltd, The Moor, Bilsthorpe, near Newark, Nottingham NG22 8TS; (0623) 870384; Telex 377291; Fax (0623) 870657

Dayville Ltd, 78/92 Stamford Road, London N15 4PQ; 081-801 7331; Telex 8955329; Fax 081-808 5728

DCA Industries Ltd, Gatehouse Road, Aylesbury, Buckinghamshire HP19 3DL; (0296) 27272; Telex 83412; Fax (0296) 27426

Deans Farm Eggs Ltd, Lower Icknield Way, Aston, Clinton, Aylesbury, Buckinghamshire HP2 5JT; (0296) 630555; Telex 837146; Fax (0296) 630870

Deans Farm, Bridgeway House, Upper Icknield Way, Tring, Hertfordshire HP23 4JX; (0442) 891811; Fax (0442) 891880

Delifrance UK Ltd, Grand Union Centre, Unit 28, Ladbroke Grove, London W10 5BX

Delifrance UK Ltd, 17 Chartwell Drive, Chartwell Industrial Estate, Wigston, Leicestershire LE18 2FL; (0533) 571871

Dell Foods (Products) Ltd, Stocks Lane, Duckmanton, near Chesterfield, Derbyshire S44 5HZ; (0246) 822161; Fax (0246) 826717

Del Monte Ltd, Del Monte House, London Road, Staines, Middlesex TW18 4JD; (0784) 61555; Telex 295991; Fax (0784) 65301

Derbyshire Chilled Foods Ltd, Dunston Trading Estate, Foxwood Road, Chesterfield, Derbyshire S41 9RF

Derbyshire Chilled Foods Ltd, Dukeries Industrial Estate, Claylands Close, Worksop S81 7AJ; (0909) 482604; Fax (0909) 772797

De Roma Ice Cream Ltd, Chapel Lane, Wigan, Lancs WN3 4HJ

Derwent Lynton Co. Ltd, Siddals Road, Derby DE1 2QD; (0332) 365121; Telex 377281; Fax (0332) 43173

Derwent Valley Foods Ltd, Medomsley Road, Consett, County Durham DH8 6SU; (0207) 506416 or 501503; Telex 537543

Devon Crest Poultry Ltd, Blundell's Road, Tiverton, Devon EX16 4DF; (0884) 254211; Fax (0884) 252707

Devon Desserts Ltd, Minerva Way, Brunel Way, Newton Abbot; (0626) 334334; Fax (0626) 334255

Devon Quality Foods Ltd, 2 Week Lane, Shutterton Industrial Estate, Dawlish, Devon EX7 0PJ; (0626) 866405; Fax (0626) 888144

Devro Ltd, Moodiesburn, Chryston, Glasgow G69 0JE; (0236) 872261; Telex 778828; Fax (0236) 872557

Dewhurst Ltd, Dewhurst House, 24–30 West Smithfield, London EC1A 9EL

Dixon Pork & Bacon Co. Ltd, Conduit Street, Ashton-under-Lyne OL6 6UU; 061-339 6000

DMJ Ltd, Lidgate Crescent, Langthwaite Grange Industrial Estate, South Kirkby, near Pontefract, West Yorkshire WF9 3AP; (0977) 640355; Fax (0977) 649062

Alex S. Donald Ltd, Woodside Road, Bridge of Don, Aberdeen AB2 8EF; (0224) 703681

D'Or Produce Ltd, Dorstone Court, Dorstone, Herefordshire HR3 6AW; (0981) 550122; Telex 35309; Fax (0981) 550143

Dorset Cereals, Romans Building, Grove Trading Estate, Dorchester DT1 1ST; (0305) 269966; Fax (0305) 250065

Dorset Chilled Foods, Longmead Industrial Estate, Shaftesbury, Dorset SP7 8PL; (0747) 51511; Telex 418213; Fax (0747) 3401

Douwe Egberts UK Ltd, Douwe Egberts House, Manor Way, Borehamwood, Hertfordshire WD6 1QQ; 081-207 4400; Telex 926472; Fax 081-207 5487

Dove Valley Poultry Ltd, Blenheim Road, Ashbourne, Derbyshire DE6 1HA; (0335) 43564; Fax (0335) 300219

E.A. Draper Ltd, Freeth Street, Meadow Lane, Nottingham NG2 3GT; (0602) 860288; Fax (0602) 863042

D.& S. (Food Products) Ltd, Union Road, Bolton, Lancashire BL2 2HQ; (0204) 35247; Telex 635007; Fax (0204) 361578

D'Silva Foods, 2 Lime Road, Stretford, Manchester M32 8HT; 061-864 2836

T.M. Duche & Sons (UK) Ltd, Ford Lane, Pendleton, Salford, Manchester M6 6PB; 061-736 3041; Telex 668834; Fax 061-745 7312

F. Duerr & Sons Ltd, 15 Prestage Street, Old Trafford, Manchester M16 9LH; 061-226 2251; Telex 668579; Fax 061-232 0678

D.T. Duggins Ltd, The Abattoir, Tufnal Lane, Broad Green, near Bromsgrove, Worcestershire (0527) 71531; Telex 339678

Duncan's Bakery, 20 St Machar Lane, Aberdeen, Grampian AB2 3YU; (0224) 481322; Telex 739248

Duncan's of Scotland Ltd, Unit 1/2, Block 23, Motherwell Food Park, Bellshill, Lanarkshire ML4 3JA; (0698) 740010; Fax (0698) 740910

Dunhills (Pontefract) Ltd, PO Box 1, 26 Front Street, Pontefract, West Yorkshire WF8 1NJ; (0977) 701437; Telex 557121

E

Eatfresh Ltd, Olive House, Magnus, off Ninian Way, Tame Valley Industrial Estate, Tamworth, Staffordshire B77 5BY

Eden Vale, Warrington Road, Cuddington, near Northwich, Cheshire CW8 2SN; (0606) 883251

Eden Vale Food Ingredients Ltd, Unit 2, The Stables, Gadbrook Park, Rudheath, Northwich, Cheshire CW9 7RA; (0606) 350474

Edmonds Eccles Cakes Ltd, 390/392 Hyde Road, Manchester M12 4RP; 061-223 0618

Elisabeth the Chef Ltd, St Mary's Road, Leamington Spa, Warwickshire, CV31 3QE; (0926) 311531

Elisabeth the Chef Ltd, Caswell Road, Sydenham Industrial Estate, Leamington Spa, Warwickshire CV31 1OF

Elkes Biscuits, Dove Valley Bakeries, Uttoxeter, Staffordshire ST14 7BT; (0889) 563131; Telex 36438; Fax (0889) 565379

David J. Ellis Ltd, Riverside House, Penarth Road, Cardiff CF1 7YS; (0222) 374646; Telex 497866; Fax (0222) 222 778

Elmdale Foods Ltd, 31 Edison Road, Aylesbury, Buckinghamshire HP19 3TE; (0296) 5712

Elsenham Quality Foods Ltd, Elsenham, Bishop's Stortford, Hertfordshire CM22 6DT; (0279) 812442; Telex 81439; Fax (0279) 812715

Emborough Farm Foods Ltd, Old Down Farm, Emborough, Somerset BA3 4SA; (0761) 232222; Fax (0761) 233193

G. Embrey, Liverpool Road, Cross Heath, Newcastle, Staffordshire ST5 9HT; (0782) 717373

Emley Farm Chilled Foods Ltd, Unit 5, Hoyle Head Mills, New Street, Earls Heaton, Dewsbury, West Yorkshire WF12 8JJ; (0924) 461442

English Provender Co., Aldreth Farm, Aldreth, Isle of Ely, Cambridgeshire CB6 3PG; (0353) 740069; Telex 32496; Fax (0353) 740151

Ensors, The Abattoir, Valley Road, Cinderford, Gloucestershire GL14 3MX; (0594) 23148

Eryri Food Products Ltd, Celtic House, Gaerwen Industrial Estate, Anglesey, Gwyndd LL60 6HR; (0248) 421050

Evedale Foods Ltd, New Mill Street, Littleborough, Lancashire OL15 8YL; (0706) 713200; Fax (0706) 713303

Evedale Foods Ltd, Harrogate Road, Apperley Bridge, Bradford BD10 0LY

Everest Frozen Foods Ltd, Heath Mill Road, Wombourne, West Midlands WV5 8E; (0902) 894002

Express Foods Ltd, Keaton House, Widewater Place, Moorhall Road, South Harefield, Middlesex UB9 6NS; (0895) 825999

F

Faccenda (Chicken) Ltd, Willow Road, Brackley, Northamptonshire NN13 5HB; (0280) 703641; Telex 837745; (0280) 705 335

FairFax Meadow Farm Ltd, 6 Newmarket Drive, Osmaston Park Industrial Estate, Derby DE2 8SW

Falconis Ltd, Unit 5, Second Way, Wembley, Middlesex HA9 0TP; 081-900 005; Telex 935347

Family Bakery (AFB Bakeries), Second Way, Avonmouth, Bristol BS11 8DQ; (0272) 825145; Fax (0272) 825635

Family Bakery (AFB Bakeries), Chertsey Road, Byfleet, Surrey KT14 7AX; (0932) 353251; Fax (0932) 342100

Family Bakery (AFB Bakeries), 88 Wycombe Lane, Wooburn Green, High Wycombe, Bucks HP10 0HH; (0628) 522061; Fax (0628) 523483

Farley Health Products Ltd, Mint Bridge Road, Kendal LA9 6NL; (0539) 723815

Farm Crest Ltd, 40 Annesbrough Industrial Estate, Lurgan, Craigavon, County Armagh BT67 9JD; (0762) 326111; Telex 747553

Farm Fed Chickens Ltd, Gateside Road, Coleraine, County Londonderry, Northern Ireland BT52 2PB; (0265) 52485; Telex 74503

Farmers Boy Ltd, Young Street, Bradford, West Yorkshire BD8 9TD; (0274) 42211

Farmer's Glory Ltd, Marylands, Newby Wiske, Northallerton, North Yorkshire DL7 9ER; (0609) 771711; Fax (0609) 780459

Farmhouse Biscuits Ltd, Pasture Lane, Barrowford, near Nelson, Lancashire BB9 6QZ; (0282) 63520; Telex 63201; Fax (0282) 411204

Faroe Seafood Co. Ltd, Estate Road 5, South Humberside Industrial Estate, Grimsby, South Humberside DN31 2TG; (0472) 361251; Telex 527220; Fax (0472) 55368

Farquhar, North & Co. Ltd, Cray Avenue, St Mary Cray, Orpington, Kent BR5 3QB; (0689) 20245

John Farrah & Harrogate Toffee Ltd, 7 Royal Parade, Harrogate, North Yorkshire HG1 2SZ; (0423) 65014

Faux Brand Foods, Luckyn Lane, Pipps Hill Industrial Estate, Basildon, Essex SS14 3AX; (0268) 293333; Fax (0268) 287950

Favour Parker Ltd, The Hall, Stoke Ferry, King's Lynn, Norfolk PE33 9SE; (0366) 500911; Telex 81135; Fax (0366) 500907 (registered office)

Favourite Food Products Ltd, 2–3 Prospect Place, East Pimbo, Skelmersdale, Lancashire; (0695) 25451; Telex 627059

Fawcetts Frozen Foods Ltd, PO Box 4, Richmond, North Yorkshire DL10 4RY; (0748) 850660; Fax (0748) 850157

Fenland Foods, Turnpike Close, Earlesfield Estate, Grantham, Lincolnshire NG31 7XU; (0476) 67733

Fenton Barns (Scotland) Ltd, Fenton Barns, North Berwick, East Lothian EH39 5AQ; (062 085) 202; Fax (062 085) 304

James Ferguson Ltd, 1/5 Telfair Street, Belfast BT1 4DY; (0232) 324871

Fermanagh Creameries Ltd, Eagle Mount, Lisnaskea, County Fermanagh, Northern Ireland (03657) 21501; Telex 747442; Fax (03657) 22329

Ferndale Foods Ltd, Thamesmead Eastern Industrial Estate, 40 Hailey Road, Erith, Kent DA18 4AR; 081-310 7991; Fax 081-311 6301

Ferrero Ltd, High Street, Rickmansworth, Herts WD3 1DF; (0923) 890501; Fax (0923) 890555

Fiddes Payne Ltd, The Spice Warehouse, Pepper Alley, Banbury, Oxford OX16 8JB; (0295) 53888

A.C. Fincken & Co. Ltd, Imperial Way, Station Estate, Waterford, Hertfordshire WD2 4JB; (0923) 34473

Findus (UK) Ltd, Benton Lane, Long Benton, Newcastle-upon-Tyne NE12 8EF; 091-266 4001; Fax 091-266 0000

Fine Lady Bakeries Ltd, Southam Road, Banbury, Oxon OX16 7RR; (0295) 250333

James Finlay plc, Finlay House, 10–14 West Nile Street, Glasgow G5 2PP; 041-204 1321; Telex 777844; Fax 041-248 4751 (registered office)

Albert Fisher Group plc, 'C' Sefton Park, Bells Hill, Stoke Poges, Bucks SL2 4HS; (0753) 677877

Fisher Frozen Foods Ltd, Focus 31 Mark Road, Hemel Hempstead, Herts. HP2 7BW; (0442) 611116; Fax (0442) 212302

Fisher Quality Foods Ltd, Timothy's Bridge Road, Stratford upon Avon, Warwickshire CV37 9YL; (0789) 299565; Fax (0789) 261413

William Fishwick & Sons Ltd, Liverpool Road, Ashton-in-Makerfield, Wigan, Lancashire

Fivemiletown & Brookborough Co-op Agricultural & Dairy Society, Main Street, Fivemiletown, County Fermanagh, Northern Ireland BT75 0RX; (03655) 21209; Telex 74296

Paul Flatman Ltd, Packards Lane, Wormingford, Colchester, Essex CO6 3AH; (0206) 240378

Fleming Poultry Ltd, 31 Ballymena Road, Ballymoney, County Antrim, Northern Ireland BT53 7EX; (02656) 62247; Telex 747250; Fax (02656) 65045

Fletchers Bakeries Ltd, Claywheels Lane, Wadsley Bridge, Sheffield S6 1LY; (0742) 348171

Food Brokers Ltd, Food Broker House, Northarbour Road, North Harbour, Portsmouth, Hampshire PO6 3TD; (0705) 219900; Telex 86779; Fax (0705) 219222

Food Enterprises Ltd, Unit 3, Grosvenor Grange, Woolston, Warrington, Cheshire; (0925) 825196; Fax (0925) 831411

Food Enterprises Ltd, Dunston Trading Estate, Foxwood Road, Chesterfield, S41 9RF; (0246) 260460; Fax (0246) 260461

Food For Thought (UK) Ltd, Cinder Hill, Horsted Keynes, West Sussex RH17 7BD; (0342) 810777

Food Ingredients Specialities, North Hyde Gardens, Hayes, Middlesex UB3 4SL; 081-569 2069; Fax 081-569 1623

Food Production Developments Ltd, 320–324 Kilburn High Road, London NW6 2QP

Forrester Foods Ltd, 4–5 Dallas Road Industrial Estate, Kempston, Bedford MK42 9EL; (0234) 217991; Fax (0234) 214200

Foston Farm Produce Ltd, Hay Lane, Foston, Derby DE65 5PJ; (0283) 585595; Fax (0283) 585313

Fountains Dairy Products Ltd, Kirkby Malzeard, Ripon, North Yorkshire HG4 3QD; (0765) 83212; Fax (0765) 83732

Four Square Division of Mars (GB) Ltd, Armstrong Road, Basingstoke, Hampshire RG24 0NU; (0256) 471500; Fax (0256) 56915

Fox's Biscuits Ltd, PO Box 10, Batley, West Yorkshire WF17 5JG; (0924) 444333; Telex 557792; Fax (0924) 470200

Fox's Biscuits Ltd, PO Box 3, Kirkham, Preston PR4 3AX; (0772) 683501

Fox's Glacier Mints Ltd, Sunningdale Road, Braunstone, Leicester (0533) 873561; Telex 341014

Framptons Ltd, Charlton Road, Shepton Mallet, Somerset BA4 5PD; (0749) 342831; Telex 44718; Fax (0749) 344997

Fraser Meats Ltd, Whitehouse Industrial Estate, Runcorn, Cheshire WA7 3BL; (0928) 712345

Frederick's Dairies Ltd, Propsect Place, East Pimbo, Skelmersdale, Lancashire WN8 9PJ; (0695) 25451; Fax (0695) 26111

Freebooter Seafoods Ltd, Gothenburg Way, Sutton Fields Industrial Estate, Hull, North Humberside HU8 0YG; (0482) 831176; Telex 527360

E. & B. Freeman (Newent) Ltd, Town Farm, Newent, Gloucestershire GL18 1HP; (0531) 820686

Freezia Foods Co. Ltd, 10 Cats St., Vauxhall Industrial Estate, Nechells, Birmingham B7 4TS; 021-359 1123; Fax 021-359 0353

Fresh Country Foods Ltd, M62 Trading Estate, Rawcliffe Road, Goole, North Humberside DN14 8JW; (0405) 768477; Telex 57629; Fax (0405) 5441

Fresha Bakeries Ltd, 480 Thurmaston Boulevard, Leicester LE4 7LN; (0533) 460800

Freshers Foods Ltd, The Kirklees Industrial Estate, Cale Lane, New Springs, Wigan WN2 1HB

Fribo Foods Ltd, Miners Road, Llay, Wrexham, Clwyd Ll12 0PB; (0978) 852161; Fax (0978) 854029

Fridays (Cranbrook) Ltd, Chequer Tree Farm, Benenden Road, Cranbrook, Kent TN17 3PN; (0580) 714014; Telex 957107; Fax (0580) 713512

Frobirch Ltd, PO Box 19, 76 Mosley Street, Burton-on-Trent, Staffordshire DE14 1DS; (0283) 61661

Frozen Quality Ltd (Albert Fisher), Fleet, Spalding, Lincolnshire PE12 8LR; (0406) 22727

J.S. Fry & Sons – see Cadbury Ltd, Bristol address

Furniss Group Ltd, Druids Road, Redruth, Cornwall TR15 3RW; (0209) 215425; Fax (0209) 215176

Fyffes Group Ltd, 12 York Gate, Regents Park, London NW1 4QJ; 071-487 4472; Fax 071-487 3644

G

Gainsborough Pork & Bacon Co., Coringham Road Industrial Estate, Gainsborough, Lincolnshire; (0427) 810100; Fax (0427) 611624

Galbani (London) Ltd, Grove House, 7 Weir Road, London SW12 0LT

Galloway Cheese Company Ltd, Commerce Road, Stranraer, DG9 7DA

J.W. Galloway Ltd, 27 Glenburn Road, College Milton North, East Kilbride G74 5BA (registered office)

Charles Gardiner Ltd, 56 Bermondsey Street, London SE1 3EU; 071-407 2295; Fax 071-407 4631

Gartner UK, Thornton Park, Somercotes, Derby DE55 4XJ; (0773) 608822

Garvin Honey Co. Ltd, 158 Twickenham Road, Isleworth, Middlesex TW7 7DL; 081-560 7171; Telex 935055; Fax 081-569 8036

Gaskell Bros. Ltd, The Bacon Factory, Woolston, Warrington, Cheshire WA1 4AY; (0925) 815000

Geest plc, White House Chambers, Albion Street, Spalding, Lincolnshire PE11 2AL; (0775) 761111; Telex 32481; Fax (0775) 710766 (registered office)

Geest Convenience Foods, Forward Drive, Christchurch Avenue, Harrow, Middlesex HA3 8NT; 081-427 7778

Gelatine Products Ltd, Sutton Weaver, Runcorn, Cheshire WA7 3EH; (0928) 716444; Telex 629303; Fax (0928) 718325

General Foods Ltd, Southan Road, Banbury, Oxon OX16 7QU; (0295) 4433; Telex 83196; Fax (0295) 59018

General Provisions Ltd, Langley Drive, Castle Bromwich, Birmingham B35 7AB; 021-748 4811

Gerber Foods International Ltd, Northway House, 1379 High Road, Whetstone, London N20 9LP; 081-446 1424; Telex 269379 or 886353; Fax 081-446 1575

Gerber Foods Manufacturing Ltd, 78 Wembdon Road, Bridgwater, Somerset TA6 7QR; (0278) 458151; Fax (0278) 426580

Gerber Frozen Foods, Pytchley Lodge Road, Kettering, Northants NN15 6JW; (0536) 511655; Fax (0536) 411115

G.F. Dietary Group of Companies Ltd, Lowther Road, Stanmore, Middlesex HA7 1EL; 081-204 6968

Gianova Foods Ltd, 5th Floor, 195 Brompton Road, London SW3 1LZ; 071-581 4911; Fax 071-581 8306

Gilbertson & Page Ltd, PO Box 321, Welwyn Garden City, Hertfordshire AL7 1LF; (0707) 371105

Giles Foods Ltd, Broadfields, Aylesbury, Buckinghamshire HP19 3AE; (0296) 395888 or 395262 (sales); Fax (0296) 395282

F.A. Gill Ltd, The Abattoir, Parkfields, Wolverhampton WV4 6EH; (0902) 331141; Fax (0902) 340772

Gill & Duffus Group plc, St Dunstan's House, 201 Borough High Street, London SE1 1HW; 071-407 7050; Telex 887162; Fax 071-403 4656

Gillard plc, Westbury Road, Walthamstow, London E17 6RJ; 081-521 8831; Fax 081-509 2689 (registered office)

S.W. Gilmour, 4 Commercial Road, Strathaven, Lanarkshire ML10 6LX; (0357) 21158

Ginsters Cornish Pasties Ltd, Callington, Cornwall PL17 7DU; (0579) 82711; Telex 45794; Fax (0579) 83001

Glass Glover Group plc, Loversall Hall, Loversall, Doncaster, South Yorkshire DN11 9DD; (0302) 850999; Telex 54325; Fax (0302) 855846

Glass Glover & Co. Ltd, Coldharbour Road, Elizabeth Way, Pinnacles Industrial Estate, Harlow, Essex CM19 5BE; (0279) 35422; Telex 817208; Fax (0279) 441509

Glenbrook Foods Ltd, 51 Seagoe Industrial Estate, Craigavon, Northern Ireland BT63 5QE; (0762) 336525

Glendale Foods Ltd, Mill Street, Pendleton, Salford; 061-736 8030 or 061-736 8052

Glengettie Tea Co. Ltd, Unit 4, Cherrycourt Way, Stanbridge Road, Leighton Buzzard, Bedfordshire LU7 8UH; (0525) 375743; Telex 826864; Fax (0525) 382302

Glengrove Ltd, South Point, 8 Paston Place, Brighton, East Sussex BN2 1HA; (0273) 693622; Fax (0273) 677986

Glisten Co. Ltd, Hill Street Works, Blackburn, Lancashire BB1 3HG; (0254) 677113; Telex 63216; Fax (0254) 682317

Global Foods Ltd, Parkchesterfield House, Chapel Road, Smallfield, Surrey; (034 284) 3596; Telex 95524

Global Group plc, Boundary House, 91–93 Charterhouse Street, London EC1M 6HR; 071-250 0388; Telex 261379; Fax 071-608 2428

Gloriette Patisserie (Knightsbridge) Ltd, 128 Brompton Road, London SW3 1JD; 071-584 1182

Gold Crown Foods Ltd, 82 Wood Street, Liverpool L1 4LU; 051-708 9030; Telex 628318

Golden Cow Dairies Ltd, Artabrackagh Road, Portadown, County Armagh BT62 4HB; (0762) 338411; Telex 747994

Golden Cross Snack Ltd, Motherwell Food Park, Block 2, Unit 2, Righead Industrial Estate, Bellshill, Lanarkshire ML4 3JU

Golden Crust (Nuneaton) Ltd, 17 Chartwell Drive, Wigston, Leicestershire LE8 2FL; (0533) 811920

Golden Crust (N. Ireland) Ltd, H. Stevenson & Co. Ltd, 95 Williams Street, Londonderry BT08 9AA; (0504) 263277

Golden Food Products Ltd, Queens Court, Alderley Edge, Cheshire SK9 7QD; (0625) 586563; Fax (0625) 586573

Golden Vale Dairies Ltd, 43 Portmanmoor Road, East Moors Industrial Estate, Cardiff

Golden Vale Dairies Ltd, Bridgend Creamery, Bridgend Industrial Estate, Bridgend, Mid Glamorgan; (0656) 653417

Golden Vale D.P.P. Ltd, Parkway House, Palatine Road, Northenden, Manchester M22 4DB; 061-945 7557; Fax 061-945 7404

Golden Vale D.P.P. Ltd, Llantrisant, Mid Glamorgan, Wales

Golden Vale plc, Charleville, Co. Cork, Ireland; 010 353 6381501; Fax 010 353 6389900

Golden West Foods Ltd, Three Cherry Trees Lane, Hemel Hempstead, Hertfordshire HP2 7HG

Golden West Foods Ltd, Genner Road, Willow Brook North Industry, Corby, Northants NN17 1FD

Golden Wonder Ltd, Edinburgh House, Abbey Street, Market Harborough, Leicestershire LE16 9BG; (0858) 410410; Telex 342791; Fax (0858) 410539

Golden Wonder Ltd, Gavin Road, Ditton, Widnes, Cheshire WA8 8RD

Goldenfry Foods Ltd, Sandbeck Way, Wetherby, West Yorkshire LS22 4DW; (0937) 63631; Fax (0937) 580024

Goldenlay Eggs Ltd, Carlton House, Sandy Walk, Wakefield, West Yorkshire WF1 2DP; (0924) 378141; Telex 556305

Golden Wonder Ltd, Abbey Street, Market Harborough, Leicestershire LE16 9BG; (0858) 410410; Fax (0858) 410539

Goldrei Bakery Services, Brookfield Drive, Aintree, Liverpool, Merseyside L9 7AW; 051-525 5969; Telex 627726

Goodlife Foods Ltd, 34 Tatton Court, Kingsland Grange, Warrington, Cheshire; (0925) 837810

Gott Foods Ltd, Westgate, Morecambe, Lancashire LA3 3DD; (0524) 36219; Telex 65213; Fax (0524) 841 687

Grampian Country Food Group Ltd, Mugiemoss Road, Bucksburn, Aberdeen AB2 9NP; (0224) 715454; Fax (0224) 714046

Grampian Country Chickens Ltd, Whytemyres Place, Mastrick, Aberdeen AB2 6HT; (0224) 662040; Telex 685853

Grampian Country Chickens Ltd, Tannery Street, Banff AB4 1ER; (02612) 5881; Fax (02612) 5655

Grampian Country Pork Ltd, Great Western Park, Buckie, Grampian AB5 1YA; (0542) 35555; Fax (0542) 31535

Grand Metropolitan plc, 11/12 Hanover Square, London W1A 1DP; 071-629 7488; Telex 299606 (registered office)

Grand Metropolitan Brewing Ltd, The Stag Brewery, Mortlake, London SW14 7ET

Grand Metropolitan Fast Food Services, Liverpool Street, Hull HU3 4HW

Granelli McDermott Ltd, Demmings Road, Cheadle, Cheshire SK8 2PE; 061-428 7595

Granose Foods Ltd, Howard Way, Newport Pagnell, Buckinghamshire MK16 9PY; (0908) 211311; Telex 826335; Fax (0908) 210514

Grant Bros. (Meat Canners) Ltd, Richmond Bridge, Galston, Ayrshire KA4 8JU; (0563) 820401; Telex 777967; Fax (0563) 821877

Gray Dunn & Co. Ltd, 75 Stanley Street, Kinning Park, Glasgow G41 1JF; 041-429 2674

Edward Gray of Dudley Ltd, 54 North Street, Dudley, Worcestershire (0384) 52139

Greco's Biscuits Ltd, Biscona Works, Craven Road, Leeds LS6 2SN; (0532) 456595; Telex 517773; Fax (0532) 429006

Green Dragon Bakery Ltd, Blenheim Road Airfield Industrial Estate, Ashbourne, Derbyshire DE6 1HA; (0335) 42373

Greenfield Bacon Co. Ltd, Brunt Cliffe Road, Morley, near Leeds; (0532) 536767

Greggs plc, 1 Lambton Road, Jesmond, Newcastle-upon-Tyne NE2 4RX; 091-281 7721; Fax 091-281 1444 (registered office)

Griffith Labs. (UK) Ltd, Cotes Park Estate, Somercotes, Derbyshire DE5 4NN; (0773) 832171; Telex 37456

Grindsted Products Ltd, Northern Way, Bury St Edmunds, Suffolk IP32 6NP; (0284) 69631; Telex 81203; Fax (0284) 60839

Grosvenor Confectionery Co. Ltd, Pleasant Road, Marine Parade, Southend-on-Sea, Essex SS1 2HJ; (0702) 466367

Guinness plc, International House, 7 High Street, London W5 5DB; 071-486 0288; Fax 081-566 3435 (registered office)

Wm. Gunstone & Sons Ltd, Stubley Lane, Dronfield, Sheffield S18 6PF; (0246) 414651; Fax (0246) 410866

H

Haagen-Dazs, Runnymede House, Heriot Road, Chertsey, Surrey KT16 9DT; (0932) 570011

Halal Meat Co. (Birmingham) Ltd, 162/190 Bishop Street, Birmingham B5 7EJ; 021-622 1497

Halal Meat Co. (Wales) Ltd, The Abattoir, Henllan, near Denbigh, Clwyd, North Wales; (0745) 812187

Haldane Foods Ltd, 25 Hayhill Industrial Estate, Sileby Road, Barrow-on-Soar, Leicestershire LE12 8LD; (0509) 816611

Haldane Foods Group, Howard Way, Newport Pagnell, Bucks MK16 9PY; (0908) 211311; Fax (0908) 210514

Hales Snails (UK) Ltd, Imperial House, Barton Dock Road Trading Estate, Stretford, Manchester M32 0ZH

Halib Foods International Ltd, 456 Antrim Road, Belfast BT15 5GD; (0232) 370903

David A. Hall Ltd, East Main Street, Broxburn, West Lothian EH52 5AW; (0506) 853300; Telex 727062; Fax (0506) 857049 (registered office)

Hall Farm Hingham Ltd, Hall Farm, Hingham, Norwich, Norfolk NR9 4PX; (0953) 850259; Fax 0953 850812

Halo Foods Ltd, Tywyn, Gwynnedd LL36 9RU; (0654) 711171; Fax (0654) 711744

Hamlyn Foods, Meadows Road, Queensway, Newport, Gwent NP9 0SS; (0633) 276276

Hammonds of Yorkshire, Harrogate Road, Apperley Bridge, Bradford, West Yorkshire BD1 0LY; (0274) 615401; Telex 517779

Handsworth Dairies Ltd, Island Road, Handsworth, Birmingham B21 8LD; 021-523 9211 (registered office)

Handsworth Dairies Ltd, Parkley Dairy, Park Avenue, Whitchurch, Shropshire SY13 1SH; (0948) 5240

Hankow Batchelors Tea Co. Ltd, Ocean Parade, Ferring, Worthing, Sussex; (0903) 503244; Telex 877110

Hannah's of Johnstone, Walkinshaw Works, Walkinshaw Street, Johnstone, Renfrewshire PA5 8AB; (0505) 21131; Telex 779885; Fax (0505) 29281

H. Hargrave (Wholesale Pork) Ltd, Horseshoe Road, Spalding, Lincolnshire PE11 3JA; (0775) 767486; Telex 32632; Fax (0775) 768101

H. Hargrave & Co. (Wholesale Pork) Ltd, 93–95 Polmadie Road, Glasgow G5 0BA; 041-429 8989; Fax 041-420 1639

H. Hargrave & Co. Ltd, Cargo Fleet Lane, Middlesbrough, Cleveland TS3 8DD; (0642) 240131; Fax (0642) 231581

Hargrave International Ltd, Eastern House, Town Drove, Quadring, Spalding, Lincolnshire PE11 4PU; (0775) 821202; Fax (0775) 821350

The Harris Pork & Bacon Group, 5 Ormond Terrace, Regent Street, Cheltenham, Gloucestershire GL50 1HR; (0242) 221727; Telex 437340; Fax (0242) 221710

The Harris Pork & Bacon Group, 114C Brettell Lane, Amblecote, Stourbridge, West Midlands DY8 4BS; (0384) 377780; Telex 338819; Fax (0384) 373601

Harris (Barnstaple), Pottington Industrial Estate, Pottington, Barnstaple, Devon EX31 1JH; (0271) 329868

Harris (Elmswell), Elmswell, Bury St Edmunds, Suffolk IP30 9HF; (0359) 40521; Telex 81527; Fax (0359) 40225

Harris (Ipswich), Hadleigh Road, Ipswich, Suffolk IP2 0HQ; (0473) 232345; Fax (0473) 252421

Harris (Leeming Bar), Leases Road, Leeming Bar, Northallerton, Yorkshire DL7 9AW; (0677) 22661; Fax (0677) 24986

Harris (Winsford), Wharton Bridge, Winsford, Cheshire CW7 3BB; (0606) 557321; Fax (0606) 55915

Harrisons & Crosfield plc, 20 St Dunstan's Hill, London EC3R 8LQ; 071-626 4333; Telex 885636; Fax 071-782 0112 (registered office)

Harrisons (Poultry) Ltd, Antrobus, near Northwich, Cheshire CW9 6JB; (0606) 89128; Telex 668574; (0606) 891705

Harvestime Ltd, Raleigh Street, Walsall, West Midlands WS2 8RB; (0922) 25566

Harvian Group, Millhouse Works, Houghton-le-Spring, Tyne & Wear DH5 8AH; 091-584 2139; Telex 537681

Haslington Cheese Co. Ltd, Lancaster Fields, Crewe Garden Farm Industrial Estate, Crewe, Cheshire CW1 1FF; (0270) 589136

Hassy Ltd, PO Box 11, Soham, Ely, Cambridgeshire CB7 5XD; (0353) 720027; Telex 81480; Fax (0353) 723112

Hatting Bageri UK Ltd, Delta 605, Delta Business Park, Welton Road, Swindon, Wiltshire SN5 7XF

Haven Foods Ltd, PO Box 139, Chesford Grange, Woolston, Warrington, Cheshire WA1 4LQ; (0925) 818101; Telex 627034

William Hay & Sons (Aberdeen) Ltd, 14 Berryden Road, Aberdeen, Grampian AB9 2WD; (0224) 636231

Hazlewood Desserts Ltd, Newmarket, Nailsworth, Stroud, Glos. GL6 0RG; (0453) 832691; Fax (0453) 834883

Hazlewood Foods plc, Empire Works, Parcel Terrace, Rowditch, Derby DE1 1NB; (0332) 295295; Telex 377872; Fax (0332) 292300

Hazlewood Frozen Foods Ltd, 36/40 Abenbury Way, Wrexham, Clwyd LL13 9UZ; (0978) 661966; Telex 617060; Fax (0978) 661856

Hazlewood & Co. (Selby) Ltd, Barlby Road, Selby, North Yorkshire YO8 7BJ; (0757) 703636; Fax (0757) 701916

Health & Diet Food Co. Ltd, Stubley Lane, Dronfield, Sheffield, South Yorkshire S18 6PF; (0246) 413294; Fax (0246) 414002

Heather Cameron Foods Ltd, Block 3, Unit 3, Etna Industrial Estate, Motherwell; (0698) 66799

Hedgehog Foods Ltd, Unit 4, Severn Farm Industrial Estate, Welshpool, Powys, Wales SY21 7DF; (0938) 5053

H.J. Heinz Co. Ltd, Hayes Park, Hayes, Middlesex UB4 8AL; 081-573 7757; Telex 261477; Fax 081-848 2325 (registered office)

H.J. Heinz Co. Ltd, Waxlow Road, Harlesden, London

Helmsley Meats Ltd, 1a Bardsley Road, Earlstree Industrial Estate, Corby, Northamptonshire; (0536) 204464; Fax 0536 400146

Hereford Herbs, Ocle Pychard, Herefordshire HR1 3RB; (0432) 78379; Telex 35332

Heritage Foods Ltd, Wharfedale Road, Ipswich, Suffolk IP1 4JP; (0473) 48391

Hero Foods (UK) Ltd, Kegworth, Derby DE74 2FJ; (0509) 674915; Fax (0509) 670198

Hewros Manufacturing Ltd, 20 Spring Croft Road, Tyseley, Birmingham B11 3EL; 021-702 2919; Fax 021-702 2995

Heygate & Sons Ltd, Bugbrooke Mills, Northampton NN7 3QH

Hey Soft Drinks (UK) Co. Ltd, Skinner Lane, Pontefract, West Yorkshire WF8 1HG; (0977) 70[illegible]161; Telex 55121; Fax (0977) 607068

Highlander Snacks Ltd, Whitehill Industrial Estate, Bathgate, Lothian EH48 2EN; (0506) 630778

Highland Venison Marketing Ltd, Grantown on Spey, Scotland PH26 3JY; (0479) 2255; Telex 75129; Fax (0479) 3055

Hill Biscuits Ltd, Smith Street, Ashton-under-Lyne, Lancashire OL7 0DB; 061-330 3617; Telex 665410; Fax 061-343 2108

Hilliers Ltd, 6 Bell Close, Newham Industrial Estate, Plympton, Plymouth PL7 4HD; (0752) 343311

Hillsdown Holdings plc, Hillsdown House, 32 Hampstead High Street, London NW3 1QD; 071-794 0677; Telex 297220; Fax 071-435 1355 (registered office)

Hillsdown Meat Group, Hillsdown House, PO Box 1, Mill Street, Calne, Wiltshire SN11 8DW; (0294) 812261; Telex 444505; Fax (0249) 821165

Hilton Meat Products Ltd, Unit 21A, Kilroot Park, Carrickfergus, County Antrim, Northern Ireland BT38 7PR; (09603) 63155; Fax (09603) 51027

Hinton Poultry, Norton St Philip, Bath, Somerset BA3 6LH; (0373) 834511; Telex 449763; Fax (0373) 834711

Hitchen Foods plc, Dobson Park Industrial Estate, Manchester Road, Wigan, Lancashire WN2 2DX; (0942) 824100

HL Foods Ltd, Bridge Road, Long Sutton, Spalding, Lincolnshire PE12 9EQ; (0406) 362101; Telex 32855; Fax (0406) 364352

Hockenhull & Hayes Ltd, Norwestern House, Staden Lane, Ashbourne Road, Buxton, Derbyshire SK17 9RZ; (0298) 26222; Telex 666608; Fax (0298) 72048

Holgran Group Ltd, Granary House, Wetmore Road, Burton-on-Trent, Staffordshire DE14 1TF; (0283) 43221

Walter Holland & Sons, Baxenden, Accrington, Lancashire BB5 2SA; (0706) 213591

Peter Hollingsworth Ltd, Horseley Fields, Wolverhampton, West Midlands WV1 3DG; (0902) 452452; Fax (0902) 870959

Holme Valley Cooked Meats, 17 Church Street, Honley, Huddersfield, Yorkshire HD7 2AH; (0484) 661542

Holmesterne Farm Co. Ltd, The Old Goods Yard, Boroughbridge, North Yorkshire YO5 9BL; (0423) 323979; Fax (0423) 324457

Homepride Foods Ltd, Compass House, 80 Newmarket Road, Cambridge CB5 8DZ; (0223) 461600; Telex 818878

Homestead Eggs Ltd, Clifton House, Glasgow; 041-332 2632; Telex 779613

Honeytop Foods Ltd, 43–45 Gransden Avenue, Hackney, London E8 3QA; 081-985 4349

Horizon Farms, Tremar, Liskeard, Cornwall Pl14 6EA; (0579) 345938

Horlicks Ltd, 11 Stoke Poges Lane, Slough, Berkshire; (0753) 33433

Horlicks Farms & Dairies Ltd, Hort Bridge, Ilminster, Somerset TA19 9PR; (0460) 52922; Fax (0460) 52985

Horsham Bacon Factory Ltd, 17 Foundry Lane, Horsham, West Sussex RH13 5PX; (0403) 52365

Hortons Ice Cream Co. Ltd, Blewshill Creamery, Hortonwood 2, Telford, Shropshire TF1 4ES; (0952) 670099; Telex 35571; Fax (0952) 603075

HP Foods Ltd, 45 Northampton Road, Market Harborough, Leicestershire LE16 9BQ; (0858) 410144; Telex 341770; Fax (0858) 410053 (registered office)

Hunters Foods, Princewood Road, Corby, Northamptonshire NN17 2JJ; (0536) 204200

Hunters the Bakers, Kingsway, Team Valley, Gateshead NE11 0LT; 091-491 0077

Huntsman Meats Ltd, Station Road, St Neots PE19 1NG; (0480) 77241; Fax (0480) 215804

Peter Hunt's Bakery Foods, The Bakery, 14 Lyon Road, Kearsley, Bolton BL4 8TX; (0204) 793446; Fax (0204) 794553

Huttons International Ltd, 633/653 Hedon Road, Hull HU9 5LQ

Hygrade Meats Ltd, Latona Road, London SE15 6SD; 071-732 8711; Telex 8955626; Fax 071-732 1221

Hygrade Foods Ltd, West Mead Lane, Chippenham, Wiltshire SN15 3HZ; (0249) 655911

Hygrade Foods Ltd, Furnham Road, Chard, Somerset TA20 1AA; (0460) 67272

Hyman Foods Ltd, Hibbert Street, Whitehill Industrial Estate, Stockport, Cheshire SK4 1NS; 061-480 0506; Telex 668307; Fax 061-480 1538

Hytech Foods, c/o Fresh-Pak Eggs Ltd, Unit 4, Jubilee Drive, Loughborough, Leicestershire LE11 0XS; (0509) 233327; Fax (0509) 232538

I

IAWS Group plc, 151 Thomas Street, Dublin 8, Eire; (01) 717131; Fax (01) 717321 or 717919

Icelandic Freezing Plants Ltd, Estate Road No. 2, South Humberside Industrial Estate, Grimsby, South Humberside DN31 2TG; (0472) 44181; Telex 527457 (registered office)

Frank Idiens & Sons Ltd, Weasenham Lane, Wisbech, Cambridgeshire PE13 2RN; (0945) 64471; Telex 32232; Fax (0945) 63878

Ilchester Cheese Co. Ltd, Free Street, Ilchester BA22 8LZ; (0935) 840531

Ingham Bros. (Sausage Manufacturers) Ltd, Ashday Lane, Southowram, Halifax, West Yorkshire HX3 9RH; (0422) 53431

Inghams (Steak Canadienne) Ltd, 192 Whalley Road, Clayton Le Moors, Accrington, Lancashire BB5 5HF; (0254) 888723

Ingle Smokehouse Ltd, Unit 8, Shore Road, Perth PH2 8BD; (0738) 30121

Inlaks Seafoods Ltd, The Harbour, Kilkeel, County Down, Northern Ireland BT34 4AX; (06937) 62473; Telex 747253; Fax (06937) 62198

Inlaks Seafoods Ltd, Unit 14, Prospect Business Centre, Dundee Technical Park, Dundee DD21 1TY; (0382) 561549; Telex 76467; Fax (0382) 561579

International Cuisine Ltd, Unit 9, No. 1 Industrial Estate, Medomsley Road, Consett, Co. Durham DH8 6ST; (0207) 502999

International Fish Canners (Scotland) Ltd, Bath Street, Fraserburgh, AB4 5DY; (0346) 513191; Fax (0346) 519247

International Flavours & Fragrances (GB) Ltd, Duddery Hill, Haverhill, Suffolk CB9 8LG; (0440) 704488; Fax (0440) 61599

Inverurie Scotch Meat Co., North Street, Inverurie, Aberdeenshire AB5 9TL; (0467) 20631; Telex 739019; Fax (0467) 24653

Invicta Lamb Ltd, Kingsley Road, The Pill, Bideford, Devon EX39 2AF

Invicta Lamb Ltd, Brewer Street, Tunbridge Wells, Kent TN3 8DP; (0892) 890678; Telex 95309; Fax (0892) 890121

H.W. Irvine, 17 Perth Street, Blairgowrie, Perthshire; (0250) 2477

W.D. Irwin & Sons Ltd, 12 Woodhouse Street, Portadown, County Armagh BT62 1JH; (0762) 332421

J

William Jackson & Son plc, 40 Derringham Street, Hull, North Humberside HU3 1EW; (0482) 224131; Telex 597553; Fax (0482) 24526

William Jackson & Son Ltd, Yarm Road, Stockton-on-Tees TS18 3RP; (0642) 675191

Jacksons Meat Products, Inglemire Lane, Cottingham, Hull, North Humberside HU16 4PJ; (0482) 841171; Telex 597553; Fax (0482) 24526

Ernest Jackson & Co. Ltd, Crediton, Devon EX17 3AP; (0363) 772251

Jacksons Foods Ltd, Madewell House, Ainley Industrial Estate, Elland, West Yorkshire HX5 9JP; (0422) 73511; Fax (0422) 74795

Jacob's Bakery Ltd, Suttons Business Park, Earley, Reading RG6 1AZ; (0734) 492000; Fax (0734) 492400

Jacobs Suchard Ltd, Miller Road, Bedford MK42 9PB; (0234) 55141; Telex 82118; Fax (0234) 57737

Jamesons Chocolates Ltd, Willoughby Lane, Tottenham, London N17 0RX; 081-807 4417; Telex 27754; Fax 081-807 6162

Alec Jarrett Ltd, High Street, Oldland Common, Bristol BS15 6TN; (0272) 322725; Telex 449982

Jenks Group, Castle House, 71–75 Desborough Road, High Wycombe, Buckinghamshire HP11 2HS; (0494) 33456; Telex 83506; Fax (0494) 24019

JLI Group plc, PO Box 54, JLI House, Guildford Street, Chertsey, Surrey KT16 9ND; (0932) 569600; Fax (0932) 569585

JL Quality Foods Ltd, Unit 48, Waverley Road, Beeches Industrial Estate, Yate, Bristol BS17 5QR; (0454) 321765; Fax (0454) 321770

Job's Dairy Ltd, Raleigh Way, Hamworth, Feltham, Middlesex TW13 7NN; 081-890 2261

J. Jolly Poultry Products Ltd, Westhoughton Industrial Estate, James Street, Westhoughton, Bolton, Lancashire BL5 3QR; (0942) 818888; Telex 377292; Fax (0942) 813485

Jon's Toffee Ltd, Victoria Road, Sowerby Bridge, West Yorkshire HX6 3AQ; (0422) 831515

W. Jordan (Cereals) Ltd, Holme Mills, Biggleswade, Bedfordshire SG18 9JX; (0767) 318222; Telex 827525; Fax (0767) 317324

Jugg Foods plc, St Lukes Yard, Bravington Road, Queens Park, London W9 3AW; 081-969 0945; Fax 081-964 0263

Jus-rol Ltd, Tweedside Trading Estate, Berwick-upon-Tweed, Northumberland TD15 2UX; (0289) 307737; Telex 537047

K

Kanes Food Products Ltd, PO Box 4, Middle Littleton, Evesham, Worcestershire WR11 5JS; (0386) 832531

Kara Foods Ltd, 15 Tanners Drive, Blakelands, Milton Keynes MK14 5BU; (0908) 616166; Telex 826893; Fax (0908) 616041

Katie's Kitchen, Forward Drive, Christchurch Avenue, Harrow, Middlesex HA3 8NT; 081-427 7778; Fax 081-861 3083

Kavli Ltd, Kingsway, Team Valley Trading Estate, Gateshead, County Durham NE11 0ST; 091-487 7146

Kays Cooked Meats Ltd, 76 Higham Hill Road, Walthamstow, London E17 6EJ

Kears Group Ltd (Bakeries), Claremont House, Yorkley, Lydney, Glos GL15 4TR

Keddie Saucemasters Ltd, Prince of Wales Industrial Estate, Abercarn, Newbridge, Gwent; (0495) 246855; Fax (0495) 244626

Kelco International Ltd, Westminster Tower, 3 Albert Embankment, London SE1 7RZ; 071-735 0333; Fax 071 735 1363

Kelco International Ltd, Applications Research & Development Dept., Pitwood Park Industrial Estate, Waterfield, Tadworth, Surrey KT20 5HQ; (0737) 362651; Fax (0737) 352003

Kellogg Company of Great Britain Ltd, The Kellogg Building, Talbot Road, Manchester M16 0PU; 061-869 2000; Fax 061-869 2100

Kennerty Farm Dairies Ltd, 9 West Bryson Road, Edinburgh EH11 1EJ; 031-337 2414; Telex 739881

Kennerty Farm Dairies Ltd, Craigshaw Drive, West Tullos Industrial Estate, Aberdeen, Grampian AB9 2XB; (0224) 890444; Telex 739881

Kent Salads Ltd, Northbourne, Deal, Kent CT14 0LW

Kentucky Fried Chicken Ltd, 88/97 High Street, Brentford, Middlesex TW8 8BG; 081-569 8733; Fax 081-569 8733

Kernels Foods plc, 1A Standard Road, London NW10 6EX; 081-965 8687

Kernel Nut Ltd, Unit Y1, Herald Way, Binley Industrial Estate, Coventry CV3 2NY

Kerry Foods Ltd, Thorpe Lea Manor, Thorpe Lea Road, Egham, Surrey TW20 8HY; (0784) 430777

Kerry Foods Ltd – Burton, 76 Mosley Street, Burton on Trent, Staffordshire DE14 1DS; (0283) 61661; Fax (0283) 511048

Kerry Group plc, Princes Street, Tralee, County Kerry, Eire; (010-353) 66 22433; Telex 73152; Fax (010-353) 66 22353

Kerrygold Co. Ltd, Leek, Staffordshire ST13 6EN; (0538) 399111; Telex 36134; Fax (0538) 399111 (registered office)

Kershaw Frozen Foods Ltd, Pool Hey Lane, Scarisbrick, Southport, Merseyside PR8 5LD; (0704) 30767; Telex 677149

Kibun Co. (UK) Ltd, Kibun House, 5–6 Bartholomew Place, London EC1A 7HH; 071-726 6801; Fax 071-726 2034

Kibun Co. (UK) Ltd, 4th Floor, Great West House, Great West Road, Brentford, Middlesex TW8 9DF; 081-568 9545; Fax 081-569 7542

Kibun Co. (UK) Ltd, Motherwell Food Park, Bellshill Industrial Estate, Bellshill ML4 3JA

Kikkoman Europe, Unit 3, 1000 North Circular Road, Staples Corner, London NW2 7JP; 081-450 4626; Fax 081-452 3734

Kilron Seafoods Ltd, Harbour Road, Fraserburgh, Grampian AB4 5TB; (0346) 25101; Telex 739583; Fax (0346) 29292

Kimpton Brothers Ltd, Berkshire House, 168–173 High Holborn Road, London WC1V 7AF; 071-379 6422; Fax 071-240 8726

Kimpton Tamlyn Ltd, Top Barn, Newton, Cambridge CB2 5PG; (0223) 872444; Fax (0223) 872512

Alexander King (Wishaw) Ltd, King Street, Wishaw, Lanarkshire ML2 8BT; (0698) 372196; Fax (0698) 351171

Kinnerton (Confectionery) Co. Ltd, 26 Nottingham Place, London; 071-486 8881

Kippax Biscuits Ltd, King Street, Colne, Lancashire BB8 9HU; (0282) 864198; Telex 635070; Fax (0282) 868891

Kitchen Range Foods Ltd, Viking Way, Bar Hill, Cambridge CB3 8EX; (0954) 780087; Telex 817672

Kitchen's of Sara Lee (UK) Ltd, Lancaster Road, Carnaby Industrial Estate, Bridlington YO15 3QY; (0262) 678212; Telex 52552

Knightsbridge Cake (Manchester) Ltd, PO Box 6, Brixham Road, Old Trafford, Manchester M16 0WT; 061-872 1613

KP Foods Group, St Helen's Court, North Street, Ashby de la Zouch, Leicestershire LE6 5UQ; (0530) 417744; Fax (0530) 417595

KP Foods Group (H.O.), Heathgate House, 57 Colne Road, Twickenham, Middlesex TW2 6QA; 081-894 5600; Fax 081-894 6715

Kraft Jacob Suchard Ltd, St George's House, Bayshill Road, Cheltenham, Gloucester GL50 3AE; (0242) 236101; Telex 43560; Fax (0242) 512084 (registered office)

Kraft Jacob Suchard Ltd, Moorgate Road, Kirkby, Liverpool L33 7XJ; 051-546 2710

Kraft Jacob Suchard Ltd, Bridge Road, Kingswood, Bristol BS15 4PT; (0272) 571859

L

Lagan Meat Co. Ltd, Duncrue Road, Belfast, Northern Ireland BT3 9BS; (0232) 749421 or 755811; Telex 74160; Fax (0232) 748153

Lambourn Food Co. Ltd, Abex Road, Newbury, Berkshire RG14 5EL; (0635) 521130; Telex 897572; Fax (0635) 38697

Lakeland Creameries/Dairies/Ice Cream, – see MD Foods plc

Lakeland Meats – see Cavaghan & Gray

Lancashire Dairies Ltd, Derby Street, Manchester M8 8LB; 061-834 8772; Telex 668715; Fax 061-833 1547

Langwood Prepared Vegetables Ltd, Gardiners Place, West Gillibrands, Skelmersdale, Lancashire WN8 9SP; (0659) 24371

Larderfresh, Potterton Way, Dartmouth Road, Smethwick, Warley, West Midlands B66 1AU

Lark Valley Foods Ltd, Eastlea Road, Bury St Edmunds, Suffolk IP32 7BY; (0284) 701461; Fax (0284) 750060

Laurel Confectionery Co., The Bridge Street, Crook, County Durham; (038 882) 2304

Laxgate Group, Group House, King Street, Middlewich, Cheshire CW10 9LY

Mr Lazenby's Sausages, Unit 7, Station Road, Stokesley, North Yorkshire TS9 7AB; (0642) 712540

LBL (Cooked Meats) Ltd, Second Avenue, Greasley Street, Bulwell, Nottingham NG6 8NE; (0602) 271166

Lea & Perrins Ltd, Midlands House, 3 Midlands Road, Worcester, Worcestershire WR5 1DT; (0905) 352770; Fax (0905) 763079

Leaf United Kingdom Ltd, Carlyle Road, Greenbank, Bristol BS5 6HR; (0272) 511122; Fax (0272) 510724

Leckpatrick Co-operative Agricultural & Dairy Society Ltd, 47 Berryhill Road, Artigarvan, Strabane, County Tyrone, Northern Ireland BT82 0HN; (0504) 44381; Telex 747504; Fax (0504) 41721

John J. Lees plc (Lee's of Scotland), Newlands Factory, Newlands Street, Coatbridge, Lanarkshire ML5 4BD; (0236) 24456; Telex 779301; Fax (0236) 33867 (registered office)

Leisure Fine Food and Drinks, Willow Road, Trent Lane, Castle Donnington, Derby DE74 2NP; (0332) 850616; Fax (0332) 850605

Leo's Home Made Sweets Ltd, 6 Silver Street, Stockton-on-Tees, Cleveland TS18 1SX; (0642) 603060

Leon Frenkel Ltd, Crabtree Manorway South, Belvedere, Kent DA17 6BB; 081-311 7171; Telex 896538

Le Pain Croustillant Ltd, Queensgate House, 18 Cookham Road, Maidenhead, Berkshire SL6 8AJ; 081-997 4487

Lesme-Callebaut Ltd, S & A, Sunya House, Daventry Road, Banbury, Oxon OX16 7UU; (0295) 257651; Telex 837382; Fax (0295) 273294

Libby, McNeill & Libby Ltd, 36 Park Street, Croydon, Surrey CR9 1TT; 081-686 8981; Telex 23117

Linnells Meat Products Ltd, Cambridge Street, Northampton NN2 6DP; (0604) 714070

Lion Confectionery Co. Ltd, Westgate, Cleckheaton, West Yorkshire BD19 5EB; (0274) 872537; Fax (0274) 873469

Lion Mark Holdings Ltd (Lion Foods), Faraday Road, Astmoor, Runcorn, Cheshire WA7 1PE; (0928) 565221; Telex 628343; Fax (0928) 561172 or 565221

Lipton Tea Co. Ltd, Station Avenue, Leighton Buzzard, Bedfordshire LU7 8QQ; (0525) 372200; Telex 825141; Fax (0525) 383866

Livera Foods plc, PO Box 100, Concorde House, Concorde Way, Preston Farm, Stockton on Tees, Cleveland TS18 3RP

Lockwoods Ltd, Ripley Road, Ambergate, Derby DE5 2JR; (0773) 856686; Fax (0773) 856687

Lockwoods (Liverpool) Ltd, Bridle Way, Netherton, Bootle, Merseyside L30 4UJ; 051-5259258

Loders Croklaan Ltd, HO and Factory, Cairn Mills, Factory Road, Silvertown, London E16 2EL; 071-476 5611

Lodge Farm Foods Ltd, Rye Hill Close, Lodge Farm Industrial Estate, Harlestone Road, Northampton NN5 7UA; (0604) 56621; Telex 312662; Fax (0604) 582762

Lofthouse of Fleetwood Ltd, Maritime Street, Fleetwood, Lancashire FY7 7LP; (03917) 2435

London Herb & Spice Co., Apothecary House, 18 Selsdon Road, South Croydon CR2 6PA; 081-680 8337; Telex 943763; Fax 081-760 0377

Londreco Ltd, Nestlé Avenue, Hayes, Middlesex UB3 4RG; 081-573 8948; Telex 8955189; Fax 081-848 3647

Howard Long International Ltd, Brandon Road, Methwold, Thetford, Norfolk IP26 4RH

Long Clawson Dairy Ltd, West End, Long Clawson, Melton Mowbray, Leicestershire LE14 4PJ; (0664) 822332; Telex 341627; Fax (0664) 823236

Longday Foods Ltd, Wensum Point, 32 Whiffler Road, Norwich NR3 2AZ; (0603) 788344; Telex 975607; Fax (0603) 789516

Lopen Group Ltd, Mill Lane, Lopen, South Petherton, Somerset TA13 5JS; (0460) 41310; Telex 46351; (0460) 41766

Lord Rayleigh's Dairies Ltd, Bury Lane, Hatfield Peverel, Chelmsford, Essex CM3 2AG; (0245) 380372; Fax (0245) 381942

Loseley Park Farms, Estate Office, Loseley Park Farms, Guildford, Surrey GU3 1HS; (0483) 571881; Telex 859972

Lowthers (Cakes) Ltd, Cale Lane, Aspull, Wigan WN2 1HE; (0942) 43801

Lovells Confectionery Ltd, Queensway Meadows, Rexville, Newport, Gwent NP9 0XA; (0633) 281777; Telex 498037

Lovell & Christmas (Ulster) Ltd, Agivey Bacon Factory, 150 Bann Road, Ballymoney, County Antrim, Northern Ireland BT53 7PD; (026 56) 62044; Telex 747880; Fax (026 56) 65965

Lovell & Christmas (Ulster) Ltd, Enkalon Industrial Estate, Unit F1, 25 Randalstown Road, Antrim, County Antrim BT41 4LJ; (08494) 65403

Lucas Ingredients, Moravian Road, Kingswood, Bristol BS15 2NG; (0272) 674681; Fax (0272) 605409

Lucas Ingredients, Draycott Mills, Dursley, Gloucestershire GL11 5NA; (0453) 3272; Telex 449825

W.H. Luke & Son Ltd, Endsleigh Road, Oreston, Plymouth, Devon PL9 7LS; (0752) 408011

Lyme Regis Fine Foods Ltd, Intercon House, 522 Fulham Road, London; 071-384 2677; Fax 071-371 0573

J. Lyons & Co. Ltd, 325 Oldfield Lane North, Greenford, Middlesex UB6 0AZ; 081-578 2040; Telex 935975; Fax 081-575 6887

Lyons Bakeries (UK) Ltd, 325 Oldfield Lane North, Greenford, Middlesex UB6 0AZ; 081-578 2040; Fax 081-575 8264

Lyons Cakes Ltd, Fish Dam Lane, Carlton, near Barnsley, South Yorkshire S71 3HQ; (0226) 286191; Telex 547479; Fax (0226) 286191

Lyons Biscuits Ltd, Devonshire Road, Blackpool FY3 7AN; (0253) 394133; Telex 67131; Fax (0253) 300238

Lyons Cakes Ltd, Thornes Road, Wakefield, West Yorkshire EWF2 8PJ; (0924) 372141; Telex 566419; Fax (0924) 291746

Lyons Patisserie Ltd, Parkview Industrial Estate, Brenda Road, Hartlepool TS25 1TE; (0429) 233622; Telex 587649; Fax (0429) 279608

Lyons Seafoods Ltd, Barrow House, Bishopstrow, Warminster, Wiltshire BA12 9HU; (0985) 214565; Telex 449357; Fax (0985) 214278

Lyons Soft Drinks Ltd, 129 Rannoch Road, London W6 9SN; 081-748 4631

Lyons Tetley Ltd, 325 Oldfield Lane, Greenford, Middlesex UB6 0AZ; 081-578 2345; Telex 935975; Fax 081-578 7483

M

Macaw (Soft Drinks) Ltd, 7 Roseway, Lomeshaye Industrial Estate, Nelson BB9 5SU

Macfish Ltd, Watermill Road, Fraserburgh, Grampian AB4 5HA; (0346) 23741; Telex 73188; Fax (0346) 25902

Macfish Ltd, St Peters Street, Peterhead, Grampian AB6 6UT; (0779) 72115; Fax (0779) 74215

M & W Mack Ltd, 43 North Street, Chichester, Sussex PO19 1NF; (0243) 787646

Mackay's of Carnoustie, 21 Thistle Street, Carnoustie, Angus DD7 7PR; (0241) 53109; Fax (0421) 52678

Mackie's Aberdeen Dairy Co. Ltd (trading as Mackie's Quality Farm Foods), Westertown, Rothienorman, Inverurie, Grampian AB5 8US; (04675) 466; Fax (04675) 448

Macphie of Glenbervie Ltd, Glenbervie, Stonehaven, Grampian AB3 2YB; (05694) 641; Telex 73589; Fax (05694) 677

D.A. Macrae Ltd, 66 St Peters Street, Peterhead, Grampian AB4 6UT; (0779) 72115; Fax (0779) 74215

Madigan Foods Ltd, Unit 8, Kenfig Industrial Estate, Margam, Kenfig Hill, Mid Glamorgan, South Wales

Magna Specialist Confectioners Ltd, Magna House, Stafford Park, Telford, Shropshire TF3 3BH; (0952) 610071; Telex 35602; Fax (0952) 290380

Magnet Pet Foods Ltd, Magnet House, Rishton, Blackburn BB1 4DX; (0254) 885484; Telex 635073

Maine Soft Drinks Ltd, 35 Ballymena Road, Ballymoney, County Antrim, Northern Ireland BT53 7EX; (02656) 62088; Telex 747173

Malton Bacon Factory, PO Box 1, Parliament Street, Norton, Malton, North Yorkshire YO17 9HG; (0653) 693031; Telex 57860; Fax (0653) 692851

Mandora (UK) Ltd, PO Box 2, Bellamy Road, Mansfield, Nottinghamshire NG18 4LW; (0623) 24661; Telex 377054; Fax (0623) 658645

Manley Ratcliffe Ltd, Tower Road, Fane Drive, Berinsfield, Oxford OX9 8LQ; (0865) 340171; Telex 83432; Fax (0865) 340738

Manor Bakeries Ltd, King Edward House, 27–30 King Edward Court, Windsor, Berkshire SL4 1TJ; (0753) 840401; Telex 849326

Manor Bakeries Ltd, Wythenshawe, Manchester M21 4TR; 061-998 0121

Manor Bakeries Ltd, Haynesway, Swift Valley Industrial Estate, Rugby, Warwickshire CV21 1HG

Manor Farm Ducklings, H.A. Long & Co., Manor Farm, Ickburgh, Thetford, Norfolk IP26 5JD; (0842) 878236

Manor Vinegar Brewery Co. Ltd, New Road, Burntwood, Walsall, West Midlands WS7 0AB; (054 36) 5555; Telex 334181; Fax (054 36) 77149

Margetts Foods Ltd, Clee Hill Road, Tenbury Wells, Worcestershire WR15 8HD; (0584) 810366; Telex 35528; Fax (0584) 810366

Marietta Frozen Foods, Nursteed Road, Devizes, Wiltshire SN10 3DY; (0380) 725161; Telex 444685; Fax (0380) 720800

Marine Harvest Ltd, Craigcrook Castle, Edinburgh EH4 3PF; 031-336 1777; Telex 727664; Fax 031-336 5534

Markit Meats Ltd, Dalrymple Street, Liverpool L5 5EN; 051-207 3439

Marks & Spencer plc, Michael House, Baker Street, London W1A 1DN; 071-935 4422; Telex 267141; Fax 071-487 2679 (registered office)

Marlow Foods Ltd, 9 Station Road, Marlow, Buckinghamshire SL7 1NF; (0628) 890850

Marlow Foods Ltd, Station Road Industrial Estate, Stokesley, North Yorkshire, TS9 7AB; (0642) 710803; Fax (0642) 712485

Marlow Foods Ltd, PO Box 2, Belassis Avenue, Billingham, Cleveland TS23 1YN

J. Marr (Seafoods) Ltd, Livingstone Road, Hessle, North Humberside HU13 0EE

Marr Fish Merchants, Herring Arm Road, Fleetwood, Lancashire FY7 6BN; (03917) 2251

Marr Foods Ltd, 102 Gillett Street, Hull HU3 4JG; (0482) 26563; Telex 52461

Mars Confectionery, Dundee Road, Slough, Berkshire SL1 4JX; (0753) 23932; Telex 848253; Fax (0753) 35396

Mars Corporate Services, 3D Dundee Road, Slough SL1 4LG (registered office)

Marshall Food Group Ltd, Newbridge, Midlothian, Scotland EH28 8SW; 031-333 3341; Telex 72561; Fax 031-333 3341 (registered office)

Marshall Food Group Ltd, George Street, Coupar Angus, Perthshire; (0828) 2295

Marshall Food Group Ltd, Apedaile, Viewfield Road, Coatbridge, Lanarkshire; (0236) 31174

James Marshall (Glasgow) Ltd, 14 Blairlinn, Cumbernauld, Glasgow G67 2XP; (02367) 28201; Telex 779712

Martlet Natural Foods, Horam Manor, Horam, Heathfield, East Sussex TN21 0JA; (04353) 2254

Master Foods, PO Box 15, Hansa Road, King's Lynn, Norfolk PE30 4JE; (0553) 692222; Telex 81198; Fax (0553) 691553

Master Meat Packers (Omagh) Ltd, 52 Doogary Road, Omagh, County Tyrone, Northern Ireland BT79 0BG; (0662) 3201

W.T. Mather Ltd, Barn Lane, Golborne, Warrington, Cheshire WA3 3NT

Mathers (Inverurie) Ltd, Harlaw Road, Inverurie, Aberdeenshire AB5 9SG; (0467) 620366; Telex 739123; Fax (0467) 625273

E.C. Matheson & Son Ltd, 93 Hutcheson Street, Aberdeen AB2 3RS; (0224) 633163; Telex 738850; Fax (0224) 633957

E.C. Matheson & Son Ltd, Montrose Road, Brechin, Angus DD9 7PL; (0356) 622876; Fax (0356) 625760

Mattessons Wall's, – see Van den Bergh Foods

Bernard Matthews plc, Great Witchingham Hall, Norwich, Norfolk NR9 5QD; (0603) 872611; Telex 97455; Fax (0603) 871118 (registered office)

Bernard Matthews plc, Great Witchingham Hall, Norwich, Norfolk NR9 5QD; (0603) 872611; Telex 975249; Fax (0603) 872421 (factory & sales)

Bernard Matthews Foods Ltd, Factory B, Upper Holton, near Halesworth, Suffolk IP19 8NJ

Matthews Foods plc, The Healey Complex, Healey Road, Ossett, West Yorkshire WF5 8NE; (0924) 272534; Fax (0924) 277071

Lloyd Maunder Ltd, Willand, Cullompton, Devon EX15 2PJ; (0884) 820534; Telex 42871; Fax (0884) 820826

Mauri Products Ltd, Stockholm Road, Sutton Fields Industrial Estate, Kingston-upon-Hull HU8 0XW; (0482) 833133; Telex 592557; Fax (0482) 838460

Mauri Products Ltd, Clarks Mill, Stallard Street, Trowbridge, Wilts BA14 8HH

W.H. Maybery Ltd, Pontardawe Industrial Estate, Pontardawe, Mid Glamorgan; (0792) 862193; Fax (0792) 830647

Maxilin Ltd, Sharples Vale, Bolton, Lancashire BL1 6NR; (0204) 57154; Telex 635109; Fax (0204) 55292

Maxons Ltd, Bradbury Street, off Valley Road, Sheffield, South Yorkshire S8 9QQ; (0742) 554216; Fax (0742) 509683

Maxway Food Products Ltd, The Industrial Estate, Wooler, Northumberland NE71 6SR; (0668) 81222; Telex 53465

Mayhew Chicken Ltd, Five Ash Down, Uckfield, East Sussex TN22 4BS; (0825) 732491

Mayhew Chicken Ltd, Trecynon, Aberdare, Mid Glamorgan CF44 8PP; (0685) 883050; Fax (0685) 874929

McCain Foods (GB) Ltd, Havers Hill, Scarborough, North Yorkshire YO11 3BS; (0723) 584141; Telex 52565; Fax (0723) 581230 (registered office)

McCain Foods (GB) Ltd, Funthams Lane, Whittlesey, Peterborough PE7 2PG; (0733) 203051; Telex 32385; Fax (0733) 205485

McConnell Salmon Ltd, Clyde House, 170 Kilbowie Road, Clydebank, Dunbartonshire G81 2UA; 041-941 3696; Telex 778883; Fax 041-952 0126

McCormick (UK) plc, Owen House, 8 Bankhead, Crossway North, Edinburgh EH1 4BR (registered office)

McCormick Foods, Dormer Road, Thame Park Industrial Estate, Thame, Oxon OX9 3SL; (0844) 215661

McCormick Glentham Ltd, Pondwood Close, Moulton Park Industrial Estate, Northampton NN3 1RT; (0604) 790220; Fax (0604) 790230

McCormick (UK) plc, Castle House, Desborough Road, High Wycombe, Bucks. HP11 2HS; (0494) 533456; Fax (0494) 463245

McCormick-Stange Ltd, Rossmore Road East, Ellesmere Road, South Wirral, Cheshire L65 3DA; 051-355 5011; Telex 628351; Fax 051-356 0469

A. McCowan & Sons Ltd, 44 Tryst Road, Stenhousemuir, Larbert, Stirlingshire FK5 4HA; (0324) 562158; Telex 778583; Fax (0324) 552667

McDougalls Catering Foods, Imperial Way, Worton Grange, Reading, Berkshire RG2 0QW; (0734) 875 383; Telex 847632; Fax (0734) 875383

McErlean Bakeries Ltd, Arizona Street, Belfast, Northern Ireland BT11 8EW; (0232) 628121

McIntosh Donald, Portlethen, Aberdeen, Grampian AB1 4QB; (0224) 780381; Telex 73662; Fax (0224) 782039

McIntosh of Dyce Ltd, Victoria Street, Dyce, Aberdeen AB2 0RA; (0224) 771414; Telex 73489; Fax (0224) 723716

McKean Foods, 84 Glentaner, Glasgow G22 7UB; 041-336 6543

McKey Food Service Ltd, Northfield Drive, Northfield, Milton Keynes MK15 0DF; (0908) 665431; Telex 826218; Fax (0908) 674621

McKey Food Service Ltd, Luneburg Way, Skippingdale, Scunthorpe, South Humberside DN15 8NN

McLaren Foods Ltd, Beaver Industrial Estate, Beaver Road, Ashford, Kent TN23 1SH; (0233) 652526; Telex 965756; Fax (0223) 613391

William McMillan Ltd, Ravenscraig Bacon Factory, Rectory Lane, Dusart, Fife KY1 2TR; (0592) 51954

McQuillan Meats Ltd, Rathenraw Industrial Estate, Greystone Road, County Antrim, Northern Ireland BT41 2SJ; (084 94) 65121; Telex 747752; Fax (084 94) 68432

McVitie's Group, Head Office, Watermans Business Park, The Causeway, Staines, Middlesex TW18 3BA; (0784) 447744; Fax (0784) 447711

McVitie's UK Ltd, Smisby Road, Ashby de la Zouch, Leicestershire LE6 5JT; (0530) 417555; Fax (0530) 416210

McVitie's UK Ltd, The Biscuit Works, 54 Church Street, Carlisle, Cumbria CA2 5TG; (0228) 24431; Fax (0228) 35900

McVitie's UK Ltd, Victoria Biscuit Works, 35 Clydeford Drive, Tollcross, Glasgow G32 8YW; 041-556 5787; Fax 041-554 5786

McVitie's UK Ltd, Waxlow Road, Harlesden, London NW10 7NY; 081-965 5787; Fax 081-965 8496

McVitie's UK Ltd, Levenshulme Biscuit Works, Crossley Road, Manchester M19 2SD; 061-432 0202; Fax 061-443 1896

MD Foods plc, Craven House, Kirkstall Road, Leeds, West Yorkshire LS3 1JE; (0532) 440141; Fax (0532) 425546

MD Foods plc, Lion Complex, Willow Street, Blackburn BB1 5NH; (0254) 697697

MD Foods plc (Lakeland Creamery), Shap Road, Kendal, Cumbria LA9 6NS; (0539) 722361

MD Foods plc (Newcastle Dairy), Whitley Road, Newcastle-upon-Tyne NE12 9SW; 091-266 9211

MD Foods plc (Lancashire Dairy), Stanley Street, Accrington, Lancashire BD5 6RJ; (0254) 391 311

Meadowbrook Foods Ltd, – see Laxgate

Meadow Farm Produce, 74 Alston Drive, Bradwell Abbey, Milton Keynes MK13 9HG; (0908) 311660; Telex 827521; Fax (0908) 319781

Meadow Vale Foods Ltd, Wrexham Road, Malpas, Cheshire SY14 7EJ; (0948) 860123; Fax (0948) 860140

Meadow Vale Food Group, Morton, Oswestry, Shropshire; (0691) 830741

Medeol UK Ltd, Cremanaze Division, 2 Centurion Way, Erith, Kent DA18 4AF; 081-311 777; Fax 081-311 6488

Meltis Confectionery Ltd, Millar Road, Bedford, MK42 9BP; (0234) 269200; Fax (0234) 269300

Memory Lane Cakes Ltd (Brossard), Maes-y-Coed Road, Cardiff, South Glamorgan CF4 4XR; (0222) 62351; Telex 497452

Mendip Foods Ltd, St Olave's House, 19 Wooley Street, Bradford-on-Avon, Wiltshire EA15 1AD; (022 16) 3939; Telex 444353

Meridian Foods Ltd, Unit 13, WDA Advanced Factory, Clawdd Poncen, Corwen, Clwyd Ll21 9RT

Meri-Mate Ltd, Block 10, Wester Gourdie Industrial Estate, Dundee (0382) 623131; Telex 76416

Merretts, Sunblest Bakery, Maes-y-Coed Raod, Cardiff CF4 4UZ; (0222) 623391

Merrydown plc, Horam Manor, Horam, Heathfield, East Sussex TN21 0JA; (043 52) 2254; Telex 957266; Fax (043 53) 3218 (registered office)

Middlebrook Mushrooms, Stock Lane, Langford, Avon BS18 7ES; (0934) 852751; Telex 449305

Midland Meat Packers Ltd, The Abattoir, Crick, Northampton NN6 7TZ; (0788) 823711; Telex 311328; Fax (0788) 822960

Milanda Bakeries (Derry) Ltd, Glen Road, Londonderry, County Londonderry, Northern Ireland BT48 0DA; (0504) 264122; Fax (0504) 264122

Milk Marketing Board of England & Wales, Thames Ditton, Surrey KT7 0EL; 081-398 4101

Milk Marketing Board for Northern Ireland, 456 Antrim Road, Belfast BT15 5GD; (0232) 770123; Telex 747136; Fax (0232) 771442

Milk Marketing Board for Northern Ireland, Bangor Dairies, Shaftesbury Road, Bangor, County Down, Northern Ireland; (0247) 467131

Milk Marketing Board for Northern Ireland, Dunmanbridge Factory, 139 Moneymore Road, Cookstown, County Tyrone, Northern Ireland; (06487) 63755

Milk Marketing Board for Northern Ireland, Dromona Creamery, 75 Dunminning Road, Cullybackey, Ballymena, County Antrim, Northern Ireland; (0266) 880253

John Millar & Sons (1844) Ltd, Thistle Industrial Estate, 155 East Main Street, Broxburn, West Lothian EH52 5BB; (0506) 855252; Telex 72465; Fax (0506) 856969

W.L. Miller & Sons Ltd, Sterte Avenue, Poole, Dorset BH15 2AS; (0202) 673281; Telex 418359

Millstore Ltd, Foxwood Road, Dunston Trading Estate, Chesterfield S41 9RF; (0246) 454085

Millway Foods Ltd, Colston Lane, Harby, Leicestershire LE14 4BE; (0949) 61371; Telex 378231; Fax (0949) 60030

Milupa Ltd, Milupa House, Uxbridge Road, Hillingdon, Middlesex UB10 0NE; 081-573 9966; Fax 081-573 2247

Joseph Mitchell (Letham) Ltd, 2 Woodside Road, Letham, by Forfar, Angus DD8 2QD; (030 781) 220; Fax (030 781) 540

Modern Health Products Ltd, Davis Road, Chessington, Surrey KT9 1TN; 081-397 4361; Fax 081-971 455

Mollington Farmhouse Cheese Co. Ltd, Mollington Grange, Chester CH1 6NP; (0244) 851226; Telex 61252; Fax (0244) 851141

Moore Fine Foods Ltd, Office 8, Hainton House, Hainton Square, Grimsby, Humberside DN32 9AQ; (0472) 240704; Telex 56261; Fax (0472) 240538

Samuel Moore Foods Ltd, Jam Factory, Easterton, Devizes, Wiltshire SN10 4PN; (0380) 812556; Telex 449554; Fax (0380) 812680

S. Moores, The Biscuit Bakery, Morcombelake, Bridport, Dorset DT6 6ES; (0297) 89253; Fax (0297) 89753

Moorland Foods Ltd, Wansford Road, Driffield, Yorkshire YO25 7NJ; (0377) 42651

Moorland Poultry Ltd, The Moor, Dalton, Thirsk, North Yorkshire YO7 3JE; (0845) 577485; Fax (0845) 577001

Moray Seafoods, 9–11 Low Street, Buckie, Grampian AB5 1UZ; (0542) 831296; Telex 73287; Fax (0542) 834859

Morisis Products Ltd, Blewshill Creamery, Halesford, Telford, Shropshire TF7 4QH; (0952) 584009; Telex 35191

Morning Foods Ltd, North Western Mills, Crewe, Cheshire CW2 6HP; (0270) 213261; Telex 36548; Fax (0270) 500291 (registered office)

Morton Foods Ltd, Wedgewood Road, Bicester, Oxon OX6 7UL; (0869) 244704; Fax (0869) 320178

Moruzzi Foods Ltd, Enterprise House, Springfield Industrial Estate, Newport, Telford, Shropshire TF10 7NB; (0952) 820289; Fax (0952) 813209

Moy Park Ltd, The Food Park, 39 Seagoe Industrial Estate, Craigavon, Northern Ireland BT63 5QE; (0762) 352233; Telex 747966; Fax (0762) 352233 (registered office)

Moy Park Ltd, Royal Oak Buildings, High Street, Crewe, Cheshire CW2 7BJ; (0270) 257722; Fax (0270) 215881

Mudies Frozen Food Products Ltd, Norwich Road, Barham, near Ipswich, Suffolk; (0473) 830096

Müller Dairy (UK) Ltd, Tern Valley Business Park, Shrewsbury Road, Market Drayton, Shropshire TF9 3SQ; (0630) 692000

Munchie Foods Ltd, House of Munchie, Duncrue Road, Belfast, Northern Ireland BT15 5AB; (0232) 757113; Telex 747666; Fax (0232) 351477

N

Napolina Ltd, Claygate House, Esher, Surrey KT10 9PN; (0372) 62181; Fax (0372) 68775

National Starch & Chemical Co., Ashburton Road East, Trafford Park, Manchester M17 1BJ; 061-872 6161; Telex 669414

Needlers plc, Sculcoates Lane, Hull, North Humberside HU5 1EH; (0482) 42435; Telex 597029; Fax (0482) 447646

Nelsons of Aintree, Long Lane, Liverpool L9 7BN; 051-525 1606; Telex 627082; Fax 051-521 8510

Neptune Fish Export Ltd, North Esplanade East, Aberdeen, Grampian AB1 2QY; (0224) 211777; Telex 739158; Fax (0224) 211668

Nestlé UK Ltd, St George's House, Croydon, Surrey CR9 1NR; 081-686 3333; Telex 23117; Fax 081-686 6072

Nestlé Ice Creams, Units E & F, Hortonwood 2, Telford, Shropshire TF1 4GW; (0952) 676654

Nestlé Rowntree, York YO1 1XY; (0904) 653071; Telex 57846; Fax (0904) 22467 (registered office)

Nestlé Rowntree, Albion Mills, Halifax, Yorkshire HX3 9XT

H.W. Nevill (Sunblest), South Access Road, Walthamstow, London E17 8AX; 081-556 1031

New Covent Garden Soup Co., 35 Hythe Road, London NW10 6RS; 081-960 2229

New Primebake Ltd, Unit D, Hortonwood 31, Telford, Shropshire TF1 4EX; (0952) 608121; Fax (0952) 670381

New Vita Biscuit Co. Ltd, Sewell Street Industrial Colony, Plaistow, London E13 8AT; 081-472 5501

Henry Newbould Ltd, 2 Startforth Road, Riverside Park Industrial Estate, Middlesborough, Cleveland TS2 1PT; (0642) 225315; Fax (0642) 230307

Newforge Foods Ltd, Belle Vale Road, Gateacre, Liverpool, Merseyside L25 2PG; 051-487 7255 or 051-487 7345; Telex 628264; Fax 051-487 0768

NewMarket Foods Ltd, Little Wratting, Haverhill, Suffolk CB9 7TD; Tel/Fax (0440) 704444

Nibb-It International (UK) Ltd, Euroway Trading Estate, 32 Wharfedale Road, Bradford BD4 6SH; (0274) 684388

J.N. Nichols (Vimto) plc, Ledson Road, Wythenshawe, Manchester M23 9NL; 061-998 8801; Fax 061-998 9446

Nimmo (Holdings) Ltd, Littlemill Road, Drongan, Ayr KA6 7BP; (0292) 590444

Nimmo Quality Meat Produce (Aberdeen) Ltd, Scotstown Road, Bridge of Don, Aberdeen AB2 8HG; (0224) 822222; Fax (0224) 704229

L. Noel & Sons Ltd, Stanhill Works, Stanhill Village, Oswaldwistle, Accrington, Lancashire BB5 4PR; (0254) 31411

Noon Products plc, 4 Wellington Terrace, Bayswater Road, London W2

Nor Sea Foods Ltd, Broadford Road, Bridge of Don, Aberdeen, Grampian AB2 8EE; (0224) 703222; Telex 73311; Fax (0224) 822802

Normeat (UK) Ltd, Royal Dane House, Rutherford Drive, Wellingborough, Northamptonshire NN8 3UT; (0933) 676363; Telex 31415; Fax (0933) 674462

NORCO, – see Northern Co-operative Society Ltd

North Country Poultry Ltd, Green Lane West, Garstang, Preston PR3 1NJ; (099 52) 6474; Fax (099 52) 5435

North Downs Dairy Co. Ltd, Grove Dairy Farm, Bobbing Hill, Sittingbourne, Kent ME9 8NY (0795) 844444; Telex 966370; Fax (0795) 844029

North Downs Dairy Co. Ltd, Saxon Way, Wincanton Business Park, Lawrence Hill, Wincanton, Somerset BA9 8AB; (0963) 34422

North of Scotland Milk Marketing Board, Claymore House, 29 Ardconnel Terrace, Inverness IV2 3AF; (0463) 232611; Telex 75254; Fax (0463) 710923

North Shore Meats, Killroot Port Industrial Estate, Unit 190, Larne Road, Carrickfergus, County Antrim, Northern Ireland; (09603) 69300

Northern Co-operative Society Ltd, Millbank House, Berryden Road, Aberdeen AB9 2WE; (0224) 634121; Telex 739921; Fax (0224) 641976 (registered office)

Northern Dairies Ltd, Raines House, Denby Dale Road, Wakefield, West Yorkshire WF1 1HR; (0924) 290808; Telex 527149; Fax (0924) 290557

Northern Dairies, Paston Road, Wythenshaw, Manchester M22 4UA; 061-998 5331; Fax 061-945 5147

Northern Foods plc, Beverley House, St Stephens Square, Hull, North Humberside HU1 3XG; (0482) 25432; Telex 597149; Fax (0482) 226136 (registered office)

Northumbrian Fine Foods plc, Dukesway, Team Valley, Gateshead, Tyne & Wear NE11 0QP; 091-487 0070; Telex 53346; Fax 091-491 0826 (registered office)

O

Oak Forest Foods Ltd, Cleton Street Business Park, Tipton, West Midlands

Oakland Fast Food, Liverpool Street, Hull HU3 4HW; (0482) 26234; Fax (0482) 210375

Oakland Fast Food, Telford Place, Crawley, West Sussex RH10 1SY; (0293) 533222

Oakland UK Ltd, Moor Monkton, York YO5 8LA

Oak Lodge Salmon Ltd, The Lodge, 24–26 Eastbury Road, London Industrial Park, Beckton, London E6 4LP

Oasters, Chequer Tree Farm, Benenden Road, Cranbrook, Kent TN17 3PN; (0580) 714014; Fax (0580) 713512

Oceancrest Foods Ltd, Station Road, Heckington, Sleaford, Lincolnshire NG34 9JH; (0529) 60123; Fax (0529) 60630

O'Hara's Bakeries Ltd, 182 Cambrai Street, Belfast, Northern Ireland BT13 3JH; (0232) 751611; Telex 748069

Jesse Oldfield Ltd, Knightbridge Bakery, Brixham Road, Old Trafford, Manchester M16 0WT; 061-872 1613; Fax 061-872 9542

OP Chocolate Ltd, High Street, Dowlais, Merthyr Tydfil, Mid Glamorgan CF48 3TB; (0685) 723291; Telex 497394; Fax (0685) 723291

Orange Crush Co. Ltd, Denington Estate, Wellingborough, Northamptonshire NN8 2QJ; (0933) 226337; Telex 311342

Orchard House Foods Ltd, Fleming Road, Corby, Northants; (0536) 61599

Ormeau Bakery Ltd, 307 Ormeau Road, Belfast, Northern Ireland BT7 3GN; (0232) 491001

Overseal Foods Ltd, Swains Park, Park Road, Overseal, Burton-on-Trent, Staffordshire DE12 6JX; (0283) 224221; Telex 341345; Fax (0283) 222006

P

G.W. Padley (Poultry) Ltd, Anwick, near Sleaford, Lincolnshire NG34 9SL; (0526) 832661; Telex 56160; Fax (0526) 833180 (registered office)

G.W. Padley (Poultry) Ltd, Gonerby Foot Hill, Grantham, Lincolnshire NG31 8HZ; (0476) 71015; Fax (0476) 79713

Palethorpes, PO Box 1, Maer Lane, Market Drayton, Shropshire TF9 3AW; (0630) 652271

Palmer Sweets Ltd, 649a Anlaby Road, Hull HU3 5QD; (0482) 506955

Panda Drinks Ltd, The Brewery, Blandford Forum, Dorset DT11 9LR; (0258) 452231, 456511 or 450415; Telex 41556; Fax (0258) 454700

Pandora Pickle Co. Ltd, Moston Road, Greengate, Middleton Junction, Manchester M24 1SE; 061-653 0223; Fax 061-655 3936

Panificio Italiano, Second Way, Wembley, Middlesex HA9 0TP; 081-902 9722; Fax 081-902 1252

Paradise Cuisine Ltd, Unit 17, Bennets Field Trading Estate, Wincanton, Somerset BA9 9DT; (0963) 34610; Fax (0985) 846309

Paradise Cuisine Ltd, Woodcock Road Industrial Estate, Warminster, Wiltshire BA12 9DX; (0985) 214121; Fax (0985) 846309

Park Cake Bakeries Ltd, Ashton Road, Oldham, Lancashire OL8 2ND; 061-633 1181 or 061-652 5411; Telex 668738; Fax 061-626 6199

Park Food Group plc, Valley Road, Birkenhead, Merseyside L41 7ED; 051-653 7377

W. & J. Parker (Wholesale) Ltd, The Cattle Market, Freeman's Common Road, Leicester LE2 7LU; (0533) 548484; Telex 342439; Fax (0533) 558579

T.N. Parr Ltd, Lilac Grove, Beeston, Nottingham NG9 1PJ; (0602) 258261

Parrs Quality Confectionery Ltd, 26 Alder Road, Parkstone, Poole, Dorset BH12 2AQ; (0202) 748650

Parslows of Reading Ltd, Lodden Bridge Road, Woodley, Reading, Berkshire RG5 4BJ; (0734) 692641

The Pasta Company, Ardent Road, Barton-upon-Humber, South Humberside, DN18 5RT; (0652) 660060; Fax (0652) 34697

Pasta Food Holdings Ltd, 224 London Road, St Albans, Hertfordshire AL1 1JF; (0727) 60461; Telex 28443; Fax (0727) 64143

Pasta Foods Ltd, High Mill Road, Great Yarmouth, Norfolk NR31 0DW; (0493) 656071; Telex 975257

Pasta Reale Ltd, 53/55 Gatwick Road, Crawley, Sussex RH10 2RD; (0293) 517974

Patak's (Spices) Ltd, 252 Goswell Road, London EC1V 7EB

Patak's (Spices) Ltd, Kiriana House, Haydock Lane, Haydock, Lancashire WA11 9UZ

Paterson-Bronte Ltd, Houston Industrial Estate, Nettlehill Road, Livingston, West Lothian EH54 5DN; (0506) 31031 or 37429 (sales); Telex 72643

Paulig Ltd, 22 Deer Park Road, Lombard Trading Estate, Wimbledon, London SW19 3UN; 081-542 9541; Telex 915642; Fax 081-540 8111

Pauls plc, 47 Key Street, Ipswich, Suffolk IP4 1BX; (0473) 232222; Fax (0473) 230509

George Payne & Co. Ltd, PO Box 113, Croydon Road, Croydon, Surrey CR9 4BU; 081-688 7744; Telex 261361; Fax 081-686 0082

Pearce Duff & Co. Ltd, Humphrys Road, Woodside Estate, Dunstable, Bedfordshire LU5 4TE; (0582) 603191; Telex 825277

Peartree Farms Foods, High Birch, Wealey Heath, Clacton-on-Sea, Essex CO16 9BU; (0255) 830921

Pedigree Petfoods, National Office, Waltham-on-the-Wolds, Melton Mowbray, Leicestershire LE14 4RS

Pedigree Petfoods, Melton Mowbray, Leicestershire LE13 1BB; (0664) 411111; Telex 34675; Fax (0664) 410131

Peerless Refining Co. Liverpool Ltd, Dunnings Bridge Road, Bootle, Merseyside L30 6TJ; 051-525 5151; Telex 628784

Pelham Venison Co. Ltd, The Estate Office, Furneaux Pelham Hall, Buntingford, Hertfordshire SG4 0JP; (027 978) 348; Fax (027 978) 421

Pells Quality Confectionery Co. Ltd, 8/10 Conway Road, Newport, Gwent NP9 8PA; (0633) 271273; Fax (0633) 270398

Pendletons (Sweets) Ltd, Lees Road, Kirkby Trading Estate, Liverpool L33 7SA; 051-546 6784; Telex 628102; Fax 051-549 1740

Penguin Confectionery Co. Ltd, Penguin House, Westmorland Street, Carlisle, Cumbria CA2 5HL; (0228) 26262; Telex 64115; Fax (0228) 41753

Pennine Foods Ltd, Drakehouse, Sheffield S19 6HT; (0742) 476864

Perimax (Scotland) Ltd, Kirkton Industrial Estate, Sir William Smith Road, Arbroath, Angus DD11 4HR; (0241) 70551

Perimax Meat Co. Ltd, Crowne Trading Estate, Shepton Mallet, Somerset BA4 5QQ; (0749) 344141; Telex 46404; Fax (0749) 345333

Perkins Foods plc, Trinity Court, Trinity Street, Peterborough PE1 1DA; (0733) 555706; Fax (0733) 558499 (registered office)

J. Perkins (Wholesale Meat Packers) Ltd, Cornishway, Taunton, Somerset; (0823) 71011

Ken Perrett (Evesham) Ltd, PO Box 4, Middle Littleton, Evesham, Worcestershire WR11 5JS

Pershore Poultry Ltd, Pershore Trading Estate, Station Road, Pershore, Worcestershire WR10 2DD; (0386) 552838; Fax (0386) 556171

Peter's Savoury Products Ltd, Bedwas House Industrial Estate, Bedwas, Gwent NP1 8XP; (0222) 869721; Telex 497971

Philpot Dairy Products Ltd, Philpot House, Rayleigh, Essex SS6 7HH; (0268) 775522; Telex 99166; Fax (0268) 747666

Phoenix Foods Manufacturers Ltd, Brakey Road, Corby, Northants NN17 1LU; (0536) 200101; Fax (0536) 202218

Pigs Marketing Board (N. Ireland), Royal Ulster Agricultural Society Showgrounds, Balmoral, Belfast BT9 6GP; (0232) 381888; Fax (0232) 668511 (registered office)

Pillsbury UK Ltd, 315 Portland Road, Hove, East Sussex BN3 5ST; (0273) 417275; Telex 87582

Pinegain Ltd, 102 Gillett Street, Hull HU3 4JG

Pin-it Pastry Ltd, Abergarw Industrial Estate, Brynmenyn, Bridgend, Mid Glamorgan CF32 9LW

Pinneys of Scotland Ltd, Brydekirk, Annan, Dumfriesshire, Scotland DG12 5LP; (05763) 401; Telex 779151; Fax (05763) 466

The Pizza Factory, Finch Close, Lenton Lane, Nottingham NG7 2NN; (0602) 868204

Playtime Foods Ltd, 39 Ings Road, Leeds, West Yorkshire LS9 9EW; (0532) 504339; Telex 557822; Fax (0532) 481240

Plumrose Ltd, PO Box No. 25, Willerby, Hull HU10 6DR; (0482) 659391; Fax (0482) 659399

Plumtree Farms Ltd, 88–92 Wallis Road, Hackney, London E9 5LN; 081-986 9222; Fax 081-986 3879

Poachers Pies Ltd, 5 Marston Road, Cromwell Road Industrial Estate, St Neots, Cambridgeshire PE19 2HB; (0480) 407777; Fax (0480) 407788

Polar Fish, South Esplanade West, Aberdeen, Grampian AB1 3AA; (0224) 878891; Fax (0224) 879575

Pork Farms Group, Queens Drive, Nottingham NG2 1LU; (0602) 866541; Telex 37132; Fax (0602) 861236

Portfolio Foods Ltd, Pasadena Close, Hayes, Middlesex UB3 3NQ; 081-573 1010; Fax 081-848 4282

John Potter Ltd, 2–5 Roman Way, Coleshill Industrial Estate, Coleshill, Birmingham B46 1HG; (0675) 465323

Pourshins plc, 4 Girling Way, Great South West Road, Feltham, Middlesex TW14 0PH; 081-844 1274; Telex 919760; Fax 081-751 4397

Premier Beverages (Divn of Premier Brands) Ltd, 19 Highfield Road, Edgbaston, Birmingham B15 3DU

Premier Beverages (Divn of Premier Brands) Ltd, Knighton Adbaston, Stafford ST20 0QJ; (0785) 280215; Fax (0785) 280359

Premier Biscuits Ltd, Pasture Road, Moreton, Wirral, Merseyside L46 8SE; 051-678 8888; Telex 629659; Fax 051-473 1675

Premier Brands Ltd, PO Box 171, Franklin House, Bournville, Birmingham B30 2NA; 021-459 1199; Telex 335895; Fax 021-433 3337

Premier Brands Ltd, 22 Ainslie Place, Edinburgh EH3 6AJ

Premier Brands, The Orchard, Histon, Cambridge; (022 023) 3333; Telex 81276

Premier Poultry Ltd, 38–40 Northgate, Newark, Nottingham NG24 1EZ; (0636) 611100

Premier Poultry Ltd, The Broadway, Woodhall Spa, Lincs LN10 6PS

Premier Poultry Ltd, Ram Boulevard, Foxhills Industrial Estate, Scunthorpe DN15 8QW; (0724) 281081; Fax (0724) 281082

William Price & Sons Ltd, 38 Raleigh Street, Walsall, West Midlands; (0922) 25566

J.L. Priestley & Co. Ltd, Station Road, Heckington, Sleaford, Lincolnshire NG34 9NF; (0529) 60751; Telex 377189; Fax (0529) 60603

Prime Seafoods Ltd, West Shore Road, Fraserburgh, Grampian AB4 5LG; (0346) 26549; Fax (0346) 24067

Princes Ltd, Royal Liver Building, Liverpool L3 1JH; 051-236 9282; Telex 628095

L.E. Pritchitt & Co. Ltd, Comelle House, Elmfield Road, Bromley, Kent BR1 1LT; 081-464 3399; Telex 262895

Pritchitt Foods Ltd, 46 Belfast Road, Newtownards, County Down, Northern Ireland BT23 4TT; (0247) 814309; Telex 74575; Fax (0247) 813538

Prosper de Mulder Ltd, Ings Road, Doncaster DN5 9SW; (0302) 390900; Fax (0302) 390048

Protein Foods (UK) Ltd, Ramsay Road, Barnfield Industrial Estate, Tipton, West Midlands DY4 9DJ; 021-557 7653; Telex 337096; Fax 021-520 7926

Protein Technologies International (UK) Ltd, 16A Princewood Road, Earlstree Industrial Estate, Corby, Northamptonshire NN17 2AP; (0536) 67325; Telex 342546; Fax (0536) 61147

Proten Co. Ltd, The Old Rectory, Banbury Lane, Cold Higham, Towcester NN12 8LR; (0327) 830888; Telex 311759; Fax (0327) 830868

Pukka Pies Ltd, The Halfcroft, Syston, Leicestershire LE7 8LD; (0533) 609755; Telex 341995

Pullman Foods Ltd, Longdon Road, Prince of Wales Dock, Swansea SA1 8QY

Pullman Foods Ltd, Booker Regional Office, Stover Road, Yate, Bristol BS17 5JP

Pullman Foods Ltd, PO Box 47, Garonor Way, Portbury, Bristol BS20 9NR

Pura Foods Ltd, Orchard Place, London E14 0JH; 071-987 2066; Fax 071-538 3338

Puredrive Ltd, 36A Causeway Road, Earlstree Industrial Estate, Corby, Northamptonshire NN17 2DU; (0536) 402122; Fax (0536) 60616

Puritan Maid Ltd, Unit 13, Humphrys Road, Woodside Estate, Dunstable, Bedfordshire LU5 4TP; (0582) 609111; Telex 825207

Puritan Maid Ltd, Poynters Road, Dunstable, Bedfordshire LU5 4SJ

Puritan Maid Ltd, St Martins House, 20 Queensmere, Slough, Berkshire SL1 1YY; (0753) 73266; Telex 847836

Q

Q Peanuts Ltd, Camford Way, Sundon Park, Luton, Bedfordshire LU3 3AN

Quaker Oats Ltd, PO Box 24, Bridge Road, Southall, Middlesex UB2 4AG; 081-574 2388; Telex 934740; Fax 081-574 6615 (registered office)

Quality Food Products Ltd, Craigshaw Place, West Tullos Industrial Estate, Aberdeen, Grampian AB1 4AH; (0224) 872189

J.G. Quicke & Partners, Home Farm, Newton St Cyres, Exeter, Devon EX5 5BT; (0932) 851222; Fax (0932) 851382

Quintins Snack Foods Ltd, 15 Glebe Road, East Gillibrands, Skelmersdale WN8 9JP

R

R & B (Bristol) Ltd, 11–12 Ashley Trading Estate, Ashley Parade, Bristol BS2 9XS; (0272) 553551/4; Fax (0272) 558159

Rahbek-Food Ltd, Oxleasow Road, Redditch, Worcestershire B98 0RE; (0527) 501012

Raines Dairy Foods Ltd, Raine House, Crown Road, Enfield, Middlesex EN1 1TX; 081-804 8151; Telex 265327; Fax 081-805 4211; factory 081-804 1653

Rakusen's Ltd, Rakusen House, Clayton Wood Rise, Ring Road, West Park, Leeds LS16 6QN; (0532) 784821; Fax (0532) 784064

Rank Hovis Ltd, King Edward House, PO Box 527; 27/30 King Edward Court, Windsor, Berkshire SL4 1TJ

Ranks Hovis McDougall plc, RHM Centre, PO Box 178, 67 Alma Road, Windsor, Berkshire SL4 3ST; (0753) 857123; Fax (0753) 846537

John Rannoch Ltd, Haughley Park, Stowmarket, Suffolk IP14 3JZ; (0359) 40472; Telex 817265; Fax (0359) 40546

Rathbone Bros. Ltd, Central Bakery, Newtown, Wigan, Lancashire WN5 9AP; (0942) 41825

Rawmarsh Foods, Claypit Lane, Rawmarsh, Rotherham, South Yorkshire S62 5DU; (0709) 522292

Rayner & Co. Ltd, 4 Bull Lane, Edmonton, London N18 1TQ; 081-807 3080; Telex 265828; Fax 081-807 9205

The Realeat Co. Ltd, Howard Way, Newport Pagnell, Bucks MK16 9PY; (0908) 211311; Fax (0908) 210514

Rea Valley (Canned Meat) Ltd, Station Road, Minsterley, Shrewsbury, Shropshire SY5 0BD; (0743) 791232; Telex 35393

The Real Meat Co., East Hill Farm, Heytesbury, Warminster, Wiltshire BA12 0HR; (0985) 40436

Reckitt & Colman plc, 1 Burlington Lane, Chiswick, London W4 2RW; 081-994 6464; Telex 21268; Fax 081-994 8920 (registered office)

Redmill Snack Foods Ltd, Globe Street, Wednesbury, West Midlands WS10 0NN; 021-505 1500; Fax 021-505 2992

Reids the Bakers, 2 Wheatsheaf Road, Coleraine, County Londonderry BT51 3NP; (0265) 44326

J.F. Renshaw & Co. Ltd, Holden Street, Liverpool L8 7RF; 051-709 6133

Rhins Bacon Co. (1953) Ltd, Commerce Road, Stranraer DG9 7DE; (0776) 3437

RHM Foods Ltd, 10 Victoria Road, Willesden, London NW10 6NU; 081-965 6565; Telex 24202

RHM Ingredient Supplies Ltd, Owl Lane, Ossett, Wakefield WF5 9AX; (0924) 280444; Telex 557491; Fax (0924) 381042

RHM Research & Engineering Ltd, The Lord Rank Research Centre, Lincoln Road, High Wycombe, Buckinghamshire HP12 3QR; (0494) 26191

RHM Retail Ltd, Clark's Bakery, Restmor Way, Hackbridge Road, Wallington, Surrey SM6 7AF; 081-669 4455

RHM Retail Ltd, Wimbush Bakery, 6 Green Lane, Small Heath, Birmingham B9 5DA; 021-772 2151

Richlea Eggs Ltd, 221 Hillhall Road, Lisburn, County Antrim, Northern Ireland BT27 5JD; (0846) 661089; Fax (0846) 660486

Rich Products Ltd, Falstaff House, 33 Birmingham Road, Stratford upon Avon, Warwickshire; (0789) 414389

Ridge Farm Foods, Midland Road, Hunslet, Leeds, West Yorkshire LS10 2RJ; (0532) 702080; Fax (0532) 701838

Ridgways, Speke Hall Road, Speke, Liverpool L24 9NT; 051-486 3221; Telex 629613; Fax 051-486 8781

Ridpath Pek Ltd, 1000 Newham Way, London E6 4JN; 071-474 0555; Telex 896934; Fax 071-511 1473

Percy L. Rigby & Son Ltd, Mill Bank Dairy, Thelwall New Road, Grappenhall, Warrington WA4 2NH; (0925) 61121

Ringtons Ltd, Algernon Road, Newcastle-upon-Tyne NE6 2YN; 091-469 2728; Telex 53120; Fax 091-276 3500 (registered office)

Roach Foods Ltd, Ebenezer, Bugle, St Austell, Cornwall PL26 8RR; (0726) 850701; Telex 45613; Fax (0726) 851627

Frank Roberts & Sons Ltd, Rudheath, Northwich, Cheshire CW9 7RQ; (0606) 41122

Roberts of Port Dinorwic Ltd, Griffiths Crossing Industrial Estate, Caernarfon, Gwynedd, North Wales; LL55 1TS; (0286) 76111; Fax (0286) 77669

John Robertson & Sons (Hamcurers) Ltd, Inches Road, Ardrossan, Ayrshire KA22 8DW; (0294) 63936

James Robertson & Sons (Preserve Manufacturers) Ltd, Golden Shred Works, Droylsden, Manchester M35 6DR; 061-370 6231; Telex 668937

Robertsons (Bakers) Ltd, The Bakery, Durranhill Trading Estate, Carlisle CA1 3NH; (0228) 27541

Robinski & Co. Ltd, 4–5 Hythe Road, London NW10 6RT; 081-969 9511; Telex 922135

C. Robinson & Co. Ltd – see Wells Soft Drinks

Robinson's, Carrow, Norwich NR1 2DD

Robirch Ltd, PO Box 19, 76 Mosley Street, Burton-on-Trent, Staffordshire DE14 1DS; (0283) 61661; Telex 342816; Fax (0283) 511048

Roman Delicatessen Products Ltd, Unit 3, Barton Road, Water Eaton Industrial Estate, Bletchley, Milton Keynes MK2 3HX; (0908) 70968; Fax (0908) 270496

Rombouts Coffee (GB) Ltd, Ten Acre Lane, Egham, Surrey TW20 8RJ; (0784) 37511; Telex 8954406

Romix Foods Ltd, Lockett Road, Aston-in-Makerfield, Lancashire WN4 8DE; (0942) 712054; Telex 67678

Roocrofts (1923) Ltd, Topping Street Mills, Halliwell, Bolton, Lancashire BL1 3TW; (0204) 25633; Fax (0204) 392951

Alex Ross (Fish Merchants) Ltd, Craigshaw Drive, West Tullos Industrial Estate, Aberdeen, Grampian AB1 4AE; (0224) 878099

Ross Foods Ltd, Ross House, Grimsby, South Humberside DN31 3SW; (0472) 59111; Telex 52182

Ross Produce Ltd, 72 Eye Road, Peterborough PE1 4SJ; (0733) 60681; Telex 32363

Ross's of Edinburgh, Pentland Industrial Estate, Loanhead, Midlothian EH20 9QR; 031-440 1717; Telex 727594

Ross Young's Ltd, Head Office, Ross House, Wickham Road, Grimsby, South Humberside DN31 3SW; (0472) 359111; Telex 52182; Fax (0472) 240640

Ross Young's Ltd, 60–64 Port Street, Annan, Dumfriesshire DG12 6BW; (0461) 202021; Fax (0461) 205641

Ross Young's Ltd, Holt Road, Fakenham, Norfolk NR21 4TH; (0328) 863251; Fax (0328) 51367

Ross Young's Ltd, Brighton Street, Hessle Road, Hull, North Humberside HU3 4XT; (0482) 27997; Fax (0482) 225885

Ross Young's Ltd, St Mary's Road, Leamington Spa, Warwickshire CV31 1QB; (0926) 420731; Fax (0926) 451791

Ross Young's Ltd, Unit 23, Britannia Estate, Leagrave Road, Luton, Bedfordshire LU3 1RJ; (0582) 401177; Fax (0582) 453557

Ross Young's Ltd, Autby House, North Thoresby, Grimsby, South Humberside; (0472) 840441; Fax (0472) 840151

Ross Young's Ltd, North Road, Okehampton, Devon EX20 1BP; (0837) 53261; Fax (0837) 54657

Ross Young's Ltd, Washford Industrial Estate, Hemming Road, Redditch, Worcestershire B98 0DH; (0527) 23831; Fax (0527) 23891

Ross Young's Ltd, North Walsham, Worstead, Norfolk NR28 9RY; (0692) 403211; Fax (0692) 504911

Rossfish Ltd, Hutton Road, Fish Docks, Grimsby, South Humberside DN31 3PX; (0472) 361591; Telex 527166; Fax (0472) 240121

Rowallan Creamery, Glasgow Road, Kilmarnock, Ayr KA3 6AA; (0563) 26442; Telex 779586

Rowats Ltd, 33–41 Kelvin Avenue, Hillingdon, Glasgow G52 4LQ; 041-882-3341; Fax 041-882 5854

Rowland Confectionery, 17 Old High Street, Folkestone, Kent CT20 1RL; (0303) 54723

Rowse Honey Ltd, Moreton Avenue, Wallingford, Oxfordshire OX10 9DE; (0491) 33122; Fax (0491) 25051

Royal Greenland Ltd, Albert House, 2 Station Road, Cheadle Hulme, Cheshire SK8 5AE; 061-485 8385; Fax 061-486 9106

Roy's Quality Foods Ltd, Viking Way, Winchwen Industrial Estate, Swansea, West Glamorgan SA1 7DE; (0792) 796519; Telex 48232; Fax (0792) 795628

Rubicon Products Ltd, 23 Tennyson Avenue, London NW9 9JA; 081-206 1306; Telex 923753; Fax 081-451 0344

Edward Ryde & Sons, Grange Farm Factory, Loscoe, near Heanor, Derbyshire DE7 7JT; (0773) 716312

Ryvita Co. Ltd, 4th Floor, Columbia Centre, Market Street, Bracknell, Berkshire RG12 1LF; (0344) 487848; Telex 846560; Fax (0344) 487866

Ryvita Co. Ltd, Old Wareham Road, Parkstone, Poole, Dorset BH17 7NW; (0202) 743090; Telex 417129; Fax (0202) 732125

S

S. & A. Foods Ltd, Peartree Industrial Park, Princes Street, Derby DE3 8UG; (0332) 270670

Sahib Foods, 163/165 Brent Park Industrial Estate, Southall, Middlesex UB2 5LJ; 081-574 7360; Telex 939003

J. Sainsbury plc, Stamford House, Stamford Street, London SE1 9LL; 071-921 6000; Telex 264241; Fax 071-921 6132 (registered office)

Saint Martin Food Products Ltd, Units 1–9, Lea Road, Waltham Abbey, Essex EN9 1AS; (0992) 767167; Fax (0992) 769357

Saint Martin Food Products Ltd, 14 Colville Road, Acton, London W3 8BL; 081-992 8711; Fax 081-993 6377

Christian Salveson plc, 50 East Fettes Avenue, Edinburgh EH4 1EQ; 031-552 7101; Telex 72222; Fax 031-552 5809 (registered office)

Christian Salvesen Food Services, Easton, Grantham, Lincolnshire NG33 5AU

Salvesen Food Services, Shrewsbury Avenue, Peterborough PE2 7LB

Samworth Brothers Ltd, Fields Farm, Cropwell Butler, Nottingham NG12 3AP; (06073) 5221; Telex 377627; Fax (06073) 5137

Sandyford Foods Ltd, Sandyford, Prestwick, Ayrshire KA9 2SY; (0292) 520486; Fax (0292) 520801

William Santus & Co. Ltd, Dorning Street, Wigan, Lancashire WN1 1HE; (0942) 43464

Sappa plc, Northern Way, Bury St Edmunds, Suffolk IP32 6NZ; (0284) 753161; Fax (0284) 706308

J.J. Saunders Ltd, Old Down Farm, Emborough, Bath; (0761) 232359

Saxby Bros. Ltd, PO Box 15, Melton Bakery, Brook Street East, Wellingborough, Northamptonshire NN8 1LH; (0933) 76333; Telex 317304; Fax (0933) 227949

Sayers, The Ideal Bakery, Lorenzo Drive, West Derby, Liverpool L11 1BJ; 051-226 1291

Schooner Seafoods (Processing) Ltd, 10/14 Ropery Street, Hull, North Humberside HU3 2BU; (0482) 225949; Fax (0482) 26682

Schwan's Europe Ltd, 20 Marathon Place, Moss Side Industrial Estate, Leyland, Preston PR5 3QN

Sco-Fro Foods Ltd, 30 George Square, Glasgow G2 1EG; 041-248 7717; Fax 041-221 7524

Scotbeef Ltd, 27 Gleburn Road, College Milton North, East Kilbride G74 5BA; (03552) 25381; Telex 777919; Fax (03552) 64327

Scotbeef Ltd, Longleys, Bridge of Allan, Stirlingsshire FK9 4NE; (0786) 832911; Telex 777775

Scotch Premier Meat Ltd, North Street, Inverurie, Aberdeenshire AB51 9TL; (0467) 620631; Fax (0467) 624653

Scotch Premier Meat Ltd, 2 Newmarket Road, Edinburgh EH14 1RH; 031-443 5252; Fax 031-455 7287

Scotfresh Chilled Foods Ltd, Drum Road, Cupar, Fife; (0334) 52500

Scotfruit Ltd, Drum Road, Cupar, Fife; (0334) 54047

Scotfruit Ltd, West Pitkerro, Dundee, Tayside DD5 3RZ; (0382) 737855

Scotia Haven Foods Ltd, PO Box 139, Chesford Grange, Woolston, Warrington, Cheshire WA1 4SX; (0925) 818101

R. & W. Scott Ltd, Clydesdale Preserve Works, Carluke, Lanarkshire, Scotland ML8 5BD; (0555) 70711; Telex 779079; Fax (0555) 72237

Scottish Farm Dairy Foods Ltd, 190 Helen Street, Glasgow G51 3HJ; 041-445 4551

Scottish Milk Marketing Board, Underwood Road, Paisley, Renfrewshire PA3 1TJ; 041-887 1234

Scotts Porage Oats, Uthrogle Mills, Cupar, Fife; (0334) 52961

Scotts Soft Drinks Ltd, Wareing Road, Aintree Industrial Estate, Liverpool L9 7AU; 051-523 3445; Fax 051-530 1631

Seafresh Foods Ltd, Unit 1, Parmiter Street, London E2 9HZ; 081-983 0333

Sea Products International (SPI), Harborn Road, Edgbaston, Birmingham B15 3BU; 021-454 9999; Telex 338168

Seabrook Potato Crisps Ltd, 207 Allerton Road, Allerton, Bradford, West Yorkshire; (0274) 46405

Sebon Ltd, Sebon Bakery, 15 Ellis Road, Mitcham, Surrey CR4 4HX; 081-646 5220; Fax 081-648 0951

Sefcol (Sales) Ltd, Whitehouse Industrial Estate, Runcorn, Cheshire WA7 3BJ; (0928) 713121; Telex 265871; Fax (0928) 712827

Sefton Meadow Seafoods Ltd, Sefton Lane, Maghull, Merseyside L31 8BX; 051-531 8383; Telex 628419; Fax 051-520 2324

Semmons Taylor Ingredients, Wildmere Road, Banbury, Oxfordshire OX16 7UX; (0295) 53101; Telex 837158; Fax (0295) 69834

Sharp & Nickless Ltd, 77 College Street, Long Eaton, Nottinghamshire NG10 4NN; (0602) 732169

Richard Sharrock & Sons, Sunblest Bakery, Ashton Road, Bredbury, near Stockport, Cheshire SK6 2RE; 061-430 5151

J.A. Sharwood & Co. Ltd, 10 Victoria Road, London NW10 6NU; 081-965 6565; Telex 24202

Benjamin Shaw & Sons Ltd, Willow Lane, Huddersfield, West Yorkshire HD1 5ED; (0484) 27427 (registered office)

Benjamin Shaw & Sons Ltd, Bondgate, Kottingley Road, Pontefract, West Yorkshire WF8 2XA; (0977) 707727; Fax (0977) 708447

Elizabeth Shaw Ltd, Carlyle Road, Greenbank, Bristol BS5 6HR; (0272) 511122; Telex 44748; Fax (0272) 510724

Shaw's (Huddersfield) Ltd, 139 Wakefield Road, Huddersfield, West Yorkshire HD5 9AN; (0484) 28888

Shearway Foods Ltd, Marley Farm, Headcorn Road, Smarden, near Ashford, Kent TN27 8PJ; (0233) 77661; Telex 965001; Fax (0233) 77660

Shelf Bacon Factory Ltd, Shelf Moor Road, Shelf, Halifax HX3 7TN; (0274) 676241

Shepherdboy Ltd, 1 Norman Road, Thurmaston, Leicester LE4 8EL; (0533) 602992

Sherriffs of Kingsbridge Ltd, Poplar Drive, Kingsbridge, Devon TQ7 1QY; (0548) 857373; Telex 45139; Fax (0548) 853034

Shieldness Produce Ltd, Carriden Industrial Estate, Bo'ness, West Lothian EH51 9SJ; (0506) 823491; Fax (0506) 826055

C. Shippam Ltd, PO Box 3, East Walls, Chichester, West Sussex PO19 1PQ; (0243) 788373; Telex 86763; Fax (0243) 773278 (registered office)

C. Shippam Ltd, Station Yard, Crediton, Devon EX17 3PP; (0363) 772717

G.E. Shouler & Co. Ltd, Lancaster Road, Carnaby, Bridlington, North Humberside YO15 3QX; (0262) 670251; Fax (0262) 400071

The Shredded Wheat Co., Bridge Road, Welwyn Garden City, Hertfordshire AL7 1RR; (0707) 325100

The Silver Spring Mineral Water Company Ltd, Park Farm, Folkestone, Kent CT19 5EA; (0303) 43751; Telex 966329; Fax (0303) 56524

Simmers Ltd, (Head Office) Romano House, 43 Station Road, Corstorphine, Edinburgh EH12 7AQ; 031-334 0852; Telex 727369; Fax 031-334 7753

Simmers Ltd, 7 Station Road, Hatton, Grampian AB4 7RY; (0779) 84631; Fax (0779) 84325

Simmers Ltd, Mitchelhill Biscuit Factory, 90 Peffermill Road, Edinburgh EH16 5UU; 031-661 3384; Fax 031-652 0790

A.L. Simpkin & Co. Ltd, 3 Hunter Road, Hillsborough, Sheffield, South Yorkshire S6 4LD; (0742) 348736; Fax (0742) 325635

E. Simpson Ltd, Buxton Road, Hazel Grove, Stockport, Cheshire SK7 6LZ; 061-483 1241

W. Simpson Ltd, 95–97 Wassand Street, Hull, HU3 4AL; (0482) 24701 or 213172; Fax (0482) 216424

Simpson Ready Foods Ltd, Stretford Road, Urmston, Manchester M31 1WH; 061-865 2241; Telex 668640; Fax 061-865 2893

Sims Food Group plc, Sims House, Sims Food Park, Sherborne Drive, Tilbrook, Milton Keynes MK7 8BS; (0908) 270061; Fax (0908) 270063

Singleton's Dairy Ltd, Mill Farm, Preston Road, Preston, Lancashire PR3 3AN; (0772) 782112; Fax (0772) 785373

Sitoni Foods Ltd, Sitoni House, St Mary's Street, Whitchurch, Shropshire SY13 1QY; (0948) 5432

A.C. Skelton & Sons Ltd, Lorraine Street, Stoneferry, Hull HU8 8EJ; (0482) 20052

Slaters Foods Ltd, Briscoe Lane, Newton Heath, Manchester M10 6XG; 061-231 4004; Telex 6660075

SMA Nutrition Ltd, Huntercombe Lane South, Taplow, Maidenhead, Berkshire SL6 0PH; (0628) 660633; Fax (0628) 666368

F. Smales & Sons (Fish Merchants) Ltd, West Dock Street, Hull HU3 4HL; (0482) 24997; Telex 597260; Fax (0482) 23765

Smedley's Foods Ltd, Wardentree Lane, Pinchbeck, Spalding PE11 3OY; (0775) 710789; Fax (0775) 710504

T.G. Smith (Fine Foods) Ltd, Orrell Mount Industrial Estate, Hawthorne Road, Bootle, Merseyside L20 6NS; 051-933 3616; Telex 628389

Rowland Smith & Son, St Saviour's House, 60 George Row, London SE16 4UD; 071-231 3421; Telex 888185

Walter Smith (Birmingham) Ltd, 366 Bromford Lane, Ward End, Birmingham B8 2RZ; 021-327 2201

SmithKline Beecham plc, SB House, Brentford, Middlesex TW8 9BD; 081-560 5151; Fax 081-847 0830 (registered office)

SmithKline Beecham plc, Consumer Brands, Royal Forest Factory, Coleford, Gloucestershire GL16 8JB

Smiths Flour Mills Ltd, Albion Mills, PO Box 3, Eastgate, Worksop, Nottinghamshire S80 1QY; (0909) 472216

J. Smylie (Birkenhead) Ltd, 439/447 Cleveland Street, Birkenhead L41 4JW; 051-647 4264

The Snack Factory Ltd, Pennine Place, Skelmersdale, Lancashire WN8 9QF; (0695) 50230

Sodastream Ltd, Morley Way, Woodston, Peterborough, Cambridgeshire PE2 0BS; (0733) 231121; Telex 32443; Fax (0733) 231234

Solent Canners Holdings Ltd, Stag House, Solent Road, Western Docks, Southampton SO9 5AQ; (0703) 30556

Solway Foods, 3 Godwin Road, Earlstree Industrial Estate, Corby, Northamptonshire; (0536) 401400

Somerset Creameries Ltd, Cropwell Bishop, Nottingham NG12 3BQ; (0602) 892350; Fax (0602) 899046

Sooner Foods Ltd, Cottage Beck Road, Scunthorpe, South Humberside DN16 1TT; (0724) 855001 or 281222 (sales); Telex 527336; Fax (0724) 280891

Sous Chef Ltd, 49 Parkway, Deeside Industrial Estate, Deeside, Clwyd CH5 2NF; (0244) 830880

South City Food Co. Ltd, Ventura House, 176–188 Acre Lane, London SW2 5UL

South West Meat Ltd, Millfield Industrial Estate, Chard, Somerset TA20 2BB; (0460) 65491; Fax (0460) 61092

Soutra Foods, East Mains Industrial Estate, Broxburn, Edinburgh EH52 5AU; (0506) 854721

Sovereign Food Group Ltd, Sovereign House, Eye, Suffolk IP23 7AH; (0379) 870661; Telex 97337; Fax (0379) 870661 (registered office)

Sovereign Chicken Ltd, Freebournes Road, Witham, Essex; (0376) 517523

Sovereign Chicken Ltd, Westdown Lane, Burwash Common, Etchingham, Sussex; (0435) 883401

Sovereign Chicken Ltd, New Buckenham Road, Attleborough, Norfolk; (0953) 455506

Speciality Foods Ltd, Rakusen House, Clayton Wood Rise, Ring Road, West Park, Leeds LS16 6QN; (0532) 784821; Telex 556349; Fax (0532) 784064

Speedibake Ltd, Chald Lane, Wakefield, West Yorkshire WF2 9RJ; (0924) 374201

Speedibake Ltd, Gladstone Road, Northampton NN5 7QE; (0604) 587774; Fax (0604) 583579

Keith Spicer Ltd, Ferndown Industrial Estate, Wimbourne, Dorset BH21 7PN; (0202) 875321

Spillers Foods Ltd, New Malden Hill, 1 Blagdon Road, North Malden, Surrey KT3 4TB; 081-949 6100; Telex 8952748

Spillers Foods Ltd, Glasgow Road, Barrhead G78 1TQ; 041-881 9221

Spillers Premier Products Ltd, High Street, Puckeridge, near Ware, Hertfordshire SG11 1RW; (0920) 821291; Telex 81352

R.R. Spink & Sons (Arbroath) Ltd, 33/35 Seagate, Arbroath, Tayside DD11 1BJ; (0241) 72023; Fax (0241) 75663

G.R. Spinks & Co. Ltd, Exeter Road, Okehampton, Devon EX20 1QA; (0837) 54292; Fax (0837) 53592

Squires & Kintons Ltd, Humberstone Lane, Leicester LE4 7JS; (0533) 693921

St Edmunds Bacon Factory Ltd – see Harris (Elmswell)

St Giles Foods Ltd, St Giles House, Sandhurst Road, Sidcup, Kent DA15 7HL

St Ivel Ltd, St Ivel House, Interface Business Park, Wootton Bassett, Swindon SN4 8QF; (0793) 848444

St Ivel Ltd, Dorcan House, Eldene Drive, Swindon, Wiltshire SN3 3TU; (0793) 488333

St Merryn Meat Ltd, Unit 9, Cooksland Industrial Estate, Bodmin, Cornwall PL31 2TZ; (0208) 78155

St Merryn Meat Ltd, Probus, Truro, Cornwall (0726) 882405; Telex 45316

St Merryn Meat Ltd, School Lane, Torrington, Devon EX38 7EX; (0805) 23371

Stantons Bakeries, Western Road, Hednesford, Staffordshire WS12 4AT; (054 38) 2381

Starbake (Anglia) Ltd, The Anglia Bakery, Midland Road, Peterborough, Cambridgeshire; (0733) 65773

Starbake Ltd, The Star Bakery, Wern Avenue, Rogerstone, Newport, Gwent; (0633) 895797

Starbake Yorkshire Ltd, Stephenson Way, Wakefield 41 Industrial Estate, Wakefield, WF2 0XN; 0924 820936

Star Confections Ltd, Star Confectionery Works, 41–43 Maddison Street, Blyth, Northumberland NE24 1EY; (0670) 353192

Steak Express Ltd, Hibbert Street, Whitehill Industrial Estate, Stockport, Cheshire SK4 1NS; 061-480 0506; Telex 665462; Fax 061-480 1538

Stella Products Ltd, Barlows Lane, Fazakerley, Liverpool L9 9EH; 051-523 6000; Fax 051-530 1769

Stevens & Brotherton Ltd, S & B House, 2 Vinson Close, Knoll Rise, Orpington, Kent BR6 0XG; (0689) 877799

D. & R. Stevenson Ltd, Glen Street, Barrhead, Glasgow G78 1QA; 041-881 3218

William Stewart & Arnold Ltd, Denmark Street, High Wycombe, Buckinghamshire HP11 2ES; (0494) 25291; Telex 837562

A.G. Stewart & Co., 36 Commercial Road, Buckie, Grampian AB5 1TX; (0542) 31320; Fax (0542) 834156

Stewarts Supermarkets Ltd, Apollo Road, Belfast, Northern Ireland BRT12 6HP; (0232) 668 222

Stocks Lovell Ltd, Mill Street, Evesham, Worcestershire WR11 4PR; (0386) 49433; Telex 338329; Fax (0386) 49803

A.K. Stoddart Ltd, Gorgie Abattoir, Edinburgh EH14 1RH; 031-443 4676 or 031-443 6317; Telex 72470; Fax 031-444 0734

Stokes Bomford Ltd, Martley Road, Lower Broadheath, Worcestershire WR2 6RF; (0905) 640521; Telex 338541

Stonegate Farmers Ltd, 15 North Street, Hailsham, Sussex BN27 1DH; (0323) 846565; Telex 877883; Fax (0323) 840439

Strathmore Foods Ltd, Carseview Road, Forfar, Tayside DD8 3NG; (0307) 462333

Stroud Creamery Ltd, Lansdown Road, Stroud, Gloucestershire GL5 1RA; (045 36) 2351

Studleigh–Royd Ltd, Ainleys Industrial Estate, Elland, Halifax HX5 9JP; (0422) 310668; Fax (0422) 310695

Studleigh–Royd Ltd, Ogden Road, Shaw Lane Industrial Estate, Doncaster, DN2 4SQ; (0302) 329689; Fax (0302) 329882

Stute Foods Ltd, Clifton Heights, Triangle West, Bristol BS8 1EG; (0272) 291406; Telex 449449; Fax (0272) 226212

Suffolk Foods Ltd, Rendlesham Hall, Rendlesham, Woodbridge, Suffolk IP12 2RG; (0394) 460816

Sunbird Foods Ltd, Index House, London Road, Ascot SL5 7EU; (0990) 27323; Telex 848752

Sunblest Bakeries (Sheffield), 710A Penistone Road, Sheffield S6 2DF; (0742) 344261

Sunblest Bakeries (Yorkshire), PO Box No. 220, Dick Lane, Bradford BD4 8PP; (0274) 664311

Sunblest Bakeries (Brighton), Falmer Road, Woodingdean, Brighton, Sussex BN2 6QR; (0273) 304091

Sunblest Bakeries (Stevenage), Cartwright Road, Stevenage, Hertfordshire SG1 4QA; (0438) 3596111

Sunblest Bakeries (Aberdeen), Northfield Bakery, Granitehill Road, Aberdeen AB9 2AX; (0224) 694058; Fax (0224) 693459

Sunblest Bakeries (Dundee), 5 Lytton Street, Dundee DD2 1UE; (0382) 69292

Sunblest Bakeries (Edinburgh), Lochend Road, Edinburgh EH7 6DB; 031-554 2666

Suncharm Ltd, Queens Square, Honley, Huddersfield, West Yorkshire HD7 2QZ; (0484) 663916; Telex 517616; Fax (0484) 666975

Sun Cream Dairies Ltd, Glascote Road, Tamworth, Staffordshire B77 2BS; (0827) 54196 or 58028; Fax (0827) 58028

Sundora Foods Ltd, Burnby Lane, Pocklington, Yorkshire YO4 2QB; (075 92) 2365; Telex 57758

Sunmalt, The Bakery, Littleborough, Lancashire OL15 8LA; (0706) 72211

Sun Valley Ltd, Marion Street, Birkenhead, Merseyside L41 6JY; 051-647 5356; Fax 051-666 1683

Sun Valley Foods Ltd, The Hardwick, Abergavenny; (0873) 859494; Fax (0873) 852918

Sun Valley Foods Ltd, Unit 7, Gilchrist Thomas Industrial Estate, Blaenavon; (0495) 790798

Sun Valley Poultry Ltd, Grandstand Road, Hereford HR4 9PB; (0432) 352400; Telex 35258; Fax (0432) 342270

Sunwheel Foods, Granary House, Wetmore Road, Burton-upon-Trent, Staffordshire DE14 1TE; (0283) 43221; Telex 341929

Supreme Salt Co. Ltd, Spa Street, Cobridge, Stoke-on-Trent ST6 2LN; (0782) 25411; Telex 36554

Surrey Hills Foods Ltd, 208 Durnsford Road, Wimbledon Park, London SW19 8DR; 081-946 9619; Fax 081-946 0460

Sussex Dairies Ltd, 96 Southview Road, Southwick, Brighton, East Sussex BN4 4TT; (0273) 593174; Telex 877980

E.T. Sutherland & Son Ltd, Waleswood Factory, Kiveton, Sheffield S31 8PF; (0909) 482604; Fax (0909) 500884

Suttons Bakeries, Maudslay Road, Coventry CV5 8EH; (0203) 73351

Sweetmate Ltd, Pasadena Close, Hayes, Middlesex UB3 3NQ; 081-848 9749; Fax 081-569 1664

Sweet 'n Savoury Frozen Foods Ltd, 2/32 Oakesway, Hartlepool Industrial Estate, Cleveland TS24 0RE; (0429) 221621

Swizzels Matlow Ltd, Carlton House, New Mills, Stockport, Cheshire SK12 3HA; (0663) 744144; Telex 669685; Fax (0663) 742800 (registered office)

Symbol Biscuits Ltd – see Lyons Biscuits

W. Symington & Co., Springfield Street, Market Harborough, Leicestershire; (0858) 66511

T

Tara Meats (UK) Ltd, 7/15 Welbeck Road, Glasgow G53 7RD; 041-848-5348; Fax 041-848-6848

Tate & Lyle plc, Sugar Quay, Lower Thames Street, London EC3R 6DQ; 071-626 6525; Telex 884084; Fax 071-623 5213 (registered office)

Tate & Lyle, Enterprise House, 45 Homesdale Road, Bromley, Kent BR2 9TE; 081-464 6556; Telex 91784; Fax 081-290 0721

Tate & Lyle Sugars, Thames Refinery, Silvertown, London; 071-476 4455

Tate & Lyle Sugars, Westburn Refinery, 150 Lynedock Street, Greenock PA15 4BE; (0475) 21241

Taveners plc, Beech Street, Liverpool L7 0HA; 051-263 3451; Telex 627275; Fax 051-264 8951 (registered office)

Tay Valley Patisserie Ltd, 119 Clepington Road, Dundee; (0382) 458666

Taylors Bacon Packers (Yorkshire) Ltd, Greenhill Mills, Mytholmroyd, Hebden Bridge, West Yorkshire HX7 5QF; (0422) 885094

Henry Telfer Ltd, Redhouse Road, Moulton Park, Northampton NN3 1NP; (0604) 790666; Telex 311332; Fax (0604) 491281

Telford Foods Ltd, Halesfield 10, Telford, Shropshire TF7 4LY; (0952) 680426

Telmark Wholesale Meats Ltd, Unit J13, Halesfield, Telford, Shropshire TF7 4QT; (0952) 680323; Fax (0952) 587612

Tendafrost Frozen Foods Ltd, 49 Winsover Road, Spalding, Lincolnshire PE11 1EG; (0775) 67571; Telex 32551; Fax (0775) 69515

Tendafrost Frozen Foods Ltd (Scotfresh), Industrial Estate, Eyemouth, Berwickshire TD14 5AN; (0390) 50656; Telex 727295

Tenderchoice Meats Ltd, Fuldene House, Ring Road, Seacroft, Leeds LS14 1NH; (0532) 323420; Fax (0532) 323394

Terry's Suchard Ltd, The Chocolate Works, Bishopthorpe Road, York YO1 1YE; (0904) 653090; Telex 57981; Fax (0904) 623434

Terry's Suchard Ltd, Cowley Lodge, Guildford Street, Chertsey, Surrey KT16 9NY; (0932) 574000; Fax (0932) 574299

Terry's Suchard Ltd (Callard & Bowser), Waterton, Bridgend, Mid Glamorgan CF31 3DJ; (0656) 658011; Fax (0656) 660215

Tesco Stores plc, PO Box 18, Cheshunt, Hertfordshire EN8 9SL; (0992) 32222; Fax (0992) 30794

Thames Valley Eggs Ltd, Membury, Lambourn Woodlands, Newbury, Berkshire RG16 7TX; (0488) 71101; Telex 848894; Fax (0488) 72969

J. Thayer & Sons Ltd, Wentloog Road, Rumney, Cardiff CF3 8ED; (0222) 778808; Telex 497992

Thomas's Petcraft, Pellon Lane, Halifax, West Yorkshire HX1 5QP; (0422) 50221

Thomas's Petcraft, Oakwell Way, Birstall, Batley, West Yorkshire WF17 9LU; (0924) 47373; Telex 556361; Fax (0924) 470961

R.H. Thompson & Co. Ltd, Wimbourne House, 136 High Street, Sevenoaks, Kent TN13 1XA; (0732) 740266; Telex 957608; Fax (0732) 740419

Thorne Poultry Ltd, Coulman Street, Thorne, near Doncaster, South Yorkshire DN8 5JT; (0405) 817200; Fax (0405) 740100

Thornhill's Country Produce Ltd, St John's House, St John Street, Ashbourne, Derbyshire DE6 1GH; (0335) 42455; Telex 377206

Thornley & Sons Ltd, Ridgewood Farm, Flag Lane, Heath Charnock, near Chorley, Lancashire PR6 9ED; (025 72) 71251

Thorntons plc, Knowle House, 4 Norfolk Park Road, Sheffield S2 3QE (registered office)

Thorntons plc, Confectionery Works, Castle Factory, Derwent Street, Belper, Derbyshire DE5 1WP; (0773) 824181; Telex 377835; Fax (0773) 824874

Thorntons plc, Thornton Park, Somercotes, Derby DE55 4XJ; (0773) 608822; Fax (0773) 540757

Ticklemore Cheese Dairy, Sharpham Barton Lane, Ashprington, Harbertonford, Totnes, Devon; (0803) 732737. Ticklemore Cheese Shop, Totnes; (0803) 865926

Tiffany Sharwood Frozen Foods Ltd, Wainwright Road, Bexhill-on-Sea, East Sussex TN39 3UR; (0424) 216704 or 730704; Telex 95489

Tiffany Sharwood Frozen Foods Ltd, Gillingham Business Park, Gillingham, Kent ME8 0PP; (0634) 387051

Tillery Valley Foods Ltd, Unit 2, Cwmtillery Industrial Estate, Abertillery, Gwent NP3 1LZ; (0495) 211555; Fax (0495) 213839

Tilley's Sweets Ltd, Springfields, Oundle Road, Thrapston, near Kettering, Northamptonshire NN14 4PQ; (0832) 732151; Fax (0832) 734091

Tilley's of Cheltenham, Churchill Road, Leckhampton, Cheltenham, Gloucestershire GL53 7EH; (0242) 520076; Fax (0242) 239545

E. Timm & Son Ltd, 145 Boothferry Road, Goole, North Humberside DN14 6AX; (0405) 4271; Telex 57646; Fax (0405) 5671

Tinsley Foods Ltd, Holbeach St Marks, Spalding, Lincolnshire PE12 8HF; (040 632) 301; Telex 32366; Fax (040 632) 214

Tinsley Foods Ltd, West Deeping, Peterborough, Lincolnshire PE6 9HR; (0778) 342217 (registered office)

Tip Top Bakeries Ltd, Cray Avenue, Orpington, Kent BR5 3RT; (0689) 25621

Emile Tissot Foods Ltd, Stafford Park 10, Telford, Shropshire TF3 3BE; (0952) 290571; Telex 35882; Fax (0952) 292575

Toffee de Beurre Ltd, Kenneth Street, Ingram Road, Leeds, West Yorkshire LS11 9RG; (0532) 456130; Fax (0532) 439798

Tolona Pizza Products Ltd – see Dansco-Tolona Pizza Products

Tooks, 731 Norwich Road, Ipswich, Suffolk IP1 6LD; (0473) 241020

Top Hat Foods Ltd, West Pitkerro, Dundee DD5 3RZ; (0382) 737855; Fax (0382) 480820

Tower Farms, Deans Cross, Lydeard St Lawrence, Taunton, Somerset TA4 3QN; (09847) 252

Towers & Co. Ltd, 15a Tottenham Lane, Hornsey, London N8 9DJ; 081-340 0710; Fax 081-340 3813

Towers Barron, 202 Parkway, Worle, Weston-super-Mare, Avon BS22 0WA; (0934) 520420; Fax (0934) 520909

Towers Tendercut, Brickfield Lane, Chandler's Ford, Hants SO5 3DP; (0703) 252444; Fax (0703) 268120

Town & Country Petfoods Ltd, 26 Asfordby Road, Melton Mowbray, Leicestershire LE13 0HR; (0664) 63209; Telex 342147; Fax (0664) 62149

Barbara Townsend Ltd, The Warehouse, Lower Broadoak Road, West Hill, Ottery St Mary, Devon EX11 1UD; (040 481) 4700

Trafford Park Bakery, Ingersoll-Rand Building, Ashburton Road West, Trafford Park, Manchester M17 1QX; 061-873 7724

J. & J. Tranfield Ltd, 161 Retford Road, Sheffield S13 9LD; (0742) 691111

Treats Ice Cream Ltd, Manston Lane, Crossgates, Leeds LS15 8SX; (0532) 601111

Trebor Bassett Ltd, Hertford Place, Denham Way, Maple Cross, Hertfordshire WD3 2XB; (0923) 896565; Fax (0923) 897345

Trebor UK Ltd, Trebor House, Woodford Green, Essex 1GB 8EX; 081-550 8800; Telex 897572; Fax 081-550 8800

Trebor UK Ltd, Kreemy Works, St Peter's Street, Maidstone, Kent; (0622) 57421

Trebor UK Ltd, Birmingham Road, Chesterfield; (0732) 848599

Trebor UK Ltd, Sverals Road Industrial Estate, 2 Wymrolls Road, Colchester; (0206) 844333

Trentham Factory, Newstead Industrial Estate, Trentham, Stoke-on-Trent, Staffs ST4 8HX; (0782) 659125; Fax (0782) 641904

Tryton Foods Ltd, Trinity Street, Hull, North Humberside HU3 1EW

Tryton Inn Frozen Foods, 40 Derringham Street, Hull, North Humberside HU3 1EW; (0482) 224131; Telex 597553; Fax (0482) 24526

T.S. Confectionery, Leigh Green Industrial Estate, Unit 20, Appledore Road, Tenterden, Kent TN30 7DF; (0580) 764110; Fax (0580) 763954

Tudor Dairies Ltd, High Street, Henley in Arden, Warwickshire B95 5BT; (0564) 792141; Fax (0564) 794043

Tulip International (UK) Bacon Division Ltd, Caxton Way, Thetford, Norfolk IP24 3SB; (0842) 754521; Fax (0842) 765804

Tulip International (UK) Cooked Meat Division Ltd, Alexandra Road, New Brighton, Wallasey, Merseyside L45 0JZ; 051-639 8181

Tunnel Refineries, Thames Bank House, Tunnel Avenue, Greenwich, London SE10 0PA; 081-858 3271; Fax 081-305 0981

Thomas Tunnock Ltd, 34 Old Mill Road, Uddingston, Glasgow G71 7HH; (0698) 813551; Telex 779453; Fax (0698) 815691 (registered office)

W.A. Turner Ltd, Broadwater Lane, Tunbridge Wells, Kent TN2 5RD; (0892) 515215; Fax (0892) 510028

Turner & Son, Sunblest Bakery, Hucknall Road, Nottingham NG5 1FF; (0602) 606691

Turners Turkeys Ltd, Spalding Drove, Clay Lake, Spalding, Lincolnshire PE12 6BL; (0775) 767444; Telex 32716; Fax (0775) 712231

R. Twining & Co. Ltd, South Way, Walworth Industrial Estate, Andover, Hampshire SP10 5AQ; (0264) 62161; Telex 47218; Fax (0264) 51978

U

ULN (Union Laitiere Normande) UK Ltd, The Ham, Westbury, Wiltshire

Ulster Farmers Bacon Co. (Newry) Ltd, 92 Armagh Road, County Down, Northern Ireland BT35 6PR; (0693) 62702 or 63171; Telex 747473; Fax (0693) 67742

Ultrafresh Foods, Unit 32, Fourth Way, Wembley, Middx HA9 0LH

Unicopa U.K. Ltd, 119 Alma Road, Bournemouth, BH9 1AE; (0202) 535770; Fax (0202) 535768

Unigate Dairies Ltd, 14–40 Victoria Road, Aldershot, Hampshire GU11 1TH; (0252) 24522

Unigate Dairies (Midlands) Ltd, PO Box 99, 74 New Town Row, Birmingham, West Midlands B6 4HX; 021-359 6121

Unigate (Humberside) Ltd, Ram Boulevard, Foxhills Industrial Park, Scunthorpe DN15 8QW

Unigate Dairies Ltd, Unigate House, Western Avenue, London W3 0SH; 071-992 3400

Unigate plc, Unigate House, Wood Lane, London W12 7RP; 081-749 8888; Fax 081-749 7166

Unilever Research, Colworth Laboratory, Sharnbrook, Bedford MK44 1LQ; (0234) 222792; Telex 82229; Fax (0234) 222757

Unilever plc, Unilever House, Blackfriars, London EC4P 4BQ; 071-822 5252; Telex 28395 (head office)

Unilever plc, Port Sunlight, Wirral, Merseyside L62 4ZA (registered office)

Unimerchants Ltd, Riverview Lodge, Raymead Road, Maidenhead, Berkshire SL6 8N; (0628) 76677; Telex 847712

Unimerchants Ltd, Lancaster Way Business Park, Ely, Cambridge CB6 3NP; (0353) 661999; Fax (0353) 666199

Union International plc, 14 West Smithfield, London EC1A 9JL; 071-248 1212; Telex 883291; Fax 071-236 7627

Union International Consultants, 14 West Smithfield, London EC1A 9JN; 071-248 1212; Telex 883291; Fax 071-236 7627

Union International Technology Centre, 24–30 West Smithfield, London EC1A 9DL; 071-248 1212; Telex 883291; Fax 071-236 7627

Unipork Ltd, 70 Molesworth Road, Cookstown, County Tyrone BT80 8PJ; (064 87) 63321 or 62405; Telex 744376; Fax (064 87) 63195

Unipork Ltd, Cornagrade Road, Enniskillen, County Fermanagh, Northern Ireland BT74 6DZ; (0365) 22761 or 322820

Unique Cuisine, 900 Stinchar Road, Prestwick, Ayrshire KA9 2PE; (0292) 75640; Fax (0292) 74012

United Biscuits (Holdings) plc, Church Road, West Drayton, Middlesex UB7 7PR; (0895) 432100; Fax (0895) 448848

United Molasses Co., Sugar Quay, Lower Thames Street, London EC3R 6DQ; 071-626 8700; Telex 8951976

United Yeast Co. Ltd (UYC), Unecol House, 819 London Road, North Cheam, Sutton, Surrey SM3 9AT; 081-337 6633

V

Vale of Mowbray Bacon Co. Ltd – see Harris (Leeming Bar)

Van den Bergh Foods Ltd, PO Box 8, Godley, Hyde, Cheshire SK14 3BR; 061-368 4080

Van den Bergh Foods Ltd, 173 Manor Road, West Ham, London E15 3DA; 081-534 4255; Telex 897721

Van den Bergh Foods Ltd, South Bar, Banbury, Oxfordshire OX16 8QL; (0295) 272121; Telex 837636; Fax (0295) 273796

Van den Bergh Foods Ltd, Rennys Lane, Gilesgate Moor, Durham, County Durham DH1 2QX

Van den Berghs & Jurgens Ltd, Sussex House, Burgess Hill, West Sussex RH15 9AW; (044 46) 46300; Telex 877381

Van den Bergh Oils Ltd, London Road, Purfleet, Essex RM16 1SD; (0708) 863300; Fax (0708) 860757

Van der Laan (UK) Ltd – see Rea Valley

Van Heyningen Bros. Ltd, Toddington Lane, Littlehampton, West Sussex BN17 7PP; (0903) 717125; Fax (0903) 731342

Van Melle Ltd, Western Way, Bury St Edmunds, Suffolk IP33 3SY; (0284) 63431; Telex 81596; Fax (0284) 63435

Peter Vassallo Ltd, Unit 1, St Nicholas Hospital Grounds, Jubilee Road, Gosforth, Newcastle-upon-Tyne, Tyne & Wear NW3 3XT; 091-285 5999; Fax 091-284 0183

Veeraswamy's (Food Products) Ltd, 20 Swallow Street, Regent Street, London W1R 7HE; 071-437 1414; Sales office (0444) 414208; Fax (0444) 441424

Vegetarian Feasts Ltd, 21 Carnwath Road, Fulham, London SW6 3HR; 071-731 1425; Telex 295655

Vico Foods (UK) Ltd, 3A Bridgwater Close, Reading, Berkshire RG3 1JT; (0734) 580163; Telex 849946

Viota Foods Ltd, Port Causeway, Bromborough, Wirral, Merseyside L62 4TH; 051-334 6261; Telex 627032; Fax 051-334 9962

Viskase Ltd, Kingsway, Fforestfach, Swansea, West Glamorgan SA5 4DL; (0792) 582585; Telex 48237; Fax (0792) 582574

G.D. Vivers & Sons, Scotch Lamb Specialists, Dornocktown, Annan, Dumfriesshire DG12 6SU; (0461) 40241; Telex 778854

W

Waitrose Ltd, Southern Industrial Area, Bracknell, Berkshire RG12 4YA; (0344) 424680

Waldens Wiltshire Foods Ltd, Canal Road Trading Estate, Ladydown, Trowbridge, Wiltshire BA14 8RJ; (0225) 753636; Telex 449906

Walker & Son (Leicester) Ltd, Charnwood Bakery, Madeline Road, Bursom Industrial Estate, Beaumont Leys, Leicester LE4 1EX; (0533) 340033; Fax (0533) 340084

Matthew Walker (Derby) Ltd, Heanor Gate Road, Heanor, Derbyshire DE7 7RB; (0773) 760121; Telex 378379; Fax (0773) 530367

Walkers Crisps Ltd, Cobden Street, Leicester, Leicestershire LE1 2LB; (0533) 691691

Walkers Crisps Ltd, Feature Road, Thurmaston, Leicester, Leicestershire LE4 8BS; (0533) 691691; Fax (0533) 691776

Walkers Crisps Ltd, 11 Bursom Road, Bursom Industrial Estate, Beaumont Leys, Leicester LE4 1BS; (0533) 357922

Walkers' Nonsuch Ltd, Calverley Street, Longton, Stoke-on-Trent, Staffordshire ST3 1QS; (0782) 321525; Telex 367157; Fax (0782) 599449

Walkers Shortbread Ltd, Aberlour-on-Spey, Grampian, Scotland AB38 9PB; (0340) 871555; Fax (0340) 871355

Wander Ltd, Station Road, Kings Langley, Hertfordshire WD4 8LJ; (0923) 266122; Telex 922747

Warburtons Ltd, Back o' th' Bank House, Blackburn Road, Bolton, Lancashire BL1 8HJ; (0204) 23551 (registered office)

Warburtons Ltd, Hereford Street, Bolton BL1 8JB; (0204) 23551

Warburtons (Tyne & Wear) Ltd, 26/28 Shelley Road, Newburn Industrial Estate, Newburn, Newcastle-upon-Tyne NE15 9RT; 091-229 0166

Warburtons (Yorkshire) Ltd, Westgate End, Wakefield, West Yorkshire WF2 9NH; (0924) 384090

Warburtons (Colne Valley) Ltd, Billington Road, off Rossendale Road, Burnley BB11 5BX; (0282) 56311

Warburtons (Colne Valley) Ltd, Brinwell Road, Blackpool, Lancashire (0253) 697118

Warburtons (Pennine) Ltd, Glebe Street, Shaw, Oldham, Lancashire OL2 7SF; (0706) 847744

Warburtons (Soreen) Ltd, Taylor Road, Urmston, Manchester M31 2JQ; 061-748 1235

Warner Lambert Confectionery Ltd, Dumers Lane, Radcliffe, Manchester M26 9QT; 061-766 5471; Telex 667074; Fax 061-796 0670 (registered office)

Waterdene Frozen Foods Ltd, Unit B, Chiltern Trading Estate, Grovebury Road, Leighton Buzzard, Bedfordshire LU7 8TU; (0525) 372367

Waterford Dairies Ltd (Heald's), Elm Grove, Didsbury, Manchester M20 0RL; 061-445 8841; Fax 061-434 5602

Waterford Foods UK Ltd, Unit 12, Hambridge Lane, Newbury, Berkshire RG13 1HZ; (0635) 33556

Waterford Foods UK Ltd (Heald's Foods), 2 Broadway, Hyde, Cheshire SK14 4QQ

Waterford Foods UK Ltd (Heald's Juices), Grimshaw Lane, Newton Heath, Manchester M10 6BB; 061-202 1044; Fax 061-205 7741

Webbs Country Foods Ltd, Halnaker, Chichester, West Sussex PO18 0NF; (0243) 773491

Webbs Country Foods Ltd, Moorside Road, Winchester, Hampshire SO23 7RX; (0962) 773491; Telex 477487; Fax (0962) 68421

Webbs Country Foods Ltd, Bridge Road, Lymington, Hampshire SO41 9YW; (0590) 75280; Telex 477495; Fax (0509) 78139

Webbs Country Foods Ltd, Okeford Fitzpaine, Blandford Forum, Dorset DT11 0RQ; (0258) 860304

Webbs Country Foods Ltd, Sutton Benger, Chippenham, Wiltshire SN15 4RS; (0249) 720733; Telex 44787; Fax (0249) 720958

Webbs Poultry & Meat Group (Holdings) Ltd, Station Road, Cullingworth, Bradford, West Yorkshire BD13 5HN; (0535) 45232; Telex 51485; Fax (0535) 46741

Webbs Poultry & Meat Group Ltd, Bridgehouse Mills, Haworth, West Yorkshire BD22 8PA; (0535) 654232

E.W. Webster Ltd, Wash Farm, Rainford Road, Bickerstaffe, Ormskirk, Lancashire L39 0HG; (0695) 24322; Telex 627401; Fax (0695) 50363

Weddel Swift Ltd, Burlington House, Crosby Road North, Waterloo, Liverpool L22 0PT; 051-920 7288; Telex 627066; Fax 051-928 9013

Weetabix Ltd, Weetabix Mills, Burton Latimer, Kettering, Northamptonshire NN15 5JR; (0536) 722181; Telex 34685; Fax (0536) 726148 (registered office)

Y

Yeo Valley Farms Ltd, Lag Farm, Blagdon, Bristol, Avon BS18 6SG; (0761) 62798; Telex 444921; Fax (0761) 62181

Yoplait Dairy Crest Ltd, Eastman Way, Maylands Industrial Estate, Hemel Hempstead, Herts HP2 7EL; (0442) 61166; Fax (0442) 256927

Yoplait Dairy Crest Ltd, Kingsway Business Park, Unit F, Oldfield Road, Hampton, Middlesex TW12 2HD; 081-941 9969

York Dragee Co. Ltd, Unit 3, Block 8, Pocklington Industrial Estate, Yorkshire YO4 2NR; (0759) 305127; Fax (0759) 305128

York House Foods Ltd, York House, Belmont Road, Uxbridge, Middlesex UB8 1TD; (0895) 38250; Telex 934277

York House (Meat Products) Ltd, Shannon Place, Potton, Bedfordshire; (0767) 260114; Fax (0767) 262165

Yorkshire Biscuits Ltd, Bridgehouse Mills, Haworth, Keighley, West Yorkshire BD22 8PA; (0535) 43508

Yorkshire Dales Dairy Ice Cream, Aireside Mills, Cononley, Keighley, West Yorkshire BD20 8LW

Yorkshire Food Group, Carter Mills, 146 Cleckheaton Road, Bradford, West Yorkshire BD12 0HF; (0274) 691285; Fax (0274) 604971

Young & McMillan (Bacon) Ltd, 86 Strathmore Road, Balmore Industrial Estate, Glasgow G22 7TY; 041-336 4311

Young's Seafoods Ltd, Ixworth House, Ixworth Place, London SW3 3PU; 071-730 9942; Telex 263735

Yours Tastefully Ltd, Forge House, Collingwood Industrial Centre, Maidstone Road, Sutton Valence, Maidstone, Kent ME17 3LS; (0622) 842152; Telex 842152

Z

Zest Foods Ltd, No 44 Mochdre Estate, Newtown, Powys SY16 4LE

Part Four

SOURCES OF FURTHER INFORMATION

The following sections list some of the organizations associated with the food industry. Inevitably a handbook of this kind can only cover a small proportion of the vast amount of information available and readers may need to look for additional sources of information.

Many of the trade, educational and government organizations listed below may be able to give further information about their particular field and there are several directories which give extensive coverage of companies in the food and drinks industry.

4.1 DIRECTORIES

Yellow Pages

Many public libraries contain complete sets of Yellow Pages which can be a useful source of information on food companies. In addition all the information from Yellow Pages is also accessible via the Business Database. Companies can be located by business classification, geographical area, size, postcode, trading activity, etc.

The Business Database

Available from: Walpole House, 18/20 Bond Street, Ealing, London W5 5AA: 081-567 7300; Fax 081-840 4150.

Grocer Directories 1994

A two-volume set – *Marketing Directory* and *Food & Drink Directory* – available from: William Reed Directories Ltd, Broadfield Park, Crawley, West Sussex RH11 9RT: (0293) 613400, ext 290; Fax (0293) 613600

The Baking Industry Directory

Available from: EMAP Response, Wentworth House, Wentworth Street, Peterborough PE1 11DS.

Master Bakers' Handbook & Buyers Guide

Available from: Turret Group, Turret House, 171 High Street, Rickmansworth, Herts WD3 1SN.

Milling European Directory

Available from: Turret Group, Turret House, 171 High Street, Rickmansworth, Herts WD3 1SN.

European Food Trades Directory

A two-volume set, United Kingdom & Europe, available from: Newman Books Ltd, 32 Vauxhall Bridge Road, London SW1V 2SS: 071-973 6402; Fax 071-233 5056.

Who's Who in the Meat Industry

Available from: Yandell Publishing Ltd, 9 Vermont Place, Tongwell, Milton Keynes MK15 8JA: (0908) 613323; Fax (0908) 618529.

Frozen and Chilled Foods Yearbook

and

Fish Trader Yearbook

Available from: FMJ International Publications Ltd, Queensway House, 2 Queensway, Redhill, Surrey RH1 1QS: (0737) 768611; Fax (0737) 761685.

4.2 SMALL COMPANIES

In addition to those listed in this guide there are many other small and speciality food producers in the UK. *Food From Britain* holds information about these companies, which are generally grouped by geographical area – including *Taste of the West*, *Taste of Wales*, *Taste of Anglia*, etc. Copies of a directory produced in 1989 entitled '*Food Focus 2*' [ISBN 0-902373-08-0] are still available from: Food from Britain 301–344 Market Towers, New Covent Garden Market, London SW8 5NQ; 071-720 2144; Fax 071-627 0614.

4.3 ORGANIC FOODS

Information on these foods can be obtained from the Soil Association, UKROFS, and British Organic Farmers:

United Kingdom Register of Organic Food Standards (UKROFS)
Food from Britain, 301–344 Market Towers, New Covent Garden Market, London SW8 5NQ; 071-720 2144; Fax 071-627 0614

Soil Association
86 Colston Street, Bristol BS1 5BB; (0272) 290661; Fax (0272) 252504

Note: British Organic Farmers and the Organic Growers Association also operate from the same address as the Soil Association.

4.4 TRADE AND RESEARCH ASSOCIATIONS

Association of British Meat Processors, 217 Central Markets, London EC1A 9LH; 071-489 0005; Fax 071-248 4733

Association of Meat Inspectors, 44 Parkfield Road, Taunton, Somerset TA1 4SF; (0823) 333201

Association of Public Analysts, 30 Russell Square, London WC1B 5DT

Biscuit, Cake, Chocolate & Confectionery Alliance, 11 Green Street, London W1Y 3RF; 071-629 8971; Telex 24738; Fax 071-493 4885

British Chicken Association, High Holborn House, 52–54 High Holborn, London WC1V 6SX; 071-242 4683; Fax 071-831 0624

British Chicken Information Service, Bury House, 126–128 Cromwell Road, London SW7 4ET; 071-373 7757; Fax 071-373 3926

BFMIRA – see Leatherhead Food RA

BMMA – see British Meat Manufacturers' Association

British Meat Manufacturers' Association, 18–19 Cornwall Terrace, London NW1 4QP; 071-935 7980; Telex 262027; Fax 071-487 4734

British Frozen Food Federation, 2nd Floor, Barclays Bank Chambers, 55 High Street, Grantham, Lincolnshire NG31 6NE; (0476) 590194; Fax (0476) 590152

British Hotels, Restaurants & Caterers Association, 40 Duke Street, London W1M 6HR; 071-499 6641; Telex 296619; Fax 071-355 4596

British Refrigeration Association, Sterling House, 6 Furlong Road, Bourne End, Buckinghamshire SL8 5DG; (06285) 31186

British Sandwich Association, 29 Market Place, Wantage, Oxon OX12 8BG; (0235) 772207; Fax (0235) 769044

British Soft Drinks Association, 20–22 Stukeley Street, London WC2B 5LR; 071-430 0356; Fax 071-831 6014

British Standards Institution, British Standards House, 2 Park Street, London W1A 2BS; 071-629 9000; Fax 071-629 0506

Also at: British Standards Institution, Linford Wood, Milton Keynes MK14 6LE
Information Section on (0908) 226888; Publications sales (0908) 221166; Quality and BS 5750 (0908) 220908

British Veterinary Association, 7 Mansfield Street, London W1M 0AT; 071-636 6541; Fax 071-436 2970

Campden Food & Drink Research Association, Chipping Campden, Gloucestershire GL55 6LD; (0386) 840319; Telex 337017; Fax (0386) 841306

Consumers Association, Castlemead, Gascoigne Way, Hertford SG14 1LH; (0992) 589031; Fax (0992) 500369

Dairy Trade Federation, 19 Cornwall Terrace, London NW1 4QP; 071-486 7244; Fax 071-487 4734

Department of Health – see Government Departments, Section 4.6

Federation of Bakers, 20 Bedford Square, London WC1B 3HF; 071-580 4252; Fax 071-255 1389

Federation of Fresh Meat Wholesalers, 227 Central Markets, London EC1A 9LH; 071-329 0776

Flour Advisory Bureau, 21 Arlington Street, London SW1A 1RN; 071-493 2521; Fax 071-493 6785

Flour Milling & Baking Research Association, Chorleywood, Rickmansworth, Hertfordshire WD3 5SH; (0923) 284111

Food Commission (The), 102 Gloucester Place, London W1H 3DA; 071-935 9078

Food & Drink Federation, 6 Catherine Street, London WC2B 5JJ; 071-836 2460; Telex 299388; Fax 071-836 0580. The Food and Drink Federation is host to a number of other small trade associations which operate from the same address and telephone number and are given at the end of Section 4.6*

Food From Britain, 301–344 Market Towers, New Covent Garden, London SW8 5NQ; 071-720 2144; Fax 071-627 0614

German CMA House, 17A Church Road, Wimbledon, London SW19 5DQ; 081-944 0484; Telex 918874; Fax 081-994 0441

Hannah Research Institute, Ayr, Scotland KA6 5HL; (0292) 76013

Ice Cream Alliance, 90/94 Gray's Inn Road, London WC1X 8AH; 071-405 1996

Ice Cream Federation Ltd, 1 Green Street, Grosvenor Square, London W1Y 3RG; 071-629 0738; Fax 071-499 9095

Institute of Engineering Research, Wrest Park, Silsoe, Bedford MK45 4HS; (0525) 60000

Institute of Food Research, Norwich Laboratory, Norwich Research Park, Colney, Norwich NR4 7UA; (0603) 56122 Telex 975453; Fax (0603) 507723

Institute of Food Research, Reading Laboratory, Earley Gate, Whiteknights Road, Reading RG6 2EF; (0734) 357000; Fax (0734) 267917

Institute of Food Science & Technology, 5 Cambridge Court, 210 Shepherds Bush Road, London W6 7NL; 071-603 6316; Fax 071-602 9936; out-of-hours answerphone 071-603 6317

Institute of Grocery Distribution, Letchmore Heath, Watford, Hertfordshire WD2 8DQ; (0923) 857141; Fax (0923) 852531

Institute of Marketing, Moor Hall, Cookham, Berkshire SL6 9QH; (06285) 24922

Institute of Refrigeration, Kelvin House, Mill Lane, Carshalton, Surrey SM5 2JR; 081-647 7033

Institution of Environmental Health Officers, Chadwick Court, 15 Hatfields, London SE1 8DJ; 071-928 6006; Fax 071-261 1960

International Meat Trade Association Incorporated, 217 Central Markets, London EC1A 9LH; 071-489 0005; Fax 071-248 4733

Irish Livestock & Meat Board, 4 Tavistock Place, London WC1H 9RA; 071-833 1251; Fax 071-278 7193

Laboratory of the Government Chemist, Queens Road, Teddington, Middlesex TW11 0LY; 081-943 7000; Fax 081-943 2767

Leatherhead Food Research Association, Randalls Road, Leatherhead, Surrey KT22 7RY; (0372) 376761; Fax (0372) 386228

Long Ashton Research Station, Long Ashton, Bristol BS18 9AF; (0272) 392181

MAFF – see Ministry of Agriculture, Fisheries and Food

Meat & Livestock Commission, PO Box 44, Winterhill House, Snowdon Drive, Milton Keynes MK6 1AX; (0908) 677577; Telex 82227; Fax (0908) 609221

National Dairy Council, 5/7 John Princes Street, London W1M 0AP; 071-499 7822

National Association of Catering Butchers, 217 Central Markets, London EC1A 9LH; 071-489 0005; Fax 071-248 4733

National Federation of Meat Traders, 1 Belgrove, Tunbridge Wells, Kent TN1 1YM; (0892) 541412; Fax (0892) 535462

New Zealand Meat Producers Board, 34–36 Grays Inn Road, London EC1X 8HR; 071-405 7904; Telex 21261; Fax 071-242 6690

PA Technology, Melbourn, Royston, Hertfordshire SG8 6DP; (0763) 6122; Telex 81561

Pizza & Pasta Association, 29 Market Place, Wantage, Oxon OX12 8BG; (0235) 772207; Fax (0235) 769044

Potato Marketing Board, Broad Field House, 4 Between Towns Road, Cowley, Oxford OX4 3NA

Processing & Packaging Machinery Association, Progress House, 404 Brighton Road, Croydon, Surrey CR2 6AN; 081-681 8226; Fax 081-681 1641

PIRA, Randalls Road, Leatherhead, Surrey KT22 7RU; (0372) 376161; Telex 929810; Fax (0372) 377526

Retail Consortium, Bedford House, 69–79 Fulham High Street, London SW6 3JW; 071-371 5185

Rowett Research Institute, Greenburn Road, Bucksburn, Aberdeen, Grampian AB2 9SB; (0224) 712751; Fax (0224) 715349

Royal Society of Chemistry, Burlington House, Piccadilly, London W1V 0BN; 071-437 8656

Scotch Quality Beef & Lamb Association Ltd, 13th Avenue, Edinburgh Exhibition & Trade Centre, Edinburgh EH28 8NB; 031-333 5335; Telex 72489; Fax 031-333 2935

Sea Fish Industry Authority, Sea Fisheries House, 10 Young Street, Edinburgh EH2 4JQ; 031-225 2515; Fax 031-220 0445. *Also at*: St Andrew's Quay, Hull HU3 4QE; (0482) 27837

Sensory Research Laboratories, 4 High Street, Nailsea, Bristol BS19 1BW; (0275) 810183; Fax (0275) 810184

Snack, Nut & Crisp Manufacturers' Association, Swiss Centre, 10 Wardour Street, London W1V 3HG; 071-439 2567

Society of Chemical Industry (SCI), 14/15 Belgrave Square, London SW1X 8PS; 071-235 3681

Society of Dairy Technology, Crossley House, 72 Ermine Street, Huntingdon, Cambridgeshire PE16 6EZ; (0480) 450741; Fax (0480) 431800

Society of Food Hygiene Technology, PO Box 37, Lymington, Hants SO41 9WL; Tel/Fax (0590) 671979

The Sugar Bureau, Duncan House, Dolphin Square, London SW1V 3PW; 071-828 9465

Tea Council Ltd, Sir John Lyon House, 5 High Timber Street, Upper Thames Street, London EC4V 3NJ; 071-248 1024

Torry Research Station, PO Box 31, 135 Abbey Road, Torry, Aberdeen, Grampian AB9 8DG; (0224) 877071; Fax (0224) 874246

UK Association of Frozen Food Producers, 1 Green Street, Grosvenor Square, London W1Y 3RG; 071-629 0655; Telex 23739; Fax 071-493 2985

United Kingdom Sugar Industries Association, Duncan House, Dolphin Square, London SW1V 3PW; 071-828 9465

Welsh Lamb Enterprise, Brynawel, Great Darkgate Street, Aberystwyth, Dyfed SY23 1DR; (0970) 624011

4.5 EDUCATIONAL ESTABLISHMENTS

Cardiff Institute of Higher Education, Faculty of Tourism, Hospitality and Food, Colchester Avenue, Cardiff CF3 7XR; (0222) 551111; Fax (0222) 465057

Cranfield Institute of Technology, Silsoe Campus, Silsoe, Bedford MK45 4DT; (0525) 860428; Fax (0525) 861527

Heriot-Watt University, Department of Biological Sciences, Riccarton, Edinburgh EH14 4AS; 031-449 5111; Fax 031-451 3009

Manchester Metropolitan University, Hollings Faculty, Old Hall Lane, Fallowfield, Manchester M14 6HR; 061-247 2685

Oxford Brookes University, School of Biological and Molecular Sciences, Gipsy Lane, Oxford OX3 0BP; (0865) 819258; Fax (0865) 819242

Queens University of Belfast, Food & Agricultural Chemistry Dept., Newforge Lane, Belfast, Northern Ireland BT9 5PX; (0232) 661166; Fax (0232) 669551

Scottish Agricultural College, Food Science and Technology Department, Auchincruive, Ayr KA6 5HW; (0292) 520331; Fax (0292) 521119

Silsoe College – see Cranfield Institute

South Bank University, School of Applied Science, Borough Road, London SE1 0AA; 071-815 7900; Fax 071-815 7999

University of Bristol, Division of Food Animal Science, Langford, Bristol BS18 7DY; (0934) 852581; Fax (0934) 852741

University of Huddersfield, Department of Food, Nutrition & Hospitality Management, Queensgate, Huddersfield HD1 3DH; (0484) 472491

University of Humberside, School of Food, Fisheries and Environmental Studies, Nuns Corner, Grimsby, South Humberside DN34 5BQ; (0482) 440550

University of Leeds, Proctor Dept. of Food Science, Leeds, West Yorkshire LS2 9JT; (0532) 332959

University of Nottingham, Faculty of Agricultural and Food Sciences, Department of Applied Biochemistry and Food Science, Sutton Bonington, Loughborough, Leicestershire LE12 5RD; (0602) 484848; Fax (0509) 673917

University of Plymouth, Seale–Hayne, Faculty of Agriculture, Food and Land Use, Newton Abbot, Devon TQ12 6NQ; (0626) 325606; Fax (0626) 325605

University of Reading, Department of Food Science & Technology, Food Studies Building, Whiteknights, Reading RG6 2AP; (0734) 318700; Fax (0734) 310080

University of Surrey, School of Biological Sciences, Guildford, Surrey GU2 5XH; (0483) 300800; Fax (0483) 576978

4.6 GOVERNMENT DEPARTMENTS

MINISTRY OF AGRICULTURE, FISHERIES AND FOOD

Whitehall Place, London SW1A 2HH; Press Office: 071-270 8973; Fax 071-270 8443

At the above address are:
- Biotechnology Unit, Whitehall Place (East Block); 071-270 8481
- Food Industry, Marketing & Competition Policy Division, Whitehall Place (West Block); 071-270 8493
- MAFF Library; 071-270 8421
- National Food Survey Branch, Whitehall Place (West Block); 071-270 8562/3
- Pigs, Eggs & Poultry Division, Whitehall Place (East Block); 071-270 8462

MINISTRY OF AGRICULTURE, FISHERIES AND FOOD (MAFF)

Nobel House, 17 Smith Square, London SW1P 3JR; 071-238 3000 (MAFF switchboard); Press Office 071-270 8446

At the above address are:
- Food Science Division II
- Food Standards Division
- Chief Scientists' Group

At Ergon House, c/o Nobel House, 17 Smith Square, London SW1P 3JR:
- Advisory Committee on the Microbiological Safety of Food; 071-238 5572
- Advisory Committee on Novel Foods & Processes; 071-238 6380
- Chemical Safety of Food Division; 071-238 6380
- Consumer Protection Division; 071-238 6272 or 071-238 6763
- Consumer Panel; 071-238 5957; Fax 071-238 6330
- Consumer Helpline; 071-238 6550
- Food Advisory Committee; 071-238 6267; Fax 071-238 6263
- Food Safety Directorate of Information; 071-238 6335; Fax 071-238 6330
- Pesticides Safety Division; 071-238 6378

MINISTRY OF AGRICULTURE, FISHERIES AND FOOD

Meat Hygiene Division, Room 94, Government Buildings (Toby Jug Site), Hook Rise South, Tolworth, Surbiton, Surrey KT6 7NF; 081-330 8367; Fax 081-330 1563

MINISTRY OF AGRICULTURE, FISHERIES AND FOOD

Veterinary Medicine Services, Tolworth Tower, Surbiton, Surrey KT6 7DX; 081-399 5191; Fax 081-399 4425

DEPARTMENT OF HEALTH

Health Aspects of Environment and Food, Skipton House, 80 London Road, Elephant & Castle, London SE1 6LW; 071-972 2000 (switchboard)

* Food & Drink Federation – additional members:
Association of Cereal Food Manufacturers
Association of Fish Canners
Bakery Allied Traders Association
British Caramel Manufacturers Association
British Fruit & Vegetable Canners' Association
British Oat and Barley Millers' Association
British Pasta Products Association
British Soluble Coffee Manufacturers' Association
British Starch Industry Association
Chilled Food Association
Dessert & Cake Mixes Association
Dehydrated Potato Association
European Chilled Food Federation
European Cereal Ingredients Association
European Spice Association
Food Casings Association
Fish and Meat Spreadable Products Association
General Products Association
Infant & Dietetic Food Association
Margarine and Shortening Manufacturers' Association
National Association of Cider Makers
National Edible Oil Distributors' Association
Pet Food Manufacturers' Association
Potato Processors' Association
Pre-Packed Flour Association
Pickles & Sauces Association
Rice Association
Roast & Ground Coffee Association
Rusk Manufacturers' Association
Seed Crushers' and Oil Processors' Association
Soup and Gravy Manufacturers' Association
Seasoning and Spice Association
Table Jellies Association
Tea Packers Association
UK Association of Manufacturers of Bakers Yeast
UK Preserves Manufacturers Association
Vegetable Protein Association

INDEX